AF183151

WHICH FUELS FOR LOW CO_2 ENGINES?

IFP International Conference / Les rencontres scientifiques de l'IFP

WHICH FUELS FOR LOW CO_2 ENGINES?

Held in Rueil-Malmaison, France, September 22-23, 2004

Conference Organization

Ms Bettina CARUSO
IFP Communication Division
bettina.caruso@ifp.fr

Conference Chairman

Pierre DURET
IFP School
pierre.duret@ifp.fr

Conference Vice-Chairman

Xavier MONTAGNE
IFP Powertrain Engineering
xavier.montagne@ifp.fr

Scientific Committee

Rik BAERT, TNO
Gérard BELOT, PSA Peugeot-Citroën
Pierre DURET, IFP School
Norimasa IIDA, Keio University
Xavier MONTAGNE, IFP Powertrain Engineering
Hua ZHAO, Brunel University

Sessions Chairmen

Heinz HASS, Ford	*hhass@ford.com*
Kazumasa KATOH, Renault	*kazumasa.katoh@renault.com*
José PASTOR, CMT Valencia	*jpastor@mot.upv.es*
Salvatore CORSO, FIAT	*salvatore.corso@fiat.com*
Philippe PINCHON, IFP	*philippe.pinchon@ifp.fr*
Bruno COSTES, PSA Peugeot-Citroën	*bruno.costes@mpsa.com*

Main authors

Béatrice MAURER, PSA Peugeot-Citroën	*beatrice.maurer@mpsa.com*
Nicolas JEULAND, IFP	*nicolas.jeuland@ifp.fr*
Jean-Claude GRIESEMANN, European BTL Fuels Experts Group	*jean-claude.griesemann@renault.com*
Norimasa IIDA, Keio University	*iida@sd.keio.ac.jp*
Gen SHIBATA, Nippon Oil Corp.	*gen.shibata@eneos.co.jp*
Tom W RYAN, Southwest Research Institute	*tryan@swri.org*
Kayiu MAN, Brunel University	*kayiu.man@brunel.ac.uk*
Kevin DUFFY, Caterpillar	*duffy_kevin_p@cat.com*
Wolfgang STEIGER, Volkswagen	*wolfgang.steiger@volkswagen.de*
Rik SG BAERT, TNO	*r.s.g.baert@tue.nl*
Yukihiro TSUKASAKI, Toyota	*yukihiro@tukasaki.tec.toyota.co.jp*
Fabian FRICKE, FEV	*fricke@fev.de*
Richard TILAGONE, IFP	*richard.tilagone@ifp.fr*
Egbert LOX, UMICORE	*egbert.lox@eu.umicore.com*
Jean BOTT , Delphi Corporation	*jean.botti@delphi.com*
Deanna WANG, University of Melbourne	*deannawang@fastmail.fm*
Per TUNESTÄL, Lund University of Technology	*per.tunestal@vok.lth.se*
Michael HALLMANNSEGGER, BMW	*michael.hallmannsegger@bmw.de*
Josef AFFENZELLER, AVL List GmbH	*josef.affenzeller@avl.com*
Gabriel PLASSAT, ADEME	*gabriel.plassat@ademe.fr*
Franz-Xaver SOELDNER, European Commission	*franz-xaver.soeldner@cec.eu.int*
Bruno CELARD, Europia/Concawe	*bruno.celard@europia.com*

IFP International Conference / Les rencontres scientifiques de l'IFP

WHICH FUELS FOR LOW CO_2 ENGINES?

Proceedings of the International Conference held in Rueil-Malmaison, France, September, 22-23, 2004

Edited by
PIERRE DURET *and* XAVIER MONTAGNE
IFP

2004
Editions TECHNIP
27 rue Ginoux, 75737 PARIS Cedex 15, FRANCE

From the same publisher

- A New Generation of Engine Combustion Processes for the Future?
  Proceedings of the international congress held in Rueil-Malmaison, France, November 26-27, 2001
  P. Duret, Ed.
- A new Generation of Two-Stroke Engines for the Future?
  Proceedings of the international seminar held in Rueil-Malmaison, France, November 29-30, 1993
  P. Duret, Ed.
- Direct Numerical Simulation for Turbulent Reacting Flows
  T. Baritaud, T. Poinsot, M. Baum, Eds.
- Multi-Dimensional Simulation of Engines Internal Flows
  T. Baritaud, Ed.
- Fuels and Engines. Technology. Energy. Environment (2 Volumes)
  J.-C. Guibet
- Lubricant Properties, Analysis, and Testing
  J. Denis, J. Briant, J.-C. Hipeaux
- Rheological Properties of Lubricants
  J. Briant, J. Denis, G. Parc
- Automobiles and Pollution
  P. Degobert

All rights reserved.

No part of this publication may be reproduced or transmitted in any form or by any means, electronic or mechanical, including photocopy, recording, or any information storage and retrieval system, without the prior written permission of the publisher.

© Editions Technip, Paris, 2004.
Printed in France

ISBN 2-7108-0851-X

Table of contents

ENERGY CONVERSION IN ENGINES FROM NATURAL GAS

INTRODUCTION OF THE USE OF HYDROGEN IN IC ENGINES

WHICH FUELS FOR LOW CO_2 ENGINES?

Which Fuels For Low CO_2 Engines?
© P. Duret, X. Montagne (Eds.) and Editions Technip, Paris, 2004, pp. 1 – 3
27 rue Ginoux, 75015 Paris

OPENING ADDRESS

Pierre Duret
IFP School

INTRODUCTION

Following the success of the previous IFP International Congress on the theme of new HCCI & CAI combustion processes [1], the purpose of this new International Conference is to provide the opportunity for experts from the automotive industry, the heavy duty and small engine sectors, fuel companies, additive makers, OEM suppliers and R&D organizations to discuss the potential of this new generation of fuels and energy conversion in engines.

The key words of the title of this new conference are "Low CO_2" and these two key words are surrounded by the words "Fuels" on one side and "Engines" on the other side. It means that when we proposed this title, we made the hypothesis that any significant progress on CO_2 emissions from transportation have to be done in working in conjunction on both engines and fuels. I hope that this event will confirm that we were right in making such hypothesis and that everybody will agree.

A CONFERENCE COVERING MULTIPLE "FUEL & ENGINE" APPROACHES FOR FUTURE LOW CO_2 TRANSPORTATION

To cover the subject of fuel & engine couple for low CO_2 emissions, several approaches naturally appeared to us:

- The first one is to further improve jointly the existing engine and liquid fuel technologies, and in a second step to make some adaptation of conventional engines for their operation with new type of liquid fuels such as bio-fuels. It is the subject of the first session of this conference (and corresponding proceedings);
- The second approach goes one step further. It is to really optimize the engines with new combustion processes (allowing to overcome the "CO_2-efficiency" versus "NOx-PM" dilemma of stratified charge engines) such as HCCI (Homogeneous Charge Compression Ignition) and CAI (Controlled Auto-Ignition) and to also modify the (still liquid) fuel formulations in order to increase the benefits offered by HCCI & CAI. This is the subject extensively covered by the second session of this conference;
- Still one step further, we arrive to a third approach which is to use natural gas for the energy conversion in engines either directly in a gaseous form in a vehicle, or indirectly in a liquid form after the Gas To Liquid conversion process. Each solution presents drawbacks and advantages and it is the purpose of the third session of this conference to try to improve the knowledge and to better know the potential of this third approach;
- Finally the last approach is located at the transition between the internal combustion engine and the fuel cell. It is to use hydrogen in an internal combustion engine, either in small quantity (mixed with another main fuel) to enhance the combustion efficiency and cleanliness or as a single fuel as it is also shown in these proceedings.

To conclude the last session of this conference is an exchange of points of view concerning the future of these approaches given by different stakeholders involved in the future of low CO_2 transportation.

In the next two sections of this short paper we will examine the world-wide activities based on the analysis of the number of studies published by the researchers using appropriate requests and database [2].

THE WORLDWIDE HCCI EFFERVESCENCE

As shown by Figure 1 covering HCCI&CAI papers, their number has significantly increased since 2001, the breakthrough between 2000 and 2001 being exactly timed when the first fully HCCI&CAI-dedicated event has been organized at IFP [1].

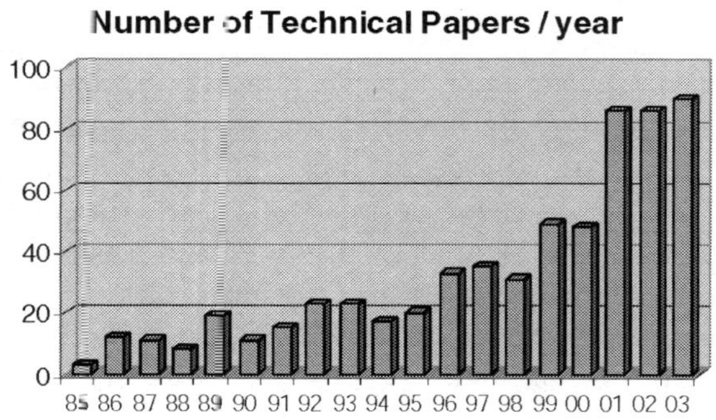

Fig. 1 Statistical analysis of the number of technical papers related to the new HCCI & CAI engine combustion processes published from 1985 to 2003

What is even more remarkable is that if we add the about 50 papers published during the first half of 2004, it means that more than 300 HCCI&CAI papers have been published within 3.5 years, which is almost the same as what has been previously published in 10 years between 1990 to 2000.

A lot of companies and institutions are now working on the subject and we can expect the spread of practical application of HCCI&CAI technologies within the next few years. Nevertheless all these first applications in production will have to be done using the conventional existing diesel and gasoline fuels. The next step will be the evolution of the fuel formulations to take increased benefit of HCCI&CAI both in terms of efficiency / low CO_2 as well as in terms of NOx / PM as shown by different authors in these Proceedings.

Now if we look to the companies and institutions which are the most deeply involved in such research, we can see in Figure 2 that several of them are present at this conference to share their respective progresses with the scientific community.

In this direction of improving the knowledge in the field of new HCCI&CAI engine combustion processes, IFP has recently set up an European Network of Excellence named "ECO-Engines" for Energy Conversion in Engines. The purpose of this network is to strengthen the European research and also to develop appropriate education & training courses in this technical field. It is interesting to notice that all the European institutions of the Figure 2 (Lund Institute of Technology, Chalmers University, Brunel university, IFP and AVL) are deeply involved in ECO-Engines and even play a major role in it.

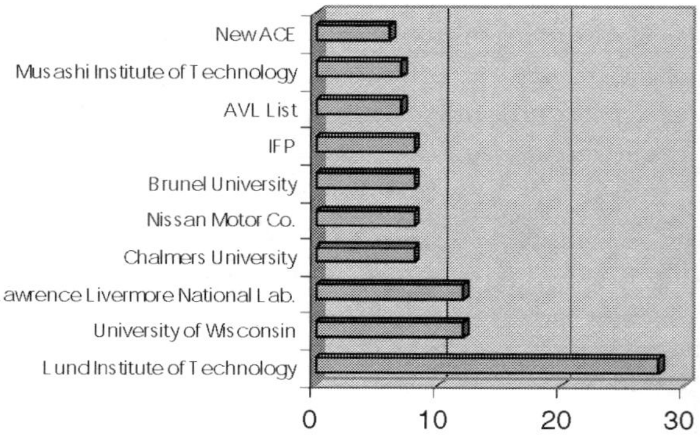

Fig. 2 Main companies and institutions (and corresponding number of papers) having published technical papers related to HCCI & CAI during the last 3.5 years

The Education and Training Part is also lead by the IFP School with the same partners. IFP School is an engineer school providing Applied graduate studies in different fields, among them the "Internal Combustion Engines" program is highly appreciated by the engine and automotive industry. It will be then of major importance for the IFP School, which celebrates its 50 years anniversary at the time of this congress, to successfully undertake this European cooperation in the development of new education and training courses and materials on HCCI&CAI related topics.

THE RELATIONS BETWEEN INTERNAL COMBUSTION ENGINES AND ALTERNATIVE FUELS

We can make similar analysis on the technical papers related to the use of alternative fuels in engines. The Figure 3 shows first the number of papers related to the use of bio-fuels.

We can first see that their number is much smaller compared with HCCI&CAI papers. This could mean that there are less activities in this field but perhaps also less need to undertake extensive research in bio-fuels and engine adaptation. This adaptation is also probably scientifically and technically less complicated than the HCCI&CAI combustion.

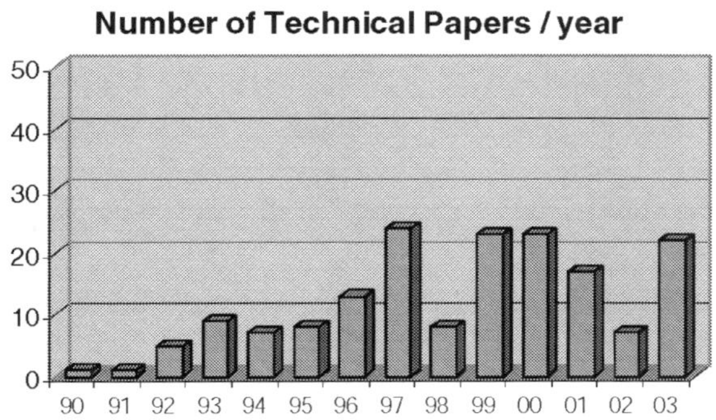

Fig. 3 Statistical analysis of the number of technical papers related to the use of bio-fuels in engines and vehicles published from 1990 to 2003

We can nevertheless observe an interesting increase of activity (with some cyclic variations) since the middle of the 1990's, probably due to the beginning of the political pressure to promote such fuels.

Regarding the number of papers related to the use of natural gas in engines and vehicles, either directly (gaseous fuels) or indirectly (GTL), the situation is much different. The number of papers is much higher with two peaks in 1995 and 2000. It asks question to us. Nevertheless, this statistical information is not sufficient to conclude that this phenomenon could be the sign of two periods of higher perspective for such fuels followed by period of reduced interest .

Finally, looking at Figure 5, we can see a relatively small number of papers on the use of hydrogen in internal combustion engines, but this number seems to grow "slowly" and "surely" since 1995. We don't know today where it will go, may be up to the industrial application of H2 in IC engines before the long term era of the fuel cells for transportation !

CONCLUSION

To conclude this opening address, I sincerely hope that this event will provide all the speakers and delegates with the opportunities to discuss in a very open and efficient manner about the future of the "engine" and "fuel" couple for low CO_2 transportation means necessary to satisfy the people growing needs of mobility all over the world.

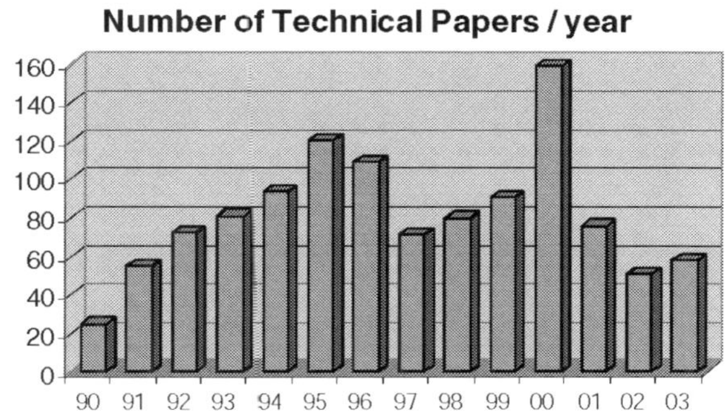

Fig. 4 Statistical analysis of the number of technical papers related to the use of natural gas in engines and vehicles

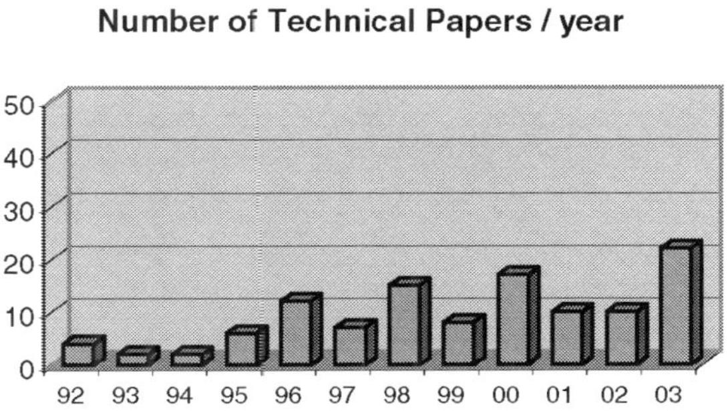

published from 1990 to 2003

Fig. 5 Statistical analysis of the number of technical papers related to the use of hydrogen in internal combustion engines published from 1992 to 2003

Acknowledgments. The author would like to deeply acknowledge Elisabeth Ubricht for her cooperation in preparing the document here above, in particular in the statistical analysis of all the published papers.

REFERENCES

1. A New Generation of Engine Combustion Processes for the Future, IFP International Congress, November 26-27, 2001; edited by Pierre Duret, Editions Technip
2. Publications database "Mobility" including all SAE papers and most of the other international technical papers in relation with the automotive domain

FUTURE OF LIQUID FUELS

Which Fuels For Low CO_2 Engines?
© P. Duret, X. Montagne (Eds.) and Editions Technip, Paris, 2004, pp. 07 – 19
27 rue Ginoux, 75015 Paris

POTENTIALITY OF ETHANOL AS A FUEL FOR DEDICATED ENGINE

Nicolas Jeuland, Xavier Montagne, Xavier Gautrot
IFP Powertrain Engineering

ABSTRACT – *One of the major challenges of the automotive industry is to reduce the greenhouse gases, and especially CO_2 emissions. Many research programs are currently being led on this subject, aiming at reducing the fuel consumption of vehicles, but also at optimizing the fuel composition. The overall process must indeed be taken into account, from the fuel production to the vehicle emissions, and a "well to wheel" balance has to be calculated for each technology.*

Ethanol seems to have important assets to comply with these new constraints: it is extracted from the biomass and consequently has a good "well to tank" CO_2 emission balance. Moreover, its properties, especially in terms of octane number and latent heat of vaporization allow a large improvement of the engine.

The present paper aims at taking stock of the ethanol path. The main production ways are tackled, especially concerning their energy balances. The main advantages and drawbacks of the use of such a fuel are then summarized. Finally, an example of the gains that can be obtained by an optimization of the engine using pure ethanol is presented. A small supercharged engine has already been modified to benefit from the potential of ethanol, especially in terms of knock resistance. The performances of this engine have been compared with those of the initial gasoline engine, showing that important gains can be obtained with such a technology.

INTRODUCTION

Ethanol as a fuel for internal combustion engines is not a new concept. Indeed, the first internal combustion engines and vehicles (N. Otto in 1877, H. Ford in 1880, in 1892 in France) were drawn to run with pure alcohol (methanol or ethanol). In the US and also in many countries such as France or United Kingdom, many studies were achieved in the 1920's and 1930's with this fuel, before the wide diffusion of leaded gasoline induced a decrease in the interest in ethanol for years.

During the last decades, a renewed interest for ethanol has grown, linked with the more and more stringent emission limits. Moreover, some economical aspects, such as agricultural development (in Brazil for instance) have also favored to the use of ethanol. Finally, the Kyoto protocol and the growing concern for greenhouse gas emissions will lead in the next coming years to an increase in biofuel productions, among which ethanol has an important role to play.

This paper presents the potentiality of the ethanol pathway as a fuel especially when used at very high concentration on a dedicated engine.

The potential of ethanol in a dedicated engine is described: the main advantages and drawbacks of this fuel are summarized and an example of application is given with the results obtained after a preliminary optimization of a small supercharged engine.

USE OF ETHANOL IN THE WORLD

Brazil is the largest user of ethanol as a fuel. It is the only country that uses ethanol blends at concentrations higher than 10%, with the exception of the use of E85 in Flexible Fuelled Vehicles (FFV). Around 20% of the Brazilian cars currently operate on 100% ethanol. The remaining cars are optimized to run on 22% blends (to meet the range of 20-24% blends). Brazil consumption of ethanol is around 15 billion liters a year.

In the United States, the US EPA regulates the use of 10% ethanol blends via the 'Substantially Similar Rule' to ensure that the use of oxygenates does not contribute to emission control system failure. The regulation prohibits the introduction, or increase in concentration of a fuel or fuel additive which is not 'substantially similar to any fuel or fuel additive already utilized'. The Substantially Similar Rule recognizes that the use of certain types of oxygenates have no adverse effect below a specified oxygen content.

The US EPA regulations also create a process by which a waiver can be granted for an oxygenated 'recipe' that is demonstrated not to cause or contribute to the failure of any emission control device or system. Under this regulation, the US EPA has granted waivers for concentrations of ethanol in petrol up to 10%. The finished product has to pass gasoline specifications as defined by ASTM D 4814-88 and US original equipment manufacturer (OEM) vehicle warranties specify that petrol must contain no more than 10% ethanol. The use of blends higher than 10% in OEM vehicles could result in the void of vehicle warranties.

The US Department of Energy (under its Clean Cities Program) and the National Corn Growers Association are cooperating to promote the development of refueling infrastructure for E85 and to encourage fleet operators to choose ethanol to meet the alternatively fuelled vehicles requirements of the Energy Policy Act. As previously stated, E85 blends require FFV.

The potential phase-out of MTBE (methyl tertiary butyl ether) and an increasing emphasis on domestic energy supplies and energy security are likely to favor in the US the increased use of ethanol as a fuel.

In Europe, the ethanol content in gasoline is limited to 5%vol. ETBE (ethyl tertiary butyl ether) is the preferred oxygenate by European refiners and car manufacturers.

ETHANOL PRODUCTION PATHWAYS

Ethanol production for fuel is a combination of biological and physical processes. The main production process is the fermentation of sugars with yeast. Ethanol is then concentrated to fuel grade by distillation. The raw materials can be obtained from various types of crops, such as corn, wheat, sugar beet...

Besides, some new ethanol production pathways appeared in the last years, producing ethanol from unconventional feedstock.

ENERGY AND GREENHOUSE GAS BALANCE

The development of biofuels, and especially ethanol, is strongly linked to their environmental performances in terms of greenhouse gas emissions and energy saving. To evaluate their efficiency, many studies focused on the "Analysis of Life Cycle" of each path (LCA).

The calculations show highly variable results, whose apparent discrepancies can be easily explained.

- As ethanol is produced from biomass, its energetic efficiency is strongly related to the raw material considered (corn, cellulose...).
- Even when considering a single raw material type, the global efficiency is still strongly linked to the development of the production process (yield, agricultural techniques, use of fertilizers...)
- Finally, the conversion rate of the raw material into ethanol can vary a lot.

The example of US corn ethanol is very symptomatic of these difficulties: before 1990, some studies showed a negative energy balance for corn ethanol. Nevertheless, the increase in yields, shown in figure 1, turned the energy balance to positive.

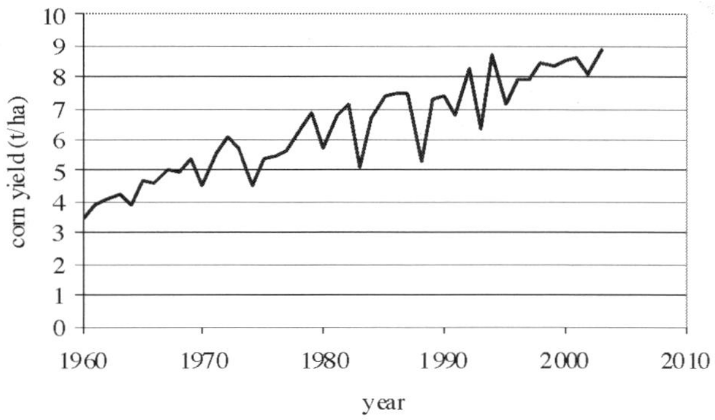

Fig. 1 US corn yield evolution [1]

Moreover, as described above, the ethanol production from the raw material is strongly linked to the raw material itself, and also to the development of the transformation industry itself. Table 1 summarizes the ethanol yields calculated in various studies.

	[9]	[10]	[5]	[6]	[7]	[8]	[11]	Mean
Barley	7.7							7.7
Corn	32.8		17.3	18.2			19.1	21.8 +/- 7
Grain Sorghum	5.4							5.4
Wheat	11.6	20.1		17.4	19.5	13.0		16.3 +/- 3
Beet		23.7		47.4		21.7		30.9 +/- 16

Table 1. Ethanol yield (hl/ha) for various crops

In Europe, corn is not used for ethanol production. The most commonly used raw materials are wheat and beet (especially in France). As for corn ethanol in the US, the energy balance and greenhouse gas balance are highly dependant on the agricultural techniques and yields. The values obtained are consequently variable as shown in the following tables, extracted from [4].

	Wheat					
Growing region	**UK**	**North France**		**UK**	**EU**	***Aver-age***
		Aver.	**Best**			
Ref./Case	[5]	[6]	[6]	[7]	[8]	
Ethanol yield t/ha	2.2	1.9	2.5	2.5	1.7	***2.1***
Energy balance	1.07	0.91	0.91	0.9	1.26	***1.04***
GHG balance	0.59	0.72	0.71	0.71	0.99	***0.75***

Table 2. Energy balance and greenhouse gas balance for ethanol produced from wheat

	Beet			
Growing region	**North France**		**EU**	***Avera ge***
	Aver.	**Best**		
Ref./Case	[6]	[6]	[8]	
Ethanol yield t/ha	5.3	6.2	3.8	***4.5***
Energy balance	0.9	0.62	0.96	***0.9***
GHG balance	0.7	0.49	0.75	***0.7***

Table 3. Energy balance and greenhouse gas balance for ethanol produced from beet

In these cases, the energy balance is defined as the ratio of the energy required to produce an amount of conventional fuel equivalent to the biofuel to the energy content of the biofuel. A figure of zero indicates a fully renewable fuel while a value of 1.0 indicates that there is no net energy saving.

The GreenHouse Gas (GHG) balance is defined as the ratio of the net GHG production, expressed as CO_2 equivalents, emitted when producing the biofuel to the amount of CO_2 equivalents emitted when producing and burning an amount of fossil fuel representing the same end-use energy. A value below one denotes a net reduction in GHG emissions.

Considering the French case, the most recently published data (ADEME, data from IFP) indicate an energy balance of 0.49 for wheat ethanol and beet ethanol, and some more prospective scenarios foresee balances down to 0.28 for wheat and 0.3 for beet [10].

The calculated energy needs for ethanol production have been shown to be highly variable, according to the raw material, the agricultural techniques and the calculation method. Nevertheless, all these calculations show a positive impact of ethanol in terms of greenhouse gas emissions, due to the fact that ethanol, as far as it is produced from biomass, should have (if ideal) a null CO_2 balance (the CO_2 emitted during the combustion should be counterbalanced by the CO_2 captured by the crop during its growth). In fact, this theoretical and ideal case is never reached, due to energy needs for the agricultural process, the fertilizer uses (nitrogen fertilizers are linked with N_2O emissions, which have a strong impact on greenhouse effect) and the transformation processes.

KEY PARAMETERS FOR THE DEVELOPMENT OF A DEDICATED ENGINE

The potential of ethanol in terms of greenhouse gas emissions, when produced from biomass, has been outlined above.

Moreover, ethanol has some very interesting physical and chemical properties that can be turned into benefits with a dedicated engine, provided some technical difficulties are overcome. The following section describes these advantages and disadvantages of ethanol and gives an example of a preliminary development with a small displacement dedicated engine.

ETHANOL PROPERTIES

The main properties of ethanol are summarized in Table 4.

	Ethanol	Gasoline (typical)
Molar mass (g/mol)	46.07	102.5
C (% wt)	52.2	86.5
H (% wt)	13.1	13.5
O (% wt)	34.7	0
Density (kg/m^3)	794	735 - 760
Latent heat of vaporization (kJ/kg)	854	289
Distillation (°C)	78.4	30 - 190
Net heating value (kJ/kg)	26805	42690
Net heating value (kJ/l)	21285	32020
Stoechiometric ratio	8.95	14.4
RON	111	95
MON	92	85

Table 4. Ethanol physical properties

This table shows that ethanol has some interesting properties to be used as a fuel for spark-ignition engines:

- A very high octane number, which induces a strong resistance to knock and consequently the ability to optimize the engine (compression ratio, spark-advance).
- A density close to the gasoline one.
- The presence of oxygen in the formula, which can provide a more homogeneous fuel / air mixing and consequently a decrease in unburned or partially burned molecule emissions (HC and CO).
- A high latent heat of vaporization enabling a "cooling effect" of air and consequently can enhance the filling efficiency.

On the opposite, some disadvantages have to be considered:

- The oxygen included in the molecule (30%wt) induces an increase in the fuel volumetric consumption.
- The high latent heat of vaporization can induce running difficulties in cold conditions, especially cold start.
- Ethanol leads to azeotropes with light hydrocarbon fractions and can lead to volatility issues.
- Ethanol is miscible with water, which can cause demixion issues when blended with hydrocarbons.
- The high oxygen content of ethanol and its ability to oxidize into acetic acid induce compatibility issues with some materials used in the engine, such as metals or polymers.
- Ethanol combustion in engines induces aldehydes emissions, which can have a negative impact on health.

These points are further detailed in the following sections.

OCTANE NUMBER

Ethanol is characterized by high octane numbers (RON and MON). A dedicated engine will consequently be less knock sensitive and its compression ratio can be increased, to enhance the engine efficiency.

This physical property is particularly useful to reduce CO_2 emissions of spark-ignition engines: yet, one of the most promising ways seems to be the downsizing of engines, with the generalization of small displacement turbocharged engines. Whereas these engines could be generally highly knock sensitive, a substantial improvement could be brought by the use of high octane fuels.

A typical correlation that is used for compression ratio calculation gives a value of 1 CR point increase for about 5 points octane [12]. When ethanol is used on a typical spark-ignition engine (CR = 9, RON 95), the compression ratio could go up to 13 - 14, inducing a substantial thermal efficiency increase. So far, most of the developments on dedicated engines selected a compression ratio around 12 - 12.5, because of they used E85 (85% ethanol mixture) instead of pure ethanol (cold start driveability improvement, see below). Moreover, such high octane numbers are subject to a high uncertainty in their measurements, and also in their physical meanings. Indeed, the ASTM D2699 method for RON values higher than 100 is based on the comparison of the tested fuel with leaded isooctane. One should assume that, even if those 2 fuels may behave the same way in a CFR engine, their radically different physical and chemical properties can induce different behaviors in real engines.

OXYGEN CONTENT

The oxygen content in ethanol is around 35%wt. The mass heating value of ethanol is consequently lower than the gasoline one (respectively 27MJ/kg and 42.7MJ/kg). Ethanol higher density tends to reduce the difference in volumic heating values, but a significant difference remains. Besides, another parameter has to be considered: the specific energy, which is the energy released for each unit mass of air consumed in stochiometric conditions. It corresponds to the ratio (heating value / air-fuel ratio). This ratio is around 3 for ethanol and 2.9 for a typical gasoline. Indeed, the oxygen in ethanol reduces the stochiometric ratio, i.e. the mass of air which is needed to get the total combustion of a given quantity of fuel. As shown in Table 4, this stochiometric ratio is around 9 for ethanol and 14 for gasoline.

As a consequence of this difference, if the engine efficiency remains similar for gasoline and ethanol, the fuel volumetric consumption is inevitably increased when running with ethanol.

Moreover, the high oxygen content of ethanol has some significant positive aspects, especially when pollutant emissions are considered. As stated above, the use of oxygenated compounds can lead to a more homogeneous combustion by bringing oxygen in the core of the fuel, where the oxygen brought by air can never arrive. A decrease in CO and HC emissions can consequently be measured, but the reduction level varies according to the vehicle type and technology. The impact on HC emissions is also important, but is highly variable because of the high number of parameters implied:

- Decrease in air / fuel local heterogeneity (decrease in HC emissions).
- Volatility increase (increase in evaporation HC emissions).
- Decrease in HC emissions when ethanol is used as a substitution for other high octane molecules such as aromatics (HC precursors).

VOLATILITY

Ethanol by itself has a moderate volatility. Table 4 shows that its boiling point is relatively high (78°C) in comparison with gasoline initial distillation point. Moreover, the vapor pressure, calculated according to Antoine's law, is low for ethanol in comparison with the light fractions of the fuel:

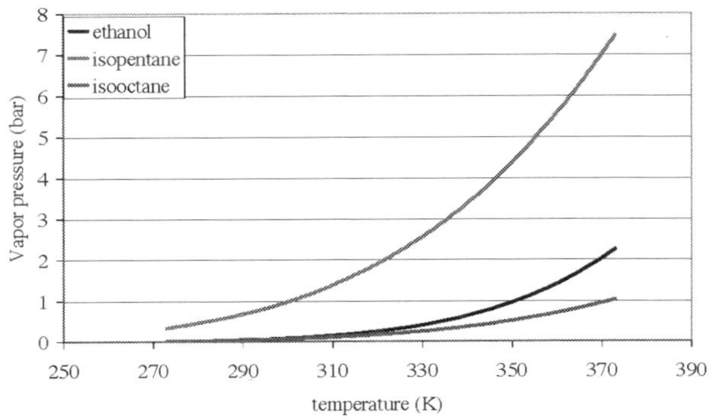

Fig. 2 Vapor pressure of ethanol - comparison with typical gasoline components

Nevertheless, another physical property must be remembered: ethanol can strongly interact with some hydrocarbons, leading to the formation of azeotropes. Azeotropes are defined as any liquid mixture having constant minimum and maximum boiling points and distilling off without decomposition and in a fixed ratio. Their main characteristic is to have a boiling point radically different from the boiling points of each component on its own, and consequently a different vapor pressure at a given temperature. Ethanol azeotropes have boiling points lower than the initial boiling points of their components.

For instance, ethanol and n-pentane can give an azeotrope which boiling point is 34°C, lower than the respective boiling points of ethanol (76°C) and n-pentane (36°C). The following table summarizes the most common azeotropes formed with ethanol.

Molecule	**Pure Boiling point (°C)**	**Azeotrope boiling point (°C) with ethanol**	**Composition of the azeotrope (%wt ethanol)**
n-pentane	36	34	5
n-hexane	69	59	21
benzene	80	68	32
cyclohexane	81	65	29.2
toluene	111	77	68
n-octane	126	77	88

Table 5. Common ethanol azeotropes characteristics [13]

The impact of these azeotropes is important when blending ethanol with fuels, by leading to an increased vapor pressure.

Of course, this phenomenon is avoided when pure ethanol is used. Nevertheless, as stated above, pure ethanol has a distillation point of 78,4°C, characteristic associated with an important latent heat of vaporization, inducing cold start difficulties. To avoid this phenomenon, E85 (ethanol blended with 15% light hydrocarbons) instead of pure ethanol is a commonly used solution. Moreover, as presented above, most countries prefer to use low ethanol concentrations (lower than 10%) in fuels. Azeotrope formation can have an important impact on these kinds of fuels: if the fuel volatility is not correctly controlled, adding ethanol can induce a strong increase in evaporative losses. For refiners, this physical phenomenon means that "splash blending" is difficult with ethanol: ethanol can only be added to selected (low volatility) base fuel.

WATER TOLERANCE

Ethanol and water are miscible in any proportions, while hydrocarbons and ethanol are not miscible. With conventional gasolines containing hydrocarbons, or ethers, the presence of water in the fuel is not a serious concern. In fact, up to 50ppm at ambient temperature (of course this value depends of the gasoline chemical composition), water remains completely soluble. Over this level, water separates without affecting the hydrocarbon portion and the water layer can be extracted if necessary.

On the opposite, in ethanol blends, traces of water can induce a total de-mixing into two phases: one phase is a mixture of ethanol and water and the other phase contains the hydrocarbons. Demixion traces are not acceptable.

Moreover, whereas ethanol can be used as an "octane enhancer" in the fuels, the demixion between the base fuel and ethanol, in presence of water leads to a decrease in the octane number of the base fuel, which could seriously damage the engine.

To avoid this phenomenon, some methods can be used, such as the use of "co solvents" (for instance TBA (tertio-butyl alcohol) in the 1980's in the US) or a careful check of fuel storing tanks water tightness.

Another way to avoid this phenomenon could be the use of pure ethanol [16]. Indeed, water is miscible in ethanol and consequently no demixion phenomenon can occur. Nevertheless, the already mentioned issues linked to engine cold start remain.

COLD START

Pure ethanol has a moderate volatility. Its high boiling point (78°C) induces difficulties of vaporization in ambient or cold conditions.

Moreover, as it has been outlined previously, ethanol has a very high latent heat of vaporization (3 times higher than typical gasoline). This property induces driveability difficulties: the vaporization of ethanol in cold conditions needs a lot of energy and consequently induces a cooling effect [17].

To lower this phenomenon, some technical solutions can be applied. The most frequently used solution is to use E85 (85% ethanol) instead of pure ethanol. The addition of 15% light hydrocarbon fraction in ethanol induces a strong increase in volatility and consequently an easier engine start. Nevertheless, this modification has to be made carefully to avoid the loss of some of the ethanol advantages (octane number, low HC and CO emissions).

Other solutions have been studied, especially in Brazil. Among these, the most widely spread are the heating of the fuel injection system (and especially the fuel-rail) or the use of a second tank, filled with highly volatile gasoline, to run the engine for a few seconds before switching to ethanol.

MATERIAL COMPATIBILITY

Polymers

Ethanol induces the swelling and weakening of rubber components, due to the absorption of fuel into rubber. Once absorbed into rubber, the oxygen of the alcohol breaks the rubber's carbon-carbon double bonds. The consequence of swelling and weakening can be a fuel leak that can endanger car users [17].

Swelling and component breakdown can all be solved by the use of compatible materials such as highly fluorinated rubbers (Viton) [18]. Nylon can also be resistant, but only at low temperature (< 30°C) [19]. This polymer can consequently be used for intake fuel line, provided the fuel temperature remains low.

Galvanic corrosion

Galvanic corrosion happens when 2 metals are in contact, generally through a conducting electrolyte. For instance, stainless steel and aluminum can undergo galvanic corrosion when placed side by side.

Typical fuels (with no additives) have an electrical conductivity between 10^{-8} to $10^{-6}\mu S/cm$. Ethanol electric conductivity is $1.35\ 10^{-3}\mu S/cm$. Moreover, the oxidation of ethanol into acetic acid induces a rapid increase in electrical conductivity ($41\mu S/cm$ for a 0.1M acetic acid solution). The presence of acetic acid can consequently enhance galvanic corrosion and chemical attack. The metals recommended for use with ethanol include carbon steel, stainless steel and bronze. Metals such as magnesium, zinc casings, brass and copper are not recommended [14].

The previous paragraph assumes that ethanol is "dry", which means it contains no water. Yet, as water is miscible in ethanol, ethanol with very high water content has been found in the past (up to 5%vol), with ion concentrations that make it much more aggressive than pure ethanol [15]. The use of anhydrous ethanol is consequently mandatory to avoid engine corrosion. This is the most important limitation to ethanol development, as far as fuel logistic and storage are concerned.

PRELIMINARY DEVELOPMENT OF A DEDICATED ENGINE

This section describes the preliminary development of a small displacement dedicated engine. Indeed, if some changes have been done on the engine, it has not been fully optimized. Nevertheless, these results show the potentiality of ethanol.

ENGINE INITIAL CHARACTERISTICS

The selected engine is a small turbocharged engine. Its main characteristics are described in Table 6.

Engine characteristic	Value
Cylinder / Valves	3 cylinders / 6 valves
Displacement	$599 cm^3$
Compression ratio	9.5 : 1
Bore	63.5mm
Stroke	63mm

Table 6. Engine initial characteristics

This engine has been fully characterized, with a RON95 EN228 gasoline, then modified as described below.

ENGINE MODIFICATIONS

To run with ethanol, the following modifications have been done:

- increase in engine compression ratio,
- modification of engine fuel system,
- modification of cylinder head.

Compression ratio

The compression ratio has been increased from 9.5 to 12.5 by modifying the piston geometry. A new piston has been designed and produced. Piston rings have also been modified to be more resistant.

The new compression ratio (12.5) was selected to optimize engine efficiency. A higher compression ratio could have been used, but tests led on natural gas engines showed that this intermediate compression ratio is a reasonable compromise between engine efficiency and mechanical constraints.

Moreover, to ensure a good resistance to pressure, the connecting rods have also been redesigned with a target of 110 bars of maximum pressure resistance. This new connecting rod design is close to diesel connecting rods.

Both pistons and connecting rods were heavier than original parts. As the engine is a 3-cylinder, the crankshaft had to be modified to be re-equilibrated.

Engine fuel system

Ethanol contains approximately 35% oxygen, which does not participate to the heat release during combustion. To maintain high and full load performances, the injection system had to be modified in a way to increase the injection flow by 35%. To reach this objective, the injector hole diameter had to be enlarged by more than 16%.

The maximum fuel flow of the initial engine was assumed to be lower than 4 kg/h, which means around 66g/min. With a maximum ethanol flow 35% higher, ethanol injectors must have a static flow of 90 g/min. Bosch LPG injectors were selected, with a static flow around 100 g/min, in line with the needs of this engine. The components of these injectors were tested according to their chemical resistance to ethanol and have shown a reasonably good behavior.

Cylinder head

Some previous tests had shown a rather poor resistance of exhaust valves and spark plug to ethanol. Indeed, due to a lack of heavy fractions, ethanol does not induce a correct cooling of the exhaust parts and thus could lead to valves fusion on very high speed / high load running conditions. Moreover, very high temperatures can lead to fusion of the spark-plug electrodes.

To avoid this phenomenon, the valves were changed to bi-material "nimonic" valves (nickel-chromium alloy with good mechanical properties and oxidation resistance at high temperatures). The spark plugs have also been changed to more temperature resistant ones.

RESULTS

Full load

The full load curve obtained with the gasoline and the ethanol versions of the engine are presented in Fig. **3**.

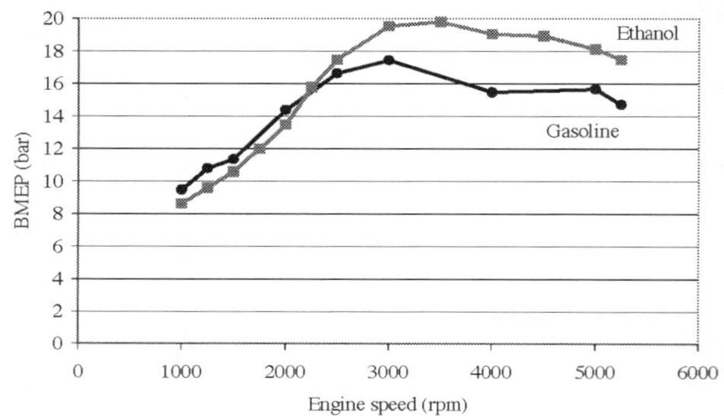

Fig. 3 Full load curve

Ethanol enables a maximum torque of 95Nm @ 3500rpm, while the gasoline engine is limited to 82Nm (+15%). The maximum power is consequently increased from 39kW for gasoline (65kW/l) to 45kW for ethanol (75kW/l). This full load curve is limited by the exhaust temperature (turbine thermal resistance) and the maximum compressor efficiency (waste gate closed).

It has to be outlined that the turbo-charging system was not changed and may not be suitable for ethanol. For instance, the Low End Torque is slightly lower with ethanol than with gasoline, which could be avoided with a more optimized compressor and an adaptation of the turbine. The study of the cylinder maximum pressure shows that an increase potential remains, as the maximum recorded pressure with ethanol is 95bars (the engine was modified to reach 110bars cylinder pressure).

Moreover, the full load curve with ethanol has been obtained in stochiometric running conditions, with an exhaust temperature (before turbine) under 950°C. Fig. 4 presents the equivalence ratio and spark advance for gasoline (full load curve) and for ethanol (same performance). The comparison of these curves shows that the spark advance can be increased by 5 to 15CA and that no mixture enrichment is necessary when using ethanol.

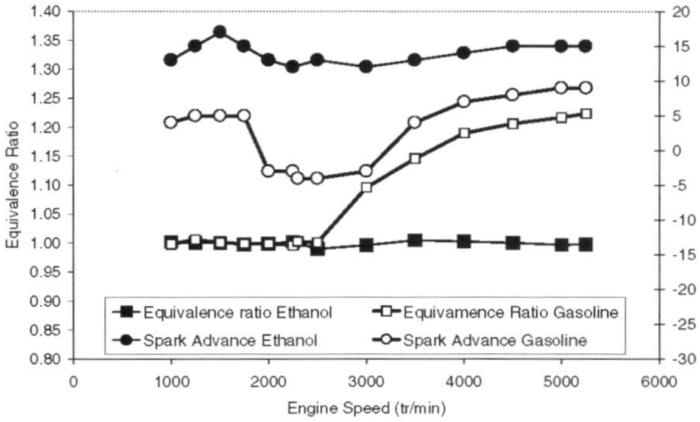

Fig. 4 Full load results: Spark advance and Equivalence ratio (iso-performances)

This figure shows that ethanol, even without taking the "biofuel" aspect into account, can provide substantial benefits for low CO_2 engines (small displacement, high compression ratio, turbocharged).

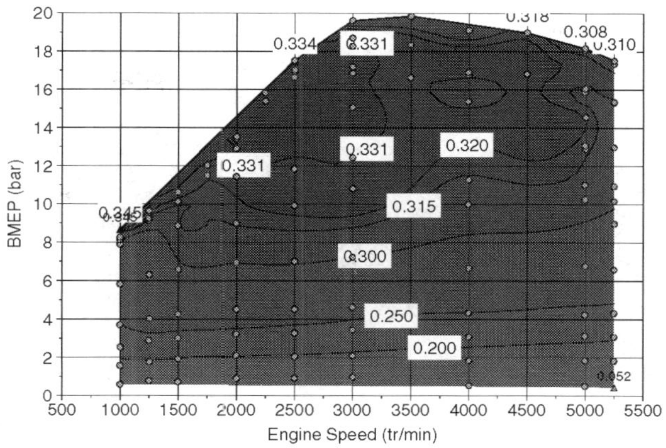

Fig. 5 Engine global efficiency – ethanol

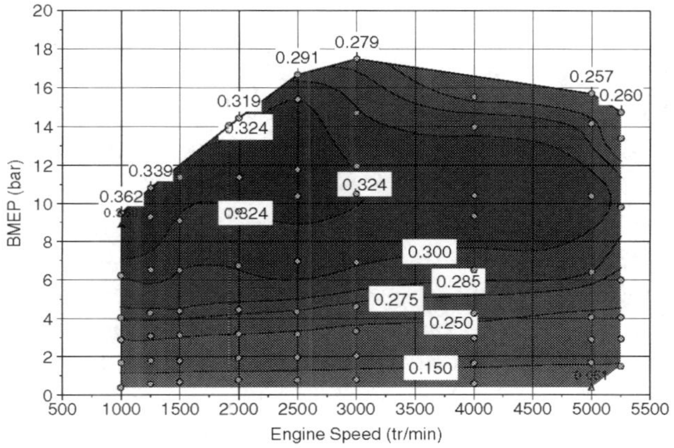

Fig. 6 Engine global efficiency - gasoline

Engine efficiency

The engine global efficiency was calculated according to the ratio released energy / maximum theoretical energy. The maximum theoretical energy was calculated according to the BSFC and the fuel net heating value (42800kJ/kg for gasoline, 26800kJ/kg for ethanol). The results obtained for this calculation are presented below:

At low speed / low load running conditions, the efficiency is higher for gasoline than for ethanol but, at high speed / high load, the impacts of the increased compression rate and of the stochiometric running are important and give the advantage to ethanol, as shown in Fig. 7.

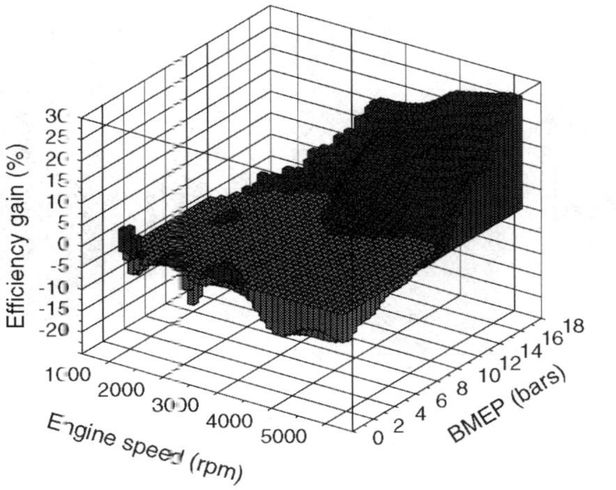

Fig. 7 Engine efficiency gain when running with ethanol

CO$_2$ emissions

In the frame of its commitment with European Union, ACEA has claimed its will to decrease CO$_2$ emissions of the fleet down to 140g/km in 2008. Thus, CO$_2$ emissions became one of the key parameters when designing an engine. Ethanol, as a biofuel, has an important potential in terms of lowering CO$_2$ emissions "from well to tank". Nevertheless, its efficiency in the engine has to be checked in order to get the CO$_2$ emissions over the full path.

The CO$_2$ emissions of gasoline and ethanol engine are presented in the following figures.

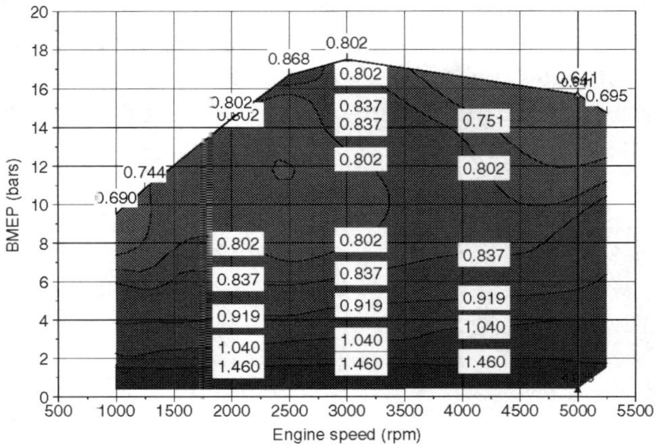

Fig. 8 CO$_2$ emissions (kg/kWh) - gasoline engine

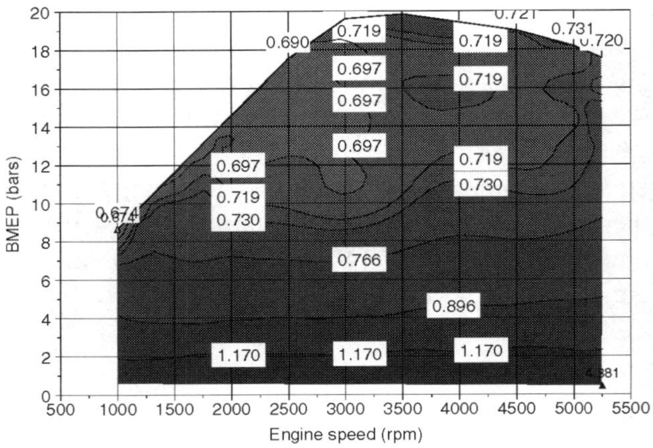

Fig. 9 CO$_2$ emissions (kg/kWh) - ethanol engine

This shows the significant benefits of ethanol on CO$_2$ emissions. An important reduction can be obtained (up to 20%), which can be explained by the high H/C ratio of ethanol (3, against 1.8 for typical gasoline). For instance, for a 2000rpm - 2bars BMEP, CO$_2$ emissions are 1.22kg/kWh for gasoline and 1.13g/kWh for ethanol (-7.4%).

To get an evaluation of the benefits on a dedicated vehicle, some simulations have been led using Amesim® [20] with the values recorded on bench tests. CO$_2$ emissions were calculated on the standard European type-approval driving cycle. The results are summarized Fig. 10.

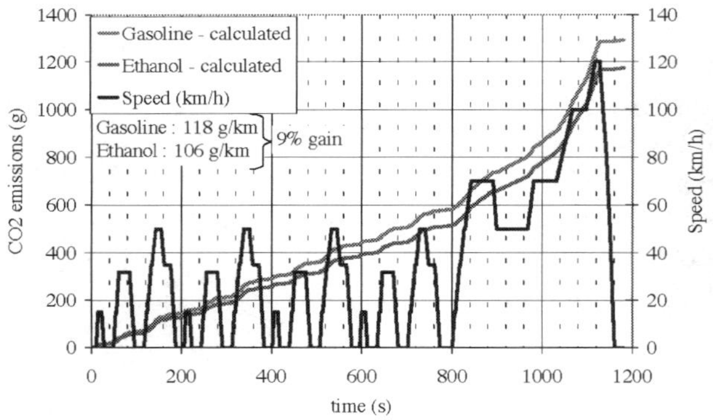

Fig. 10 Calculation of CO$_2$ emissions on the NEDC cycle

This data processing shows good results for gasoline engine in comparison with type-approval value (120g/km). The calculation on ethanol engine data shows a 9% reduction in CO$_2$ emissions.

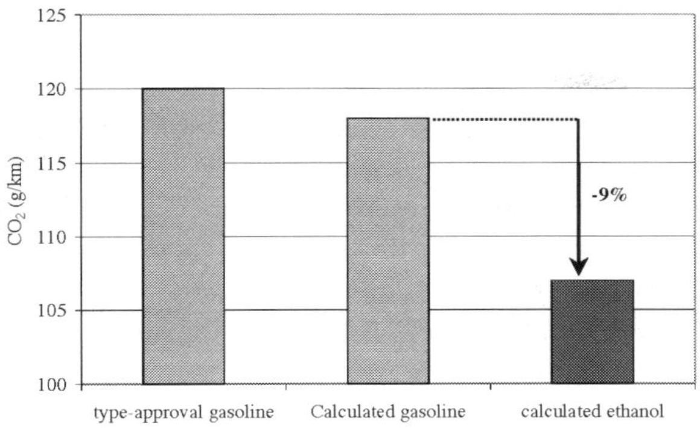

Fig. 11 Summary of global CO_2 emission on the NEDC cycle.

Besides, this assessment has been realized without any change in the vehicle design, and especially with the same gearbox ratios. As demonstrated above, ethanol allows an increase in engine maximum power and torque, enabling further CO_2 reduction via a gearbox and transmission optimization.

Pollutant emissions

HC emissions are difficult to measure accurately with the flame ionization detector typically used for engine pollutant emissions measurement. Indeed, the response factor for oxygenated compounds in these analyzers is low (for instance 1.8 instead of 2 for ethanol, which means that less than 90% of ethanol emissions are analyzed). Assuming that most of the HC emissions are made up of unburned hydrocarbons, the total HC emission level can be corrected by the response factor of ethanol. The value obtained for HC emissions in ethanol configuration is then a maximum estimate (considering that HC emissions are exclusively made up of ethanol).

To minimize this bias and obtain a more accurate HC emission characterization, chromatographic analyses have been achieved, analyzing the chemical species in HC emissions (C_1 - C_8). These analyses have shown that, as far as non-oxygenated hydrocarbons are concerned, ethylene represents more than 50% of HC emissions. This molecule comes directly from ethanol decomposition.

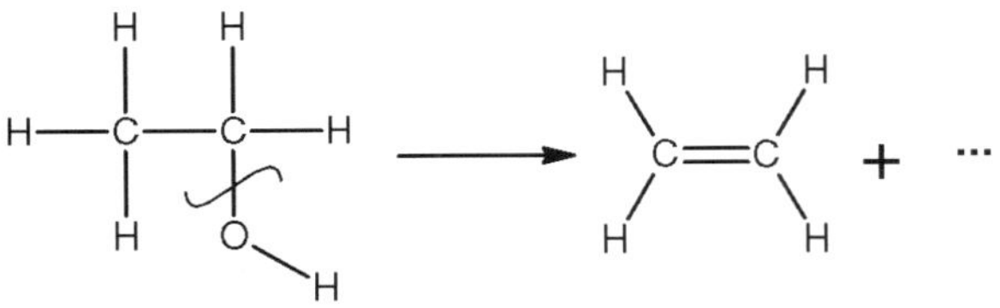

The comparison of the total HC emission measured with ionization detector and chromatography enables an evaluation of ethanol emissions. Indeed, for the 4 tested running points, the HC emissions measured by chromatography can be compared to the total HC emissions and the total amount of unburned ethanol can be calculated, as shown in Fig. 12. This calculation assumes that ethanol (or light aldehydes) is the only component of all HC that are not analyzed here in line by gas chromatography.

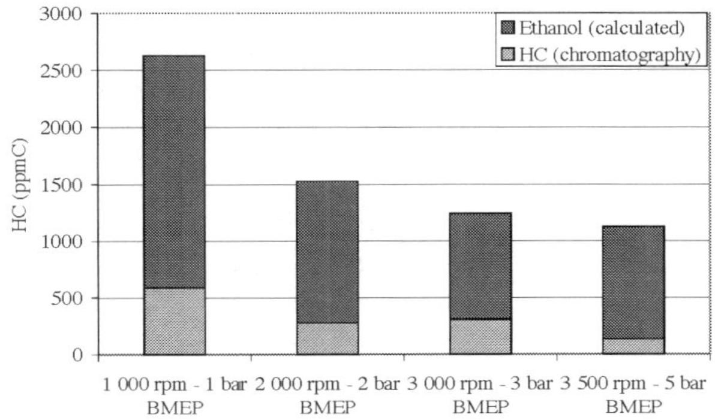

Fig. 12 Ethanol and HC emissions before catalyst

As far as CO emissions are concerned, ethanol brings a real gain. At low engine speeds, CO emissions with ethanol are approximately reduced to the half in comparison with gasoline. At very high load running conditions, ethanol induces also very low CO emissions in comparison with gasoline, because of its low exhaust temperature, enabling to run the full load curve without any mixture enrichment. For instance, at the 5250rpm - WOT running point, CO emissions before catalyst are divided by 6 when running with ethanol, despite the higher torque and power obtained with ethanol.

NOx emissions are less impacted by the use of ethanol, and the results are highly dependant on the running conditions: if for most of the running points, NOx emissions are similar with gasoline or ethanol, some significant differences can be observed for high speed running points.

- At low load, NOx emissions with ethanol are lower than with gasoline, due to differences in the latent heat of vaporization and in the combustion speed.
- At high load, the mixture enrichment in gasoline induces a decrease in the combustion

temperature and consequently in NOx emissions. As mixture enrichment is not needed with ethanol, NOx emissions are higher. On the other hand, the low exhaust temperatures when using ethanol enable a stochiometric running among the whole running range, and makes it possible to use a 3-ways catalyst to reduce NOx emissions, even at full load.

FURTHER OPTIMIZATION

These results show the true benefits brought by ethanol with only a little optimization of the engine: the increased compression ratio induces an important benefit in terms of engine efficiency and CO_2 emissions. Still, some points have to be further studied to get a fully optimized engine.

- The tested engine was a PFI engine. To get full benefits from ethanol high latent heat of vaporization via the "cooling effect" (increase in filling efficiency due to the intake air cooling when ethanol is vaporized), a direct injection engine should be used.
- No change was done on the supercharging system. The optimization of this part of the engine could lead to an increased low-end torque.
- Some critical points, such as cold start management or lubricant compatibility, have to be further studied to obtain a fully useable engine.
- Aldehyde emissions have to be considered. As we have worked on engine-out emissions and as even aldehydes have been measured, it is not realistic to conclude on these pollutants without exhaust gas after-treatment.
- The catalyst adaptation has to be checked. Indeed, the low exhaust temperature found with ethanol can induce catalyst light-off difficulties.

CONCLUSION

In the context of the reduction objective of greenhouse gas emissions, the use of biofuels in internal combustion engines worth being closely evaluated.

First, ethanol characteristics when used pure as fuel in a dedicated engine were reviewed, to determine the balance between its main advantages and disadvantages. Later, after having summarized the main production processes and their relative efficiencies, the main physical properties of ethanol have been analyzed in regard to their positive or negative impact on engine running. Finally, the results obtained on a dedicated engine are presented. The main conclusions are:

- The high diversity of the production paths and feedstock induces highly variable ethanol energy efficiency calculations. According to the raw material (sugar beet, corn, wheat...), to the agricultural yield and to the transformation process efficiency, the energy balance (energy used / energy produced) has been shown to vary from 0.3 to 1.6.
- In a dedicated engine, ethanol-fuel has many advantages, such as its high octane number or its high latent heat of vaporization. These advantages can be valorized in a dedicated engine, provided that some issues are solved (material compatibility, water tolerance, volatility). This ethanol-fuel can bring substantial benefits when used in low CO_2 engines, such as small displacement turbocharged ones.
- A preliminary optimization of a small displacement turbocharged engine has confirmed the potential benefit of this fuel, with an increase in engine efficiency and a decrease in CO_2 emissions. Some calculation using vehicle simulation models (Amesim®) have shown a net reduction of 9% in CO_2 emissions on the NEDC cycle. Moreover, a significant reduction in CO emissions has been noticed.
- Some further optimizations such as cold start strategies optimized gearbox or dedicated turbo charging system should be foreseen to get a fully useable engine.
- Ethanol-fuel can truly induce a CO_2 emission reduction in dedicated engines. Further, as this fuel can be produced from biomass, its "well to wheel" CO_2 balance appears to stand at a very interesting level.

Acknowledgments. The authors would like to thank Ph. Turbin who performed the engines tests and Ph. Kopeckny for the vehicle calculations.

REFERENCES

1. Alternatives to Transportation fuels - An Overview, Energy Information Administration, Office of Coal, Nuclear, Electric and Alternate Fuels, 1994
2. Klasson K.T., Elmore B.B., Vega J.L., Ackerson M.D., Clausen E.C., Gaddy J.L. Biological Production of Liquid and Gaseous Fuels from Synthesis Gas. *Applied Biochemistry and Bioengineering* 1990 ; 24/25 : 857-873
3. Track Records, United States Crop Production, 1866 to Date, USDA / NASS, April 2004
4. Energy and greenhouse gas balance of biofuels for Europe - an update. Concawe report no 2/02, 2002
5. Gover M.P. et al. Alternative road transport fuels – a preliminary life-cycle. 1996
6. Levy R.H. Les biocarburants. Report to the French government based on figures from the Commission Consultative pour la Production des Carburants de Substitution. 1993
7. Richards I.R. Energy balances in the growth of oilseed rape for biodiesel and of wheat for bioethanol. Report for the British Association of Bio Fuels and Oils (BABFO). Ipswich: Levington Agriculture Ltd 2000
8. E.U. Application of biologically derived products as fuels or additives in combustion engines. Publication No. EUR 15647 EN. Directorate-General XII Science, Research and Development. 1994
9. USDA website
10. Bilans Energetiques et gaz à effet de serre des filières de production de biocarburants en France, note de synthèse", ADEME, Ecobilan, DIREM, décembre 2002
11. Pimentel D. The Limits of biomass utilization. In: The Encyclopedia of Physical Science and Technology. NY Academic Press. 2001
12. Searle G.R. Octane Quality and Knock. In: Motor Gasoline. College of Petroleum and Energy Studies, Oxford, Royal Society of Chemistry publication, 1995
13. CRC Handbook of Chemistry and Physics, 74th edition
14. Bechtold R.L. Alternative Fuels Handbook. SAE press, ISBN 0-7680-0052-1, 1997
15. Mueller Associates, Inc. Status of Alcohol Fuels Utilization Technology for Highway Transportation: a 1986 Perspective, Volume 1 - Spark-Ignition Engines, Report N°ORNL/Sub/85-22007/4, National Technical Information Service, Springfield, Va.
16. Alternative Fuel in the Automotive Market. Concawe report 2/95. 1995
17. Hammel-Smith C., Fang J., Powders M., Aabakken J. Issues Associated with the Use of Higher Ethanol Blends (E17-E24), NREL technical report NREL/TP-510-32206, 2002
18. Nersasian A. The Vol. Increase of Fuel Handling Rubbers in Gasoline/Alcohol Blends. SAE Paper No. 800789, June 1980
19. Dunn J.R., Pfisterer H.A. Resistance of NBR-Based Fuel Hose Tube to Fuel-Alcohol Blends. SAE Paper No. 800856, June 1980
20. http://www.amesim.com

Contact. Nicolas Jeuland (nicolas.jeuland@ifp.fr) Fuel, Lubricant and Emission Department IFP

Which Fuels For Low CO_2 Engines?
© P. Duret, X. Montagne (Eds.) and Editions Technip, Paris, 2004, pp. 21 – 25
27 rue Ginoux, 75015 Paris

BIOMASS TO LIQUID (BTL) DIESEL FUELS. SOURCING, STAKES AND ENABLERS

Jean-Claude Griesemann
European BTL Fuels Experts Group

ABSTRACT – *The biomass based fuels, up to recently known as "biofuels" (FAME and Bioethanol), can, potentially, contribute to the supply diversification of energy sources in Europe, as well as reducing the impact of road transport on the climate equilibrium. Although real, this potential contribution remained modest until the recent focus brought onto the concept of "Biomass to Liquid" fuels.*

Promoted by Volkswagen, Daimler Chrysler, Renault and Volvo, under pilot production in Germany and Sweden, seriously considered by the DG TREN as a valuable complementary option, these fuels were investigated by a European Experts Group, assessed in terms of potential availability, energy efficiency, ecologic impact, economic credibility. Some strategic provisions were proposed in an attempt to foster the development of a production infrastructure in Europe.

INTRODUCTION

The "Biomass to Liquids" (BtL) fuels are created by thermal process; gasifying the whole undifferentiated collected biomass into "syngas" CO + H2 (step 1), then, synthesised into liquid (step 2) Fischer Tropsch Diesel fuel, Di Methyl Ether (DME) or Methanol. The feedstock considered encompass forestry and mill residues, straws and energy grasses as well as short rotation coppice and "end of life" construction residues or municipal waste. These BtL fuels are presently emerging from the status of a promising concept, under development in Europe, to that of pilot production for field tests, ready, in some case, for demonstration production. Among car manufacturers, some of these fuels are known as B.T.L. fuels or, in the case of FT-Diesel, "SunDiesel" (name agreed between Daimler Chrysler and Volkswagen) These must not be confused with the usually well-known "biofuels" (biodiesel or bioethanol), for which chemical process address only the usable part of the collected crop and let unused residues,

Once industrially produced and distributed, these B.t.L. fuels will be profitably used, in state-of-the-art or in adapted Diesel engines, either neat or blended in usual Diesel fuel. The global "Well to Wheels " assessments known to-day make them rank among the highest potentials in terms of Greenhouse effect benignity. The three automotive companies, Daimler Chrysler, Renault and Volkswagen share this vision and have agreed to discuss coordinated actions supporting this development. At longer terms, these synthetic fuels could be considered as hydrogen carriers for Fuel Cells vehicles

The technologies used for Gasification (step 1) and Syngas to Liquid (step 2) are, individually, well known. The challenge, now, is to solve the logistical, technical and economical issues raised by the integration of these processes. In Germany, Sweden, France, and UK, R&D activities are in progress. Pilot production is currently underway in Germany, soon in Sweden, in partnership with major European car manufacturers such as Daimler-Chrysler, Volkswagen, Volvo, Renault

1. SYNTHETICS FROM BIOMASS

Pilot production is currently underway in Germany, soon in Sweden, in partnership with major European car manufacturers.

1.1. THE BIOMASS TO FUEL CONVERSION: ENERGY EFFICIENCY AND YIELDS:

The most usually assessed process (as operated in CHOREN GmbH plant, Germany), is said "autothermal", using almost exclusively the biomass energy to power the process. Other processes, using externally sourced energy, or chemicals, added to the biomass as inputs for increased fuel yields, are still under investigation. European organisms, academics and industries have independently assessed the global energy balance and yields of their processes. Between Imperial College (UK), ATRAX Energi (S), CEA/IFP (F), CHOREN Industries (G) and Chemrec (S), the conclusions found a reasonable convergence in order of magnitude.

Process	Energy efficiency(GJ / GJ)	Conver. Yield (t.o.e. / ton biomass)
Autothermal	50 (+/-14) %	0,20 (+/- 0,07)
Allothermal	55 (+/-10) %	0,45 (+/-0,10)

Following the presently operated "autothermal" process. 5 tons of wood pellets (15% moisture) can produce 1 ton of Diesel equivalent BtL synthetic fuel, or SunDiesel. Such conversion process have been assessed for plants able to deliver 100 000 tons fuel / year from 1500 tons biomass per day, which is of usual logistic practice in paper mills.

1.2. FEEDSTOCK: AVAILABILITY AND YIELDS, FUEL POTENTIAL

The feedstock considered here includes any renewable carbonaceous stuff, subject it is nor necessary to human industrial activity or food, such as:

- Agricultural, crop harvest residues and manure
- Forest felling, sawmills, paper mills residues
- Municipal waste
- Energy plantations: Forests, short rotation coppice and herbaceous

Several Research organisms have, independently, undertaken to assess tentatively such figures at the European (EU 15) level: LBST (D), Kaltschmidt and IfE (D), more recently: STFI (S) and Imperial College (UK). While these figures converge in conversion energy efficiency, less homogeneous limits and range bring, for the evaluated resource, more random figures:

	Biomass yield (Mio ton / y)	Fuel Pot. (Mtoe /y)	Substituability (%)
Minimum	**70**	**13**	**4,50%**
Maximum	**170**	**32**	**10%**
Extremum	280	52	16%

Conservative evaluations predict a minimal potential of substuability of 5 % of the transport fuels in Europe, with perspectives at 10, up to a volunteer hypothesis at 15 %

1.3. EMPLOYMENT

The production and distribution of B.t.L. fuels will, not only generate local employment, but, moreover, create quality jobs where there is a great need for them, in rural areas and contribute to the protection of rural landscape against summer fires by rational exploitation and watch.

2 domains will generate jobs: Feedstock farming, felling, trading and delivery, residues collecting, conditioning, delivery and conversion to fuel. Estimations show 60 % of these jobs addressing feedstock, 40 % dedicated to the conversion, for a cumulated employment reaching 42 000 up to 90 000 jobs for 30 Mtoe fuel / year (10 % fuel substitution in EU 15 in 2010)

1.4. FINANCIAL ISSUES

Creating conversion plants sized at a production rate of 100 000 toe B.t.L. fuels per year, swapping 10 % of the transport fuels in Europe, would imply the investment of 300 plants, representing an investment of 250 to 300 Meuros each.

Depending on the B.t.L. fuel (SunDiesel, bio-DME or Methanol) and on the process, tentative costs have been calculated for these fuels, in European currency per ton of Diesel equivalent. These figures are ranging from 330 up to 800 Euros per toe, which may be seen as high, compared to crude based Diesel at 247, but fall in an order of magnitude comparable to the usually known "biofuels", while they present indisputably higher advantage in terms of Greenhouse Effect.

1.5. ATMOSPHERIC EFFECTS

The highest stake of B.t.L. fuels relies on their potential outstandingly reduced contribution to the G.H.G. emissions from road transport. This contribution can only be validly evaluated if assessed on a global “Well to Wheels” basis, including the fuel production process with the vehicle effect. On such a basis, if Natural Gas ends in a draw with Diesel, if FAME brings a reduction of 53%; B.t.L. fuels have the potential to reduce these global emissions by at least 85 %. Depending on the process assessment rules for by products, this no-contribution could reach nearby 100 %

One additional strength of this pathway is that, while “state of the to-day art” Diesel vehicles are already able to welcome some of these fuels for immediate effect, technology breakthroughs (Homogeneous combustion, Fuel cell reformers) should profitably foster this potential.

2. IMPLEMENTATION STRATEGY AND RECOMMENDATIONS

To foster a strategy towards a significant contribution of the BtL fuels to the Transport issue, some recommendations can be proposed:

1) Re-consider the Commission Strategy for Alternative fuels 2010 – 2020 as:

- Reminding the impact of the Hydrogen High Level Group on national policy makers and industry managers, create an equivalent HLG for the development of Biomass to Liquid fuels.
- Enforcing the EU target of > 20% alternative fuels in 2020, re-consider a sharing modified such as

Year	GTL, BTL, Biofuels, %	Natural gas, %	Hydrogen, %	Total, %
2020	10 - 15**	< 5	2 - 3	17 - 23

15%** among which more than 12 are BtL / Biofuels

- Having clearly in mind that measuring the G.H.G. at the vehicles tailpipe only is quite meaningless to reflect the actual benignity of the renewable fuels:

  a. consider the “Greenhouse” impact of fuels and associated powertrain, no longer on the vehicles exhaust gas basis, but on a comprehensive, commonly agreed, “Well to Wheels” assessment.

  b. Give rise to proposals targeting at the elaboration of a clear, commonly accepted technical procedure to evaluate this actual, global “Greenhouse” impact

  c. On the model of the biofuels, determine, for the BtL fuels, a fiscal frame coherent with their local emissions benefit and outstanding climate benignity potential

2) Enabling database:

Conduct, within and by the Member States and Candidate Countries a geo-economic inquiry dedicated to the local biomass resources and publish the conclusions among industry and financial decision-makers:

- Density, annual availability and nature
- Geographical locations, logistics possibilities and costs
- Employment impact

3) Enabling Regulatory corpus:

- Obtain the ECOFIN assignment on COM/2001/0547 keeping the 100 % tax exemption on biofuels, as an amendment of Directive 92/81/EWG and entitle Member States to support them through complementary taxes reduction
- Define the conditions for long term security of the legislative conditions ruling the long term investments in BtL production plants and in short rotation forestry.
- Support the new C.A.P. environmental focus, fostering rotation, non-food crops on set-aside lands
- Extend at least 10 years beyond 2006 the Council Reg 2641/1999 and amend it so as to agricultural land converted to forestry keep their “agricultural” status at long term.

4) In the domain of R&D:

Open a support frame for an ambitious “Lighthouse Project” targeting at the assessment / investigation of Biomass to renewable hydrogen, addressing the production of BtL fuels as a preparatory step:

- Agricultural modes and innovative species enhancing the feedstock production
- Residues collection, conditioning and logistics
- Conversion processes optimisation in energy intensity, specifically for conversion units of

small size (100 Ktoe fuel / year), with demonstration by, at least, 3 plants in Europe, supported by high level funding
- Conditions of profitability for the global production processes, for the same units size
- Extensive check of harmlessness for present vehicles
- Development of the "ad hoc" powertrain technologies including the HCCI combustion modes and the fuel cell reformers for the on-board production of fossil free hydrogen

5) Implementation strategy: Preliminary introductive steps, in the vision of automotive industries[1] should be:

- Intensify efforts to provide cleaner and more efficient conventional fuels supporting reduction of fuel consumption and regulated emissions.
- Utilize GTL technology to extend the energy supply and blend in increasing amounts of GTL to improve the fuel quality
- Increase progressively the phasing-in of BtL fuels as a blending component

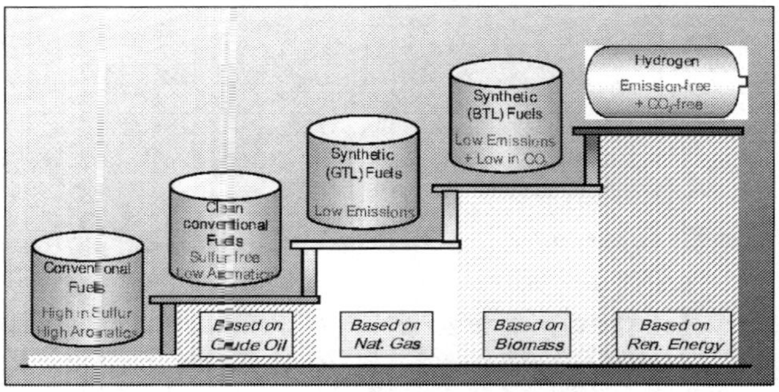

CONCLUSION

In the domain of transport fuels, if we consider: Greenhouse gases reduction, domestic sourcing, easy and fast applicability as "Benefits" and technology difficulties to be overcome, creation of infrastructures, safety regulations to be developed, delay as "Cost", the "Biomass to Liquids" (BtL) fuels have the highest potential in "Benefit to Cost" ratio.

The "primary (renewable) energy to fuel" efficiency is coherently evaluated at 50% +/- 10% and the fuel yield around 0.2 toe of fuel per ton of "dry" (15% moisture) feedstock. Subject a conversion plant is operated, on farmed wood, the yield should be about 2 toe fuel / ha.year. Promising (but not fully assessed) perspectives on the collection of unused residues, on the farming yields, on the process of conversion are under investigation.

If confirmed, these perspectives would bring the fuel potential in Europe (EU 15) from 13 Mtoe / year (conservative issues, based on minimal farmed wood footprint) up to 50 Mtoe / year, when collectable residues are accounted for (just ecologically available wheat straw in EU 15 amounts at 30 Mt/y, potentially convertible into 6 Mtoe fuel or 2% substitution)

No technology breakthrough are needed to take immediate full profit from those fuel as soon as available as, if DME needs a slightly modified, dedicated Diesel engine, F-T Diesel can power all the Diesel vehicles presently on the road, as well blended in usual fuel as well as neat. The greenhouse mitigation will be dramatically effective as each toe of fuel used will reduce by 85 to 90 % the global emission of GHG compared to usual crude based fuel.

The technologies to be implemented for the production are individually well known:

- Wood farming and / or residues collecting
- Integrated gasification of carbonaceous feedstock (IGCC) into syngas
- F-T (DME) synthesis from syngas
- Distribution of liquid fuels…

For a desirable widespread venue of these BtL fuels, the concerns which have to find a positive answer are in the domains of economy, logistics, industrial integration and profitability, communication.

The creation of the necessary production infrastructure rely (and will rely) on private initiatives. The obstacles to such private initiatives can be listed as:

- Very few decision-makers know that such production is possible and desirable.

[1] "Proposal for a European Energy strategy for Road Traffic", Pr. Dr. H.K. Kohler, VDA Conference (April 2003)

- Even less of them know where the feedstock could be made available
- The feedstock logistics remain often an issue.
- A potential remains for the improvement of agricultural yield
- The global conversion process is not yet fully optimized
- Stable and reliable political framework is still missing

The profitable economic equation remain to be consensually established

Acknowledgments. This paper endeavors to summarize the result of the cooperative work of several experts who volunteered to contribute as members of the "European BtL fuels experts Group". Among others, P.J. Danielsson and A. Röj (Volvo Bus Corp), H. Heinrich (Volkswagen), R. Maly (Daimler Chrysler), A.A.Groves and H.P. Calis (SHELL Global Solutions), Prof. L. Sjunnesson (Sydkraft AB), K. Stähl (Ducente AB), Prof. Dr. D. Murach (Tech. U. Eberswalde), J. M. Seiler (CEA Grenoble), S. His (IFP), A. Bauen and J. Woods (Imperial College), M. Rudloff (Choren GmbH). To all of them, the author of this paper is deeply indebted.

REFERENCES:

1. "Proposal for a European Energy strategy for Road Traffic", Pr. Dr. H.K. Kohler, VDA Conference (April 2003)
2. R. Edwards J.R.C. Ispra, contribution to the European Well to Wheels study 2003
3. "Feasability & market potential of Black Liquor" gasification with production of renewable motor fuel" T. Ekbom, N. Berglin, M. Lindblom at ISAF IV Conference , November 2002
4. "Bioelectricity vision: Achieving 15% of electricity from biomass in OEDC by 2020": A. Bauen, J.Woods, R.Hailes (March 2003) . (1GW installed bioelectricity corresponds to 2 million tons wood per year)
5. DME from Biomass, Atrax Energi AB for IEA, Feb. 1999
6. CHOREN Industries GmbH, under private communication to the European B.t.L. fuels Advisory Group

ENGINE AND FUEL-RELATED ISSUES IN HCCI & CAI COMBUSTION

Which Fuels For Low CO_2 Engines?
© P. Duret, X. Montagne (Eds.) and Editions Technip, Paris, 2004, pp. 29 – 44
27 rue Ginoux, 75015 Paris

THE KEY POINTS OF HCCI COMBUSTION CONTROLS

Norimasa Iida, Yudai Yamasaki, Susumu Sato

Keio University, Faculty of Science & Technology, Department of System Design Engineering

ABSTRACT – *In order to achieve a high-efficient and low emission HCCI combustion engine, They are the important points that maintenance of proper auto ignition timing of the mixture, the appropriate combustion reaction speed, the evasion of knocking and the reservation of combustion completion. The authors have carried out experimental studies on auto ignition and combustion processes in both 2 stroke ATAC and/or AR engines and a 4 stroke HCCI engine by spectroscopic observation and numerical calculation with elementary reactions, focused on the difference of molecular structure of fuels, i.e. Methane, n-Butane, DME, and so on. In this report, the key points of HCCI combustion will be discussed based on these studies.*

INTRODUCTION

Homogeneous Charge Compression Ignition; HCCI engine is an auto-ignition engine which air and fuel mixture are supplied in a combustion chamber, and makes it auto-ignition by adiabatic compression with a piston. Since low flame temperature combustion is realized under a super-lean mixture which exceeds the flammability limits, therefore, HCCI engine is observed as a next generation type internal combustion engine which can realize low emissions of NOx and PM and high thermal efficiency. [1, 2]

On the other hand, many subjects exist in introduction of HCCI combustion. In HCCI combustion, the physical ignition means of the sparks ignition and fuel injection, which are used in the conventional engine, is inapplicable. "Control of ignition timing" is mentioned as a subject. It depends for ignition of fuel-air mixture on oxidation reaction of fuel molecules. Since auto-ignition is carried out when the temperature of fuel-air mixture exceeds ignition temperature, the control is comparatively easy. Moreover, if the amount of fuel injections is increased in order to raise IMEP of HCCI engine, a combustion reaction will advance rapidly and will cause a knocking. Therefore, "Control of combustion reaction speed" has been a bigger subject than that of "control of ignition timing". Securing suitable "combustion reaction period" in the operation conditions of high load especially is required. Which combustion reaction speed may be conversely slow and the reaction after ignition may stop since the combustion temperature of HCCI combustion is low. This becomes a high CO and HC emissions. Therefore, "combustion completeness" (combustion efficiency) reservation serves as important requirements.

At this paper, reference is made about the combustion control technique of HCCI engine by showing the solution over these subjects, after clarifying positioning of HCCI combustion.

WHAT IS HCCI ENGINE

HISTORY OF HCCI ENGINE

The history of HCCI engine stems from having carried out the research report of the combustion phenomenon which the Shigeru Onishi of a Japanese clean engine research institute calls "activity heat atmosphere combustion" late in the 1970s. Professor Osamu Hirao of the University of Tokyo named this combustion ATAC (Active Thermo Atmosphere Combustion).

In Europe, it is called as "Controlled auto-ignition combustion". In the U.S., it will be PNGV (partnership for next generation vehicle) in 1997. The budget 3 million U.S. dollar of a program stuck, and systematic research was advanced.

There is a name of PREDIC, PCCI combustion, ATAC, AR combustion, etc. other than HCCI combustion.

CONVENTIONAL INTERNAL COMBUSTION ENGINE AND HCCI COMBUSTION

The place of HCCI engine can classify with the reciprocal internal combustion engine, which combined the combustion system by compression auto-ignition seen in diesel engine, and by the fuel supply system by fuel-air mixture in S.I. gasoline engine.

In a report, the S. Ashley, journalist of Scientific American, classified into four sorts, which show a reciprocated type internal combustion engine's combustion system in Table 1, and defines positioning of HCCI engine [3].

Engine is conventionally classified into four sorts according to the matrix which consists of two kinds of fuel and air supply methods, "homogeneous" and "heterogeneous", and of two kinds of ignition methods, "sparks ignition" and "compression ignition."

However, when it considers from a combustion study-viewpoint, HCCI combustion has the field that is not necessarily expressed with this diagram.

In a gasoline engine, at first, fuel-air mixture supplied in the cylinder is compressed by the piston, and compressed by heat release of flame propagating

Table 1. Classification of reciprocated engines by S. Ashley in Scientific American

Formation of Mixture / Ignition Method	Homogeneous	Heterogeneous
Spark Ignition	Gasoline Engine	Direct Injection Gasoline Engine
Compression Ignition	HCCI Engine	Diesel Engine

combustion. Auto-ignition (i.e. HCCI combustion) occur in the end gas which is a local fuel-air mixture located far away from the spark ignition point, before arrival of a flame front. The case, the strong auto-ignition combustion, is a knocking. Since gasoline engine is operated on a knock limit conditions at high load, combustion system of gasoline engine can say to have a slight HCCI combustion aspect inherently.

On the other hand, in a diesel engine, liquid fuel is injected in the air atmosphere of the high temperature and high pressure after compression and should pass the process of atomization, evaporation, and diffusion, fuel mixes with air, and forms fuel-air mixture, and is ignited. And the fuel injected continuously forms a diffusion flame after that. In a

Table 2. Classification of the engine combustion configurations

Formation of Fuel/Air Mixture		Homogeneous Mixture ⟷ Stratified Mixture / Heterogeneous Mixture ⟷ Diffusive Mixing
Engine Name		S.I. Engine / Diesel Engine
Fuel Injection	Position	In Manifold / In Cylinder
	Form	Single / Multi / Pilot + Main / Single Injection
	Timing	Intake Stroke / Compression Stroke / Around TDC / After TDC
Combustion Form		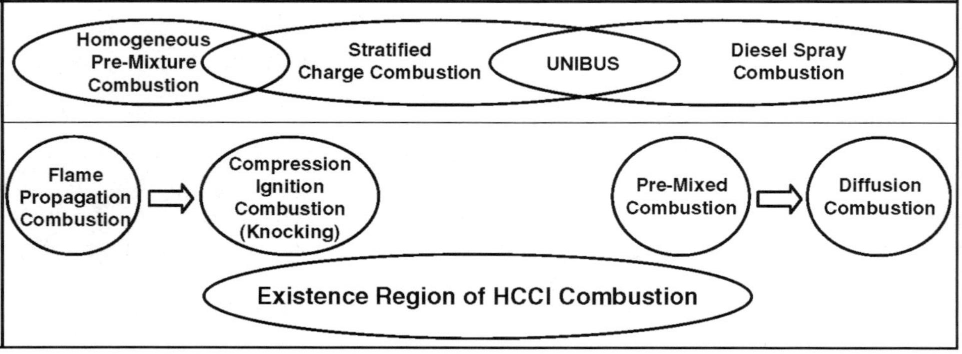

strict meaning, although premixed flame and diffusion flame in the diesel spray combustion are not clearly distinguished. The past engineer called this two-step of the combustion process as initial combustion, and main combustion process, or the premixed combustion and a diffusion combustion process. This premixed combustion is a different combustion form of flame propagation combustion. It can be called the same combustion form as HCCI combustion at the point in which fuel-air mixture carries out auto-ignition combustion.

Therefore, in a gasoline engine HCCI combustion is inherent in the final stage of the combustion, in a diesel engine it is inherent in early stages of the combustion. Thus, if it thinks, it is possible to catch other combustion continuously.

When the temperature of fuel-air mixture is reached at auto-ignition temperature (it is the peculiar value of fuel), HCCI combustion is started. In the case of a gasoline engine, it is designed so that flame propagation combustion may be completed before fuel-air mixture reaches its auto-ignition temperature. On the other hand, with a diesel engine, it is designed so that the temperature of the air exceeds the auto-ignition temperature of fuel, and combustion may be started with injection of fuel.

In the present diesel engine, if fuel supply is carried out into compression stroke, the fuel follows the process of HCCI combustion, since the fuel will be supplied before reaching auto-ignition temperature[4]. In the UNIBUS [5] diesel combustion system proposed by TOYOTA, fuel was supplied to early timing by pilot injection, it is regarded as (diesel) combustion in that the amount of premixed combustion was enlarged.

In this way, HCCI combustion realizes when fuel is supplied at the condition that the air temperature is below auto-ignition temperature, such as intake stroke and compression stroke etc. It is a combustion system that allows wide range of fuel injection timing.

The HCCI combustion existence range is shown in Table 2 classifying from the viewpoint of "mixture formation", "the position, the form, and timing of fuel injection", and "the form of combustion", of the conventional internal combustion engine.

FEATURES AND SUBJECTS OF HCCI COMBUSTION ENGINE

What are the features and the subjects of the engine that carries out HCCI combustion of all the fuel to supply?

HCCI combustion engine has the following features.

a) HCCI combustion is enabled for the ultra lean mixture beyond the flammability limits. Because the flame temperature is low, hardly produce NOx in combustion process.

b) Since fuel and air are mixed completely and the shortage of oxygen does not occur, soot does not produce at the combustion process.

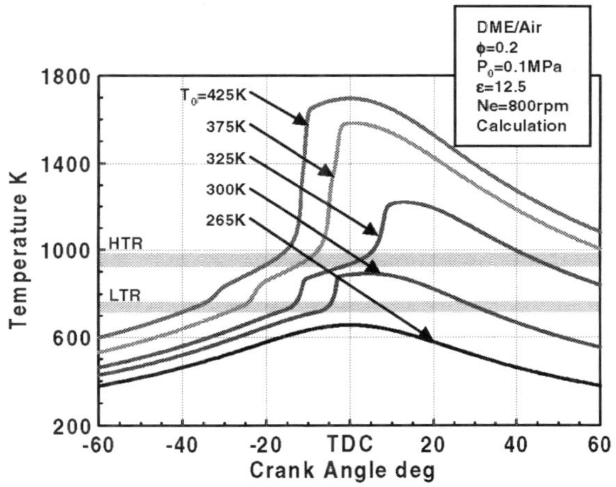

Fig. 1 Temperature histories for various initial temperatures (fuel: DME, calculation)

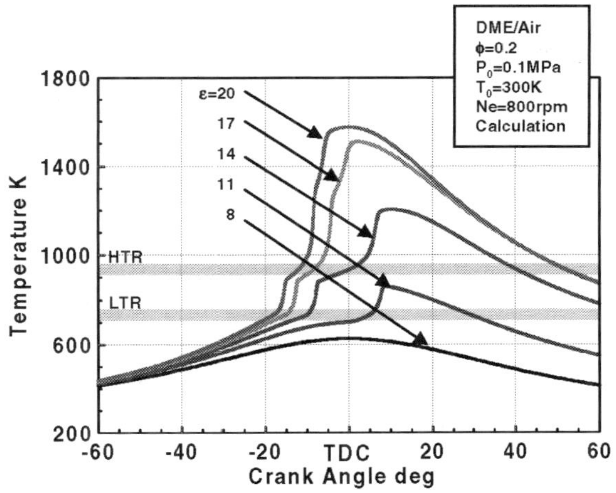

Fig. 2 Temperature histories for various compression ratios (fuel: DME, calculation)

c) Since it can operate with about the same high compression ratio as a diesel engine, thermal efficiency is high.

d) Since it can supply fuel-air mixture, without throttling the air intake, there is "no pumping loss".

As described at "INTRODUCTION", HCCI engine has the following subjects; obtaining not only control of ignition timing, but suitable combustion reaction speed, avoiding a knock, and securing-combustion completeness, are required. Here is a subsection (second level heading).

SOLUTION OVER SUBJECTS OF HCCI COMBUSTION

CONTROL OF IGNITION TIMING

Because auto-ignition is strongly depending on oxidation reaction of fuel-air mixture, temperature management of fuel-air mixture is the major control factor for auto-ignition timing.

By introducing the numerical calculation with detailed elementary reactions model, dependability of composition of the mixture that settled at the kind of fuel, equivalence ratio, rate of EGR, as well as the compression speed and ratio, on auto-ignition processes can be grasped. This will be enabled much more exact control of auto-ignition timing. Ishibashi et al reported that a 2-stoke AR engine attained operation of IMEP 0.3MPa in the large range of engine speed from idling to 12000rpm in which HCCI combustion is appropriately controlled [6].

The fuel-air mixture introduced in the combustion chamber is adiabatic compressed by a piston. When a local temperature exceeds at auto-ignition temperature, it results in auto-ignition.

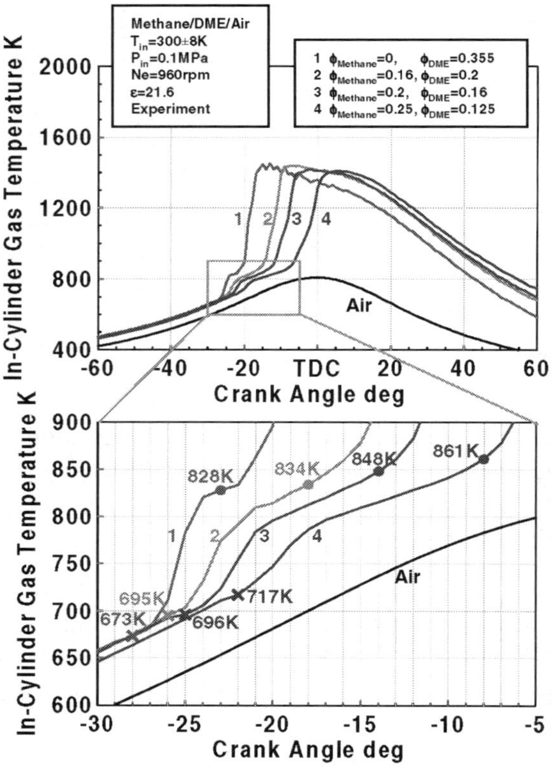

Fig. 3 Temperature histories and ignition points for various equivalence ratios of Methane and DME (experiment)

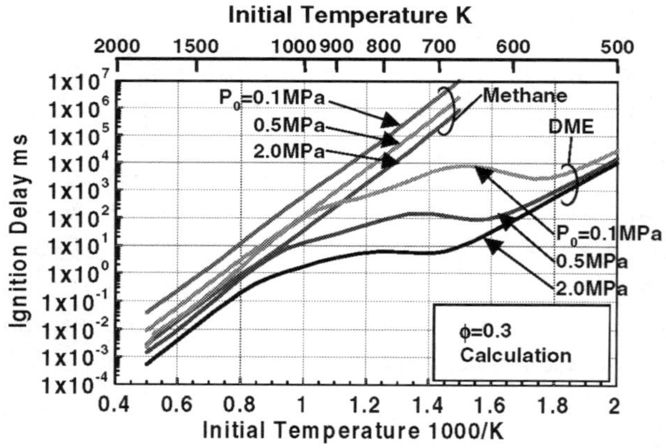

Fig. 4 Ignition delay comparison of Methane and DME (calculation)

Table 3. The oxidation process of paraffin hydrocarbon

Low Temperature Reaction	→	High Temperature Reaction	
(Cool Flame) 750K ~ 800K	850K (Blue Flame) 1050K	1100K (Hot Flame) 1500K	
1st O_2 Addition 2nd O_2 Addition	Degeneration of H_2O_2 $H_2O_2 \rightarrow 2OH$	β - Scission	Oxidation of CO
Accumulation of H_2O_2, HCHO, Ketone	Oxidation of Intermediate Products (Formaldehyde, Ketone etc.)	Generation of CO	Conclusion of Reaction
	In case of Methane, Aromatics, Isomer ↑	Taking Part in Reaction from Here	

As the direct method that controls ignition timing, the method of changing intake temperature, and the method of changing a compression ratio are mentioned. Figure 1: Temperature histories with change of intake temperature (numerical calculation with elementary reactions). Figure 2: Temperature histories with compression ratio change (numerical calculation with elementary reactions).

Ignition timing will shift early-stage the case where intake temperature as well as compression ratio is raised. However, making it a high compression ratio causes a knock.

The research of HCCI engine which carried the variable mechanisms of a compression ratio, and of the valve timing which adjusts the amount of internal EGR and enables it to adjust the temperature of fuel-air mixture arbitrarily, is also done. A difficulty of that structure is complicated and cost is high.

Moreover, the method of changing the auto-ignition temperature of the fuel is mentioned as another method for controlling ignition timing in HCCI combustion [7]. Auto-ignition temperature is changed by changing a blending ratio of two types of fuels, which one is the fuel that passes through low-temperature oxidation reaction process, and another is the fuel that does not pass through. Figure 3: Temperature histories at the time of using Methane and DME as fuel (experiment). Figure 4: Arrhenius plots of the auto ignition delay characteristics of Methane and DME. It is possible to control the ignition timing, by increasing the content (partial equivalence ratio) of DME, and decreasing auto-ignition temperature of blended fuel of Methane and DME.

In case of the paraffin hydrocarbon whose number of carbon is more than four, the temperature shown in Table 3 serves as the key point of HCCI combustion.

In the combustion of the alkane whose number of carbon is less than three (Methane, Ethane, Propane), the alkane with tertiary bonding (iso-Butane, iso-Octane), aromatics (α-methylnaphthalene), and cyclic hydrocarbon, low temperature does not exit and the auto-ignitions of these fuels occur in the range of 1000K to 1050K, in which β-scission begins and high temperature reaction starts. In the

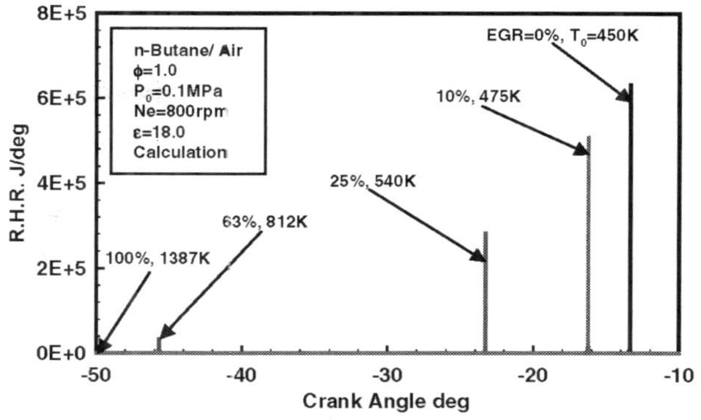

Fig. 5 R.H.R. profiles for various HOT EGR ratios (fuel: n-Butane, calculation) [8]

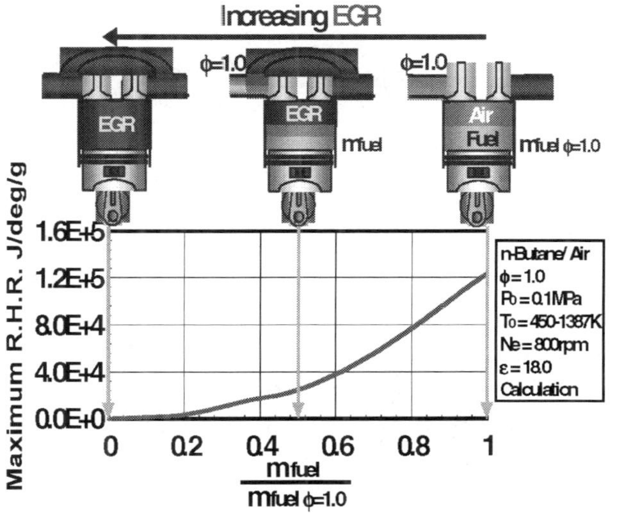

Fig. 6 Effect of dilution by HOT EGR on maximum R.H.R. (fuel: n-Butane, calculation) [8]

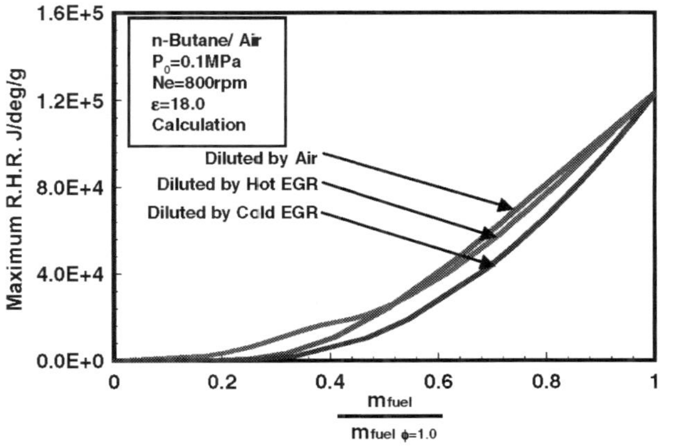

Fig. 7 Effect of dilution by HOT EGR, COLD EGR and air on maximum R.H.R. (fuel: n-Butane, calculation) [8]

case of such fuel since the auto-ignition temperature required of ignition is high, it is already over 1050K at the time of ignition, and after ignition bulk temperature exceeds 1500K in many cases easily. In other words, combustion completeness is easy to be secured.

CONTROL OF COMBUSTION REACTION SPEED

All the supplied fuel-air mixture reacts almost simultaneously by every place of a chamber, so it make a knocking with high heat release rate. Strength of the high heat release rate is almost proportional to the total amount of fuel mass supplied in the combustion chamber.

The dilution effect by EGR is mentioned as a means of control of combustion reaction speed [8]. If the rate of EGR increases, combustion reaction speed will fall as shown in Figure 5. However, if it is also observed that the same effect dilutes fuel-air mixture with air, namely, equivalence ratio decreases, reaction speed will fall (Figure 6, Figure 7). Figure 7 shows that the reduction rates of combustion speed by diluting the mixture by EGR gas and by air have almost the same values.

It turns out that the fall effect of combustion reaction speed is the same in order by the case where fuel-air mixture is diluted with burnt gas by EGR, and the case where it dilutes with air.

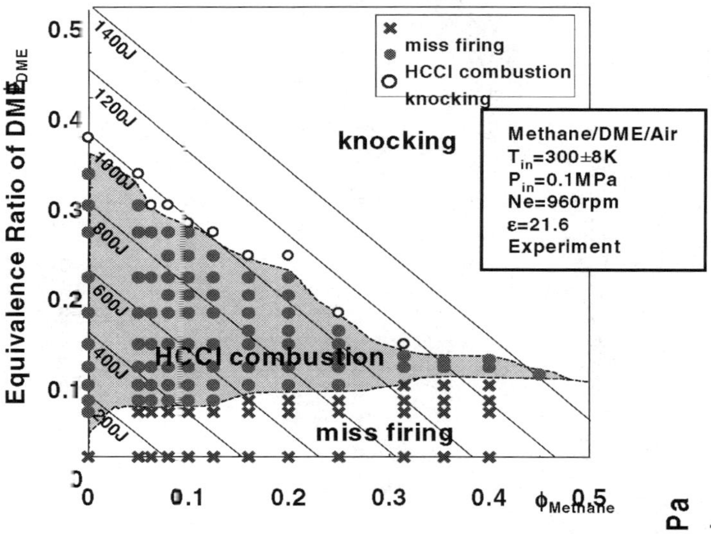

Fig. 8 Map of operation region (fuel: Methane, DME, experiment)

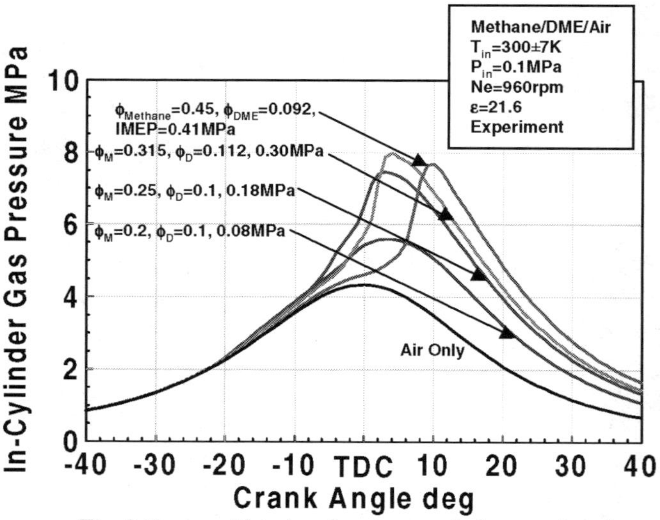

Fig. 9 Pressure histories of various equivalence ratios of Methane and DME (experiment)

The amount of fuel (absolute quantity of the fuel thrown into 1 cycle to a combustion chamber) decided the combustion reaction speed. It means that if the quantity of heat value of the fuel is increased, combustion reaction speed (maximum of the rate of heat release) increasing exponentially.

The prediction above is obtained from the numerical calculation that was analyzed by the so-called 0-dimensional model assumed to be homogeneous mixture. Generally speaking, such analysis results as which the steep combustion with higher heat release is estimated ompared with HCCI combustion of real engine. However, the EGR dependability of combustion reaction speed that shows the same tendency, although absolute values differ from the result of real engine. The value 1E+5 J/deg guessed by the 0-dimensional model is equivalent to the knock limit in real engine with our experience. In the chapter, "effect of heterogeneity of fuel/air mixture" as an effective method avoiding the knocking in HCCI engine will be discussed.

HIGH OUTPUT (HIGH IMEP) RESERVATION

Evasion of a knock is indispensable technology in order to obtain high load in HCCI combustion engine. The two next are thought to be effective as the technique of expanding a knock limit to a high load region :

- Use a heterogeneity effect.
- HCCI combustion in expansion stroke

Since the chapter of "effect of heterogeneity of fuel/air mixture" describes the former, the latter is described here.

In order to secure high IMEP, avoiding a knock, selection of auto-ignition timing becomes an important factor. When the auto-ignition occurs before the TDC, HCCI combustion is performed in compression stroke, there is a tendency that the combustion reaction after auto-ignition accelerates and a knock tends to occur.

When the auto-ignition occurs at the TDC and/or after the TDC, HCCI combustion is performed in expansion stroke, it is expected that the combustion reaction speed is reduced with temperature reduction effects. A knock tends to avoid. As a result, knocking limit will be expanded forward to high IMEP side. A high controlling technology is required to keep the auto-ignition timing at the beginning of expansion stroke after the TDC exactly.

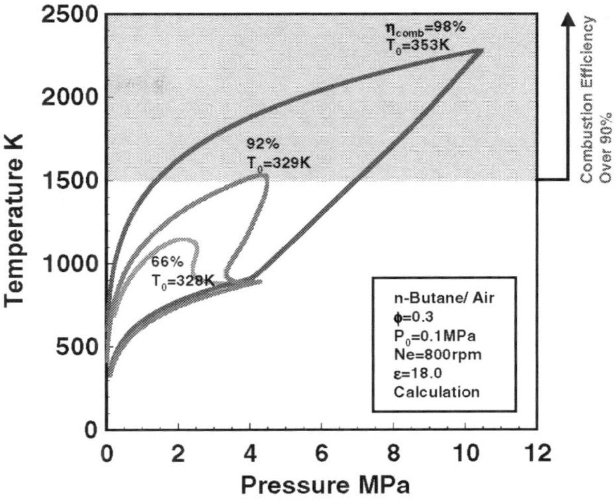

Fig. 10 T-P diagrams for various initial temperatures (fuel: n-Butane, calculation) [8]

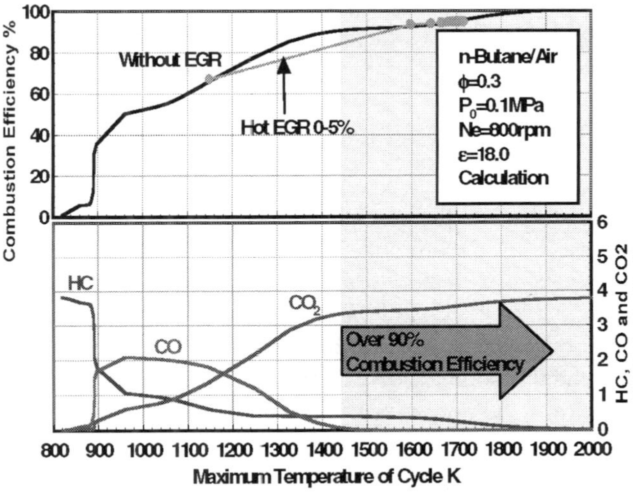

Fig. 11 Relation between maximum temperature, combustion efficiency, and emissions (fuel: n-Butane, calculation) [8]

Figure 8 shows knocking area map of a single cylinder HCCI engine operated with the mixed fuel of Methane and DME, with vertical axis for the equivalence ratio of DME and horizontal axis for equivalence ratio of Methane. Figure 8 shows that there is a domain that the knock limit has expanded like a peninsula where auto-ignition timing is after the TDC. In-cylinder gas pressure diagrams in this domain are shown in Figure 9. The maximum IMEP value at knocking limit in case of that ignition timing is after the TDC is high than that in case that ignition is before the TDC.

By the combination of these methods, a HCCI combustion engine can be operated to the range of 1.0MPa in IMEP.

RESERVATION OF COMBUSTION COMPLETENESS

If fuel-air mixture reaches to the ignition temperature, auto-ignition will take place. But having auto-ignition in HCCI combustion does not guarantee that fuel-air mixture carries out perfect combustion. In other words, combustion efficiency does not go up to 100%.

According to the result of numerical calculation, when the maximum temperature in a cycle can reach around 1500K by managing the conditions of the intake temperature and equivalence ratio of mixture, as well as a compression ratio, high combustion efficiency is given [8]. Figure 10: T-P diagrams in case of n-Butane as the fuel (calculation).

Reduction of the amount of HC discharge, the increase in the amount of CO discharge, and the increase in CO_2 emissions are seen with the rise of combustion efficiency. The CO emission will be set to about 0 if the maximum temperature in a cycle is exceeded about 1500K. Figure 11: Relation between maximum temperature, combustion efficiency, and emissions (calculation).

The relation between the exhaust emissions detected by gas chromatography and combustion efficiency was experimentally investigated using two-stroke HCCI engine [9]. With the increase in equivalence ratio, emissions of HC, which is source fuel, emissions of CO which is a intermediate products, and emissions of CO_2 of the final product were seen in order, and apparent combustion efficiency became fixed with the value of 83% (Figure 12). The portion, which had not burned, is HC of mixture in a piston crevice gap and mixture in boundary layers on the surface of a combustion chamber wall, which remained without burning. The thickness of the boundary is estimated 0.84 ~ 1.03mm (Figure 13). It turns out to be that high combustion efficiency is secured more than equivalence ratio to which the amount of CO emission decreases to almost zero. CO in an exhaust gas is the important index that

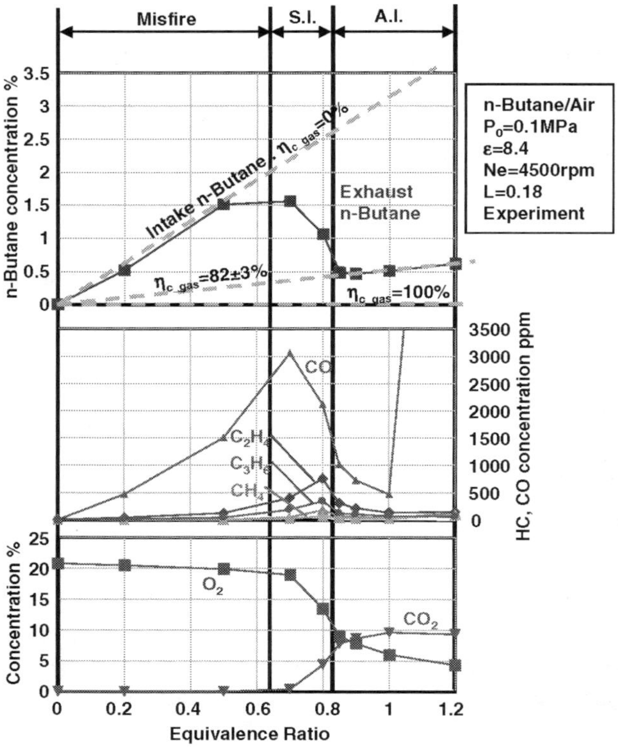

Fig. 12 Relation between equivalence ratios, n-Butane exhaust concentrations and emissions (fuel: n-Butane, experiment) [9]

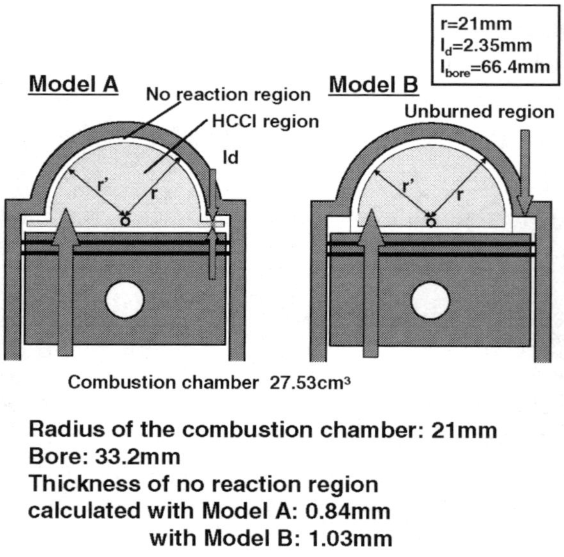

Fig. 13 The models of no reaction region in combustion chamber [9]

shows the conclusion nature of combustion. Figure 12: Relation between equivalence ratios, n-Butane exhaust concentrations, and emissions (experiment). Figure 13: The models of no reaction region in combustion chamber.

EFFECT OF HETEROGENEITY OF FUEL/AIR MIXTURE

Does the HCCI combustion include flame propagation? For checking a combustion reaction region, it is easy way to observe radical luminescence of a flame. If the windowpane made from quartz is prepared in a combustion chamber and a 2-dimensional image of luminescence is acquired, the region where the reaction has proceeded can be checked. Since a luminescence is occurred in three-dimension-space, by the 2-dimensional image, the distribution in depth direction of the optical system is inseparable. LIF is effective in grasping three-dimension-structure in a combustion chamber [11, 12].

As an object of observation, they are radicals, such as OH, CH, C_2, and HCHO. In these radicals, luminescence of OH and CH is strong and effective in the analysis of a combustion mechanism. Luminescence of OH radical is in the processes of low temperature reaction, degeneracy, and high temperature reaction (blue flame and hot flame). After a reaction is completed, luminescence continues. In case a momentary luminescence image is analyzed, it becomes impossible to distinguish the place where the combustion reaction has occurred, and the place that the reaction already completed. If continuation photography is carried out, the region of a reaction spreads out gradually will be observed[13].

By using DME as a fuel, two dimensional observation of a flame luminescence that is emitted at recombination reaction process of CO and O shows that the luminous areas on the first frame (-8 ~ -6 C.A. deg) are stopped luminosity on the second frame, to a contrary, the non-luminous areas on the first frame are started luminosity on the second frame (Figure 17). On the 3rd frame, luminous areas on the previous frames are as dark area, the new luminous areas are where non-luminous areas on previous frames. In HCCI combustion of

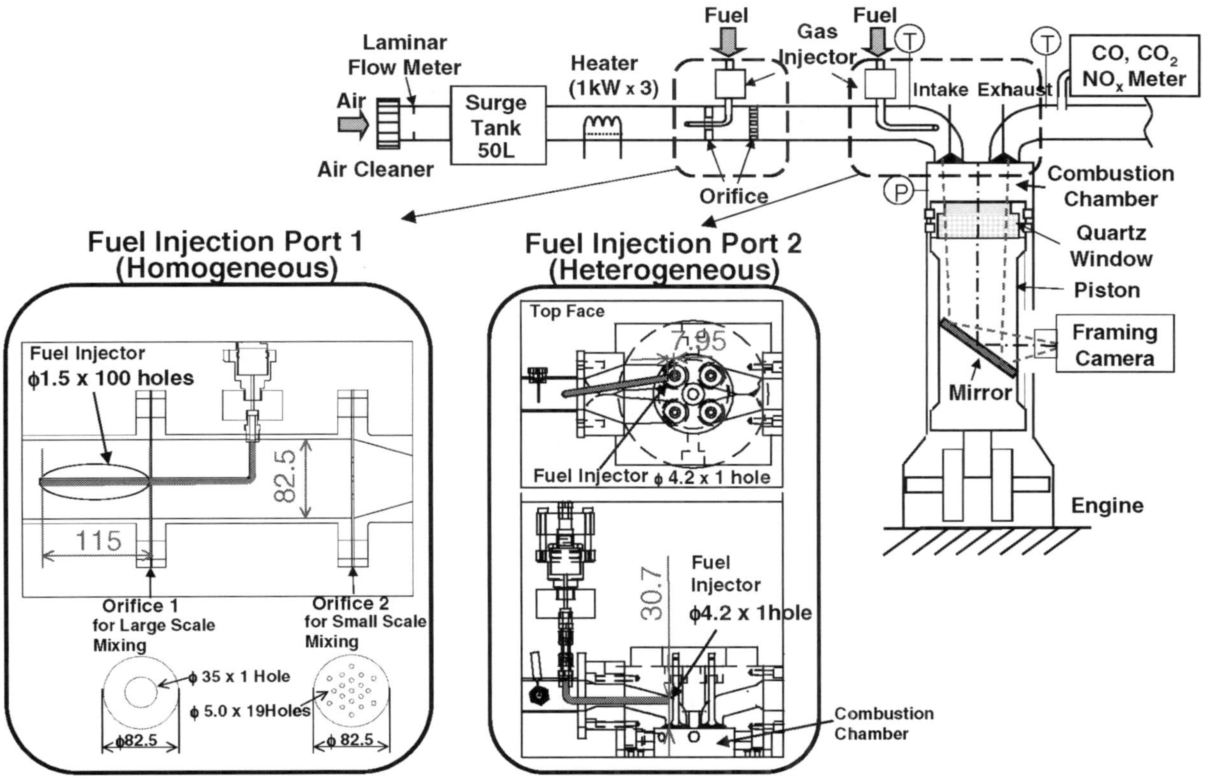

Fig. 14 Schematic of the optically accessible engine [10]

heterogeneous mixture, it is understood that combustion reaction regions locally appear and these regions spatially and in time, these regions are discovered dispersedly.

HCCI combustion is not like flame propagation combustion.

It is the effect of the heterogeneity in a combustion chamber that cannot be disregarded as a means that can solve the subject of HCCI combustion.

In the experiment of the optically accessible engine, which performed DME as fuel, it is changing a fuel injection position and experimented by producing the heterogeneity of fuel concentration in the combustion chamber [10]. When heterogeneous, it became clear that cycle variation of the pressure in a combustion chamber becomes small, and the rate of heat release of overalls is stopped low. Figure 14: Schematic of optically accessible engine. Figure15: Pressure history comparison at the time of homogeneous, and the time of heterogeneous. Figure 16: Comparison of 64 cycles averaged histories of heat release rate.

Furthermore, by 2-dimensional luminescence image measurement in a combustion chamber, when fuel concentration is homogeneous, luminescence arises at this time throughout a combustion chamber mostly, but when heterogeneous it arises with a big time lag. Figure 17: 2 dimensional image comparisons at the time of homogeneous, and the time of heterogeneous.

According to luminescence spectrum measurement of DME/Air HCCI combustion, the strong continuation spectrum applied to 300-600nm that has a peak near 430nm was checked in the duration of high temperature reaction [14]. It is known that this continuation spectrum is a thing resulting from the reaction shown below [15]. Figure 18: Luminescence spectrum in combustion chamber in HCCI combustion of DME.

$CO + O + M = CO_2^* + M$

$CO_2^* + M = CO_2 + M$

$CO_2^* = CO_2 + h\nu$

By Luminescence strength history measurement using the photo-multiplier, it is clarified that luminescence mainly arises in the second half of high temperature reaction. And by numerical calculation with elementary reactions, it is checked that the reaction of $CO + O + M = CO_2 + M$ also appears in the second half of high temperature reaction. What is said, luminescence in the HCCI combustion of DME/air mixture has a high possibility of being a thing resulting from oxidation reaction of CO. Figure 19: Histories of heat release rate and the luminescence intensity for various equivalence ratios. Figure 20: Histories of pressure, temperature, rate of heat release, and reaction rate acquired from numerical calculation with elementary reactions.

In HCCI combustion of DME/air, if fuel-air mixture is compressed and the temperature reaches 650K, a

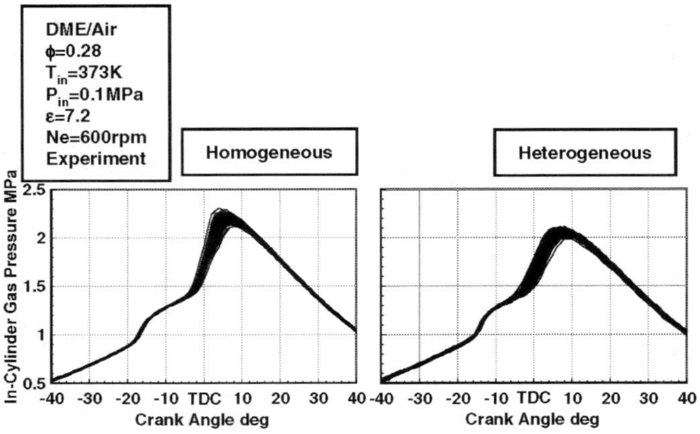

Fig. 15 In-cylinder gas pressure histories of 64 cycles (fuel: DME, experiment) [10]

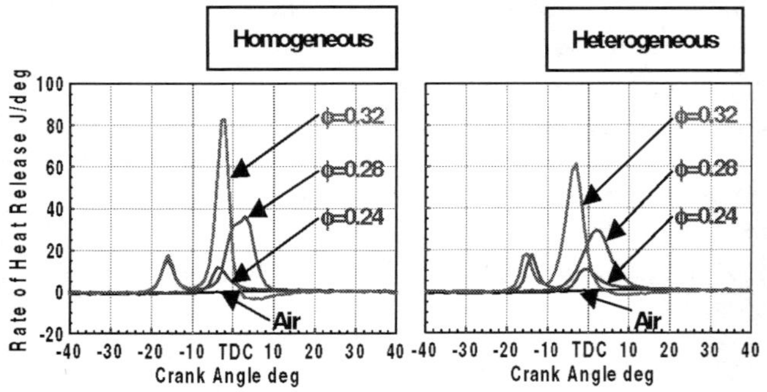

Fig. 16 64 cycles averaged histories of R.H.R. (fuel: DME, experiment) [10]

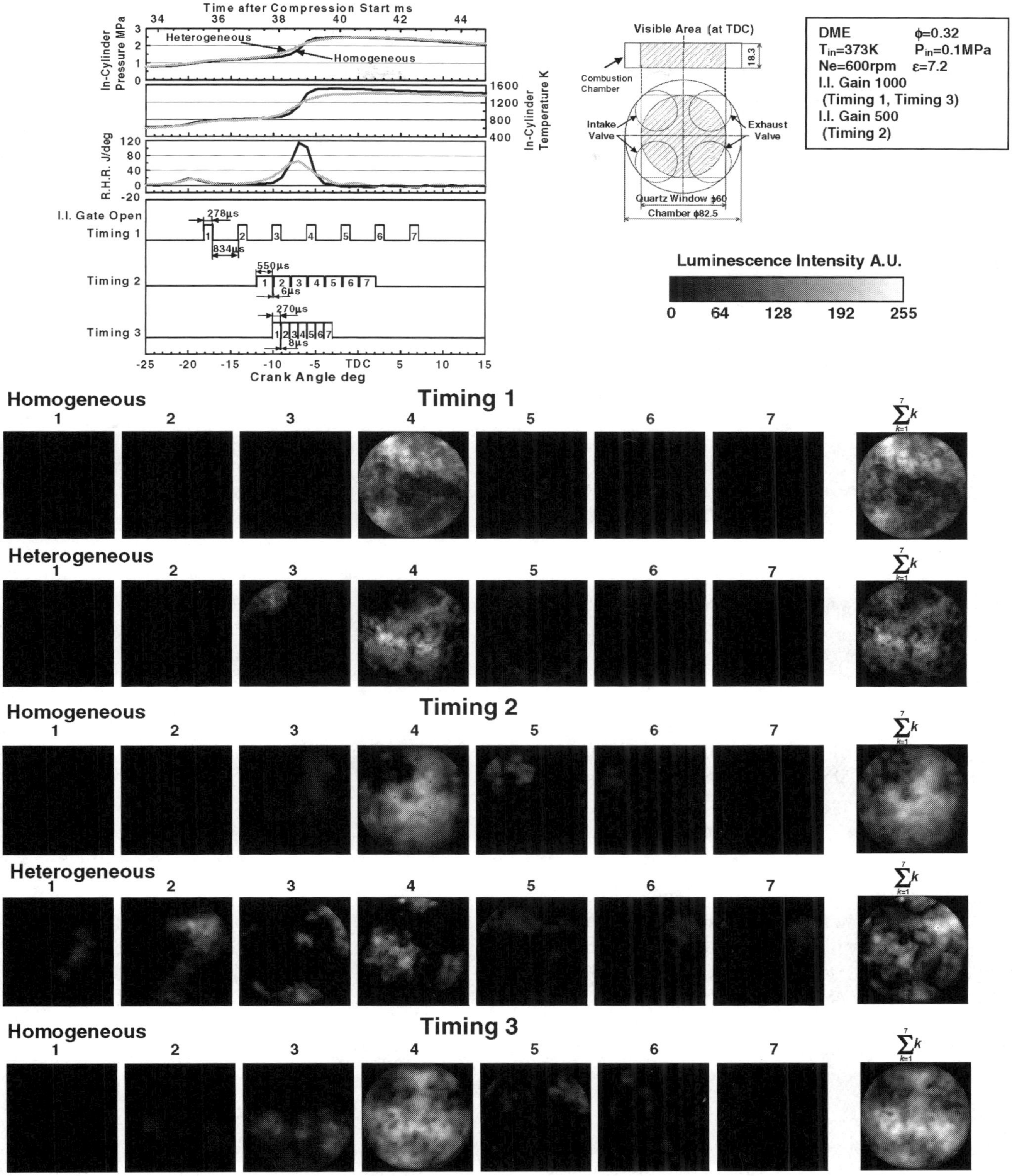

Fig. 17 Two-dimensional images of the luminescence with the histories of in-cylinder pressure, temperature, R.H.R. (fuel: DME, experiment) [10]

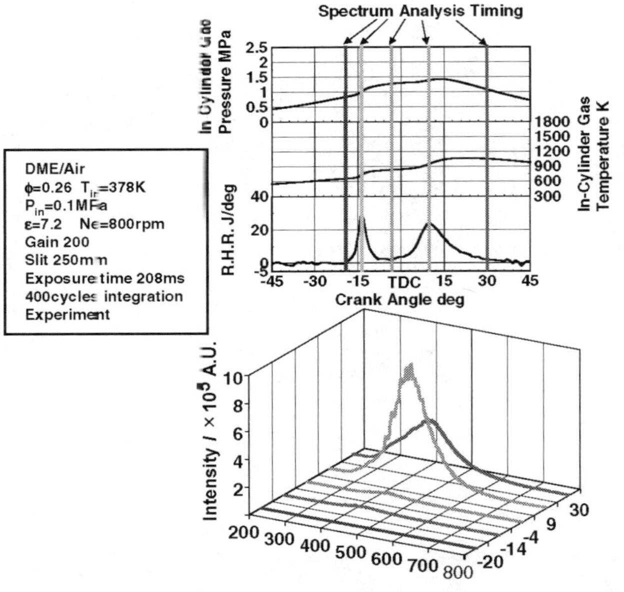

Fig. 18 Luminescence spectra under auto-ignition and combustion process (fuel: DME, experiment) [14]

part of DME reacts and HCHO (formaldehyde), H_2O_2 (hydro peroxide), etc. are generated. It is the advance of low-temperature reaction [16]. Figure 21: Mole fraction histories of main chemical species in DME HCCI combustion. Figure 22: Oxidation pass of DME [17].

According to Tezaki and others' LIF observation [18], when one third of fuel-air mixture is consumed and HCHO of the almost same number is generated, low-temperature reaction stops. The so-called degeneration starts.

When temperature in a combustion chamber reaches about 1050K by heat release of this low temperature reaction and by more compression, oxidization of HCHO and H_2O_2 will start and will shift to hot flame. However, when the bulk temperature in a combustion chamber does not exceed 1500K, a lot of CO remains. Thus, in HCCI combustion, even if fuel/air mixture ignites, a reaction does not always advance even to the final combustion products of CO_2 and H_2O.

This is a typical attitude at a low load HCCI operation, with an ultra lean fuel-air mixture and/or a low equivalence ratio, at which a knock never occur. The combustion reactions will be completed if in-cylinder gas has an additional compression until its temperature exceeds 1500K.

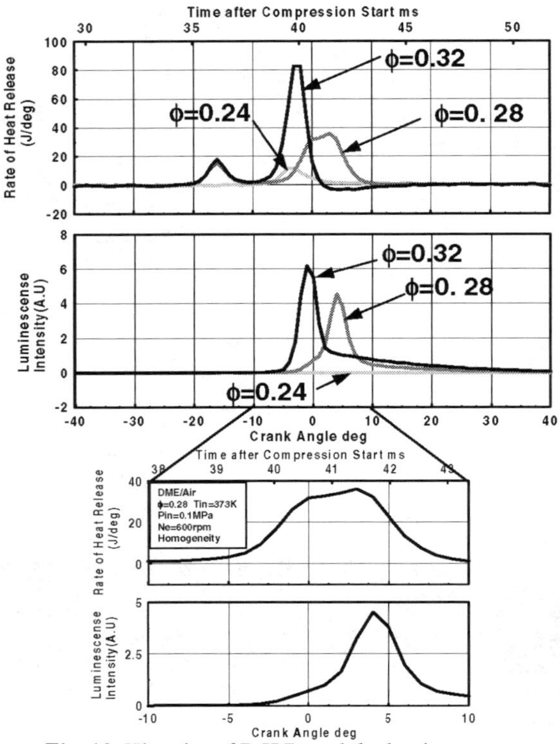

Fig. 19 Histories of R.H.R. and the luminescence intensity for various equivalence ratios (fuel: DME, experiment)

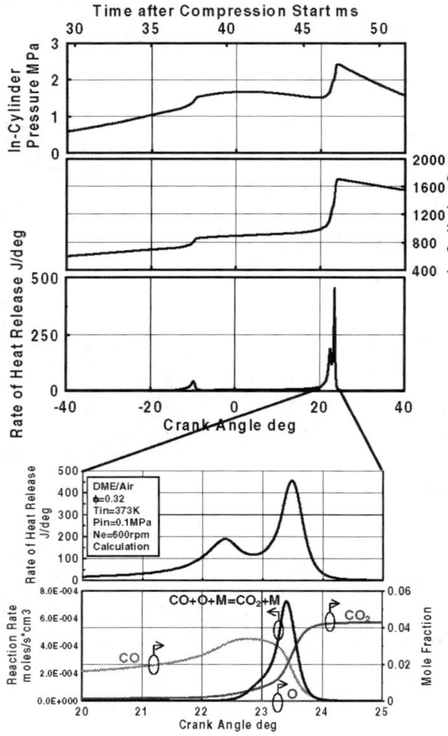

Fig. 20 Histories of pressure, temperature, and R.H.R., and reaction rate of CO oxidation (fuel: DME, calculation)

The influence affect the HCCI combustion when increasing the heterogeneity of fuel-air mixture is summarized as follows.

a) Combustion completeness: In low load, HC and CO increase, and the completeness become low.

b) Cycle variation: combustion stability of each cycle is high, i.e. the variations of ignition timing, combustion duration, peak value of rate of heat release, and etc is small. Especially, when the engine is operated near the knocking limit where the combustion efficiency is high and maximum mass averaged in-cylinder gas temperature in a cycle exceeds 1500K; the introduction of heterogeneity of mixture is effective to reduce the cycle variations.

c) Rate of heat release: Maximum value of rate of heat release decreases and combustion duration increase.

d) Mechanism and structure of HCCI combustion: The HCCI combustion domains are separated and spread over in the space of combustion chamber as like a flame let or island. Auto-ignition timing of each domain is not simultaneous but various in time. This is the reason why overall combustion duration becomes long and the knocking limit can be expanded to high load. Combustion reaction speed of each domain may be not changed.

In above discussion is made on heterogeneity of the mixture concentration; it is the important factor that controls HCCI combustion also about the heterogeneity of temperature. Practical use of the integrated effect of two kinds of heterogeneity, concentration and temperature, becomes one methods of HCCI combustion control.

In the calculation result in consideration of 0 dimensional model, rapid pressure rise and very high heat release are computed as compared with actual HCCI engine.

Not the error of a numerical analysis result but this is considered to mainly originate in heterogeneity of fuel-air mixture. By CFD calculation, Morimoto and others of Tokyo Gas Co., Ltd computed the distribution of temperature or concentration in a combustion chamber. Next, they calculated by dividing this distribution by class and giving different temperature or concentration conditions with 0 dimensional model [19]. As a result, the

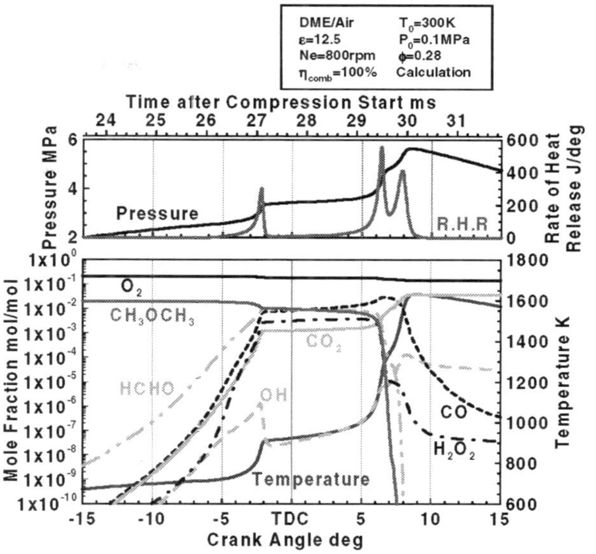

Fig. 21 Mole fraction histories of major chemical species (fuel: DME, calculation) [16]

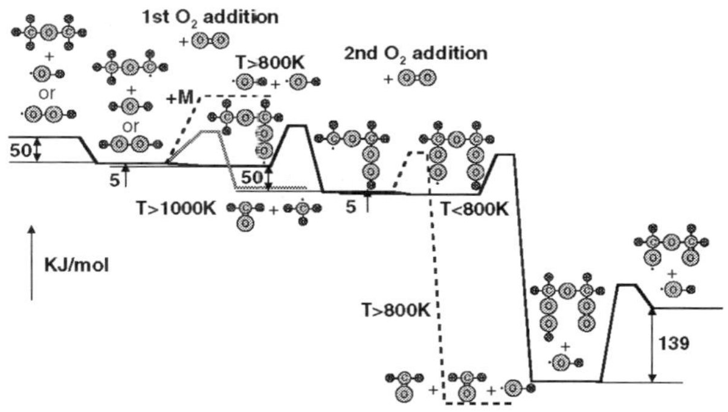

Fig. 22 Main chemical reaction pass of DME [17]

calculation result supposing heterogeneity shows the value equivalent to the result of the actual engine. Figure 23: Temperature distribution calculated from CFD, and pressure histories comparison between homogeneous condition and heterogeneous condition.

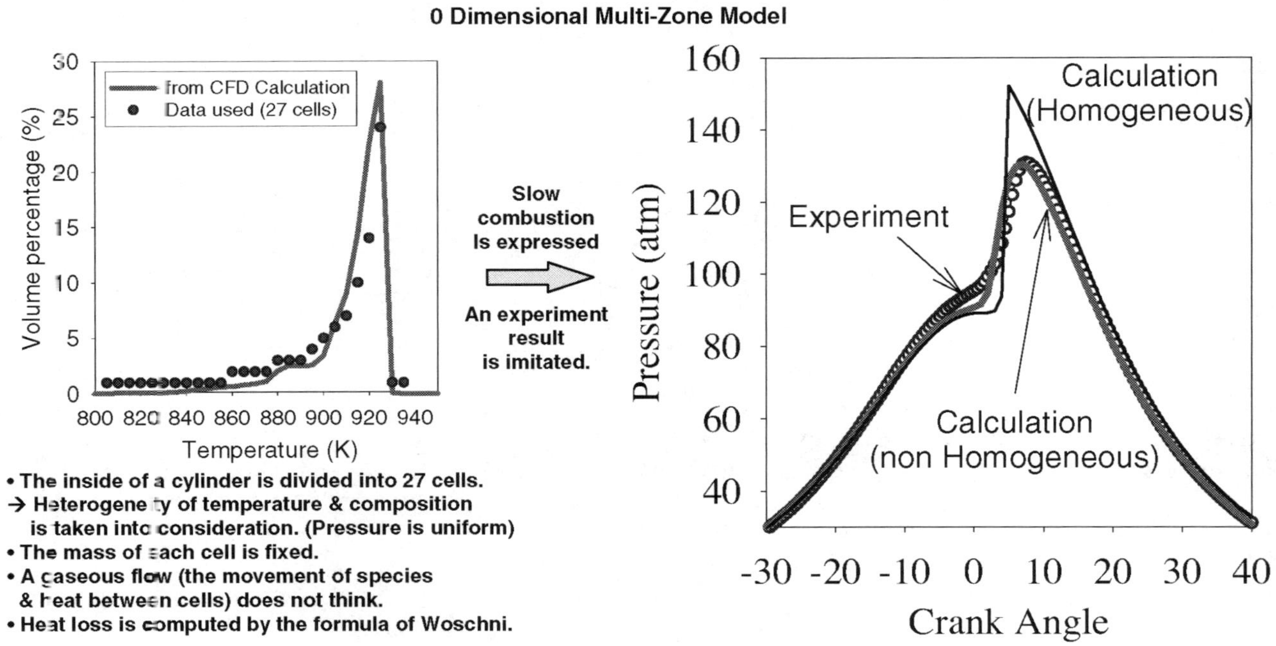

Fig. 23 Temperature distribution calculated from CFD and pressure histories comparison [19]

SUMMARY

THE KEY POINTS OF HCCI COMBUSTION CONTROL

1) The present internal combustion engines cannot be simply classified only according to the kind of "formation of fuel/air mixture" and "ignition method". When it classifies by adding the standard of a "fuel injection form" and a "combustion form" here, it turns out that the HCCI combustion is the combustion form that exists also by the gasoline engine and the diesel engine, and is crossing variably.

2) In HCCI engine, the piston compresses the fuel-air mixture introduced in the combustion chamber adiabatically, and when it exceeds auto-ignition temperature, it results in ignition. It is thought that by changing intake temperature and compression ratio that control of auto-ignition timing is possible.

3) Combustion reaction speed is depended on the fuel mass introduced in the combustion chamber. If the heat quantity introduced in the combustion chamber is increased, combustion reaction speed will increase exponentially.

4) Auto-ignition timing becomes an important factor in order to secure a high output power. It is important to retard the auto-ignition timing to after top dead center while avoiding knocking. The HCCI operation, whose load is 1.0MPa as BMEP, is possible if controllable so that auto-ignition timing and a combustion reaction arise in expansion process. In order to realize high load operation beyond this, introduction of the heterogeneity of the temperature and the concentration of fuel/air mixture becomes indispensable.

5) According to the result of numerical calculation with elementary reactions, by making intake air temperature or equivalence ratio increase, when the highest attainment temperature in a cycle rises and the temperature exceeds about 1500K, oxidization of CO can be completed, and combustion efficiency can exceed 90%. CO in exhaust gas serves as an important index that shows the completeness of combustion.

6) By introduction of the heterogeneity of fuel concentration in the combustion chamber, cycle variation of in-cylinder gas pressure becomes small and overall heat release rate is stopped low. Furthermore, by the measurement of two-dimensional image in combustion chamber, it turns out that luminescence arises almost in the combustion chamber whole region at the same time when the fuel concentration is homogeneous, and it arises, on the other hand, with a big time lag when the fuel concentration is heterogeneous.

FUTURE OF HCCI ENGINE

At the last, it is said how HCCI combustion technology is developed in the future.

At first, it is the method of using it in the short period. In other words, to a management of an exhaust catalyst, it is that only the limited short period of time uses HCCI combustion. The atmosphere of a stoichiometric ratio, reducing atmosphere, or intermediate products is supplied to a catalyst at the same time it performs temperature management of a catalyst.

Next, the method of combining the conventional combustion system (S.I. combustion, Diesel combustion) and HCCI combustion according to load is mentioned. In this case, although it will be used for a partial load domain, it is necessary to realize the smooth shift between each combustion mode.

Then, is the engine which works only by HCCI combustion realized? When using it in all load domains, the possibility of realization of 2-stroke high-speed engine is high. If it is used with 4-stroke engine, the possibility of the co-generation natural-gas engine and the heat-rejected engine can be considered. If it is made to burn by the light-oil diesel and will become, the necessity for low cetane fuel will also come out.

However, if the above-mentioned thing is realized, ascertaining of limit IMEP is needed from a combustion engineering viewpoint. Naturally, the technology of maintaining HCCI combustion stably for expansion stroke is needed. That is, introduction of the control technology of advanced ignition timing (conquest of negative feedback) or the matrix heterogeneity of temperature and composition is indispensable.

REFERENCES

1. Thring, R. H., "Homogeneous Charge Compression Ignition (HCCI) Engines", SAE Paper, 892068
2. Iida, N. and Igarashi, T., "Auto-Ignition and Combustion of n-Butane and DME/Air Mixtures in a Homogeneous Charge Compression Ignition Engine", SAE Paper, 2000-01-1832
3. Ashley, S., "A Low-pollution Engine Solution", Scientific American, June, 2001
4. Lee, J. and Iida, N., "Combustion of Diesel Spray Injected into Reacting Atmosphere of Propane-Air Homogeneous Mixture", International Journal of ENGINE RESEARCH, IMeckE and SAE, 2001 Vol.2, No.1, pp.69-80, (2001)
5. Yanagihara, H., Satou, Y. and Mizuta, J., "A Simultaneous Reduction of NOx and Soot in Diesel Engines under a New Combustion System (Uniform Bulky Combustion System -UNIBUS)", 17th Int. Vienna Motor Symposium, (1996)
6. Ishibashi, Y. and Nishida, K., "An Approach to Controlling Auto-Ignition: Two-Stroke Gas Exchanging Method", GPC congress 2002 Ann Arbor, Proceedings "Advanced Engine Design & Performance", pp.26-33, (2002)
7. Jun, D., Ishii, K. and Iida, N., "Combustion Analysis of Blend Fuel of Methane and n-Butane in an HCCI Engine Using Experiment and Elementary Reactions Calculation", Transactions of Society of Automotive Engineers of Japan, Vol.34, No.4, pp.69-74, (2003), (in Japanese)
8. Yamasaki, Y. and Iida, N., "Numerical Analysis of Auto Ignition and Combustion of n-Butane and Air Mixture in the Homogeneous Charge Compression Ignition Engine by Using Elementary Reactions", SAE Paper, 2003-01-1090
9. Kojima, Y. and Iida, N., "A Study of the Combustion Completion on the 2-Stroke HCCI Engine with n-Butane/air Mixture - Investigation of the Composition and the Exhaust Mechanism of the Exhaust Gas –", SAE Paper, 2004-01-1978
10. Kumano, K. and Iida, N., "Analysis of the Effect of Charge Inhomogeneity on HCCI Combustion by Chemiluminescence Measurement", SAE Paper, 2004-01-1902
11. Collin, R., Nygren, J., Richter, M., Alden, M., Hildingsson, L. and Johansson, B., "Simultaneous OH- and Formaldehyde-LIF Measurements in an HCCI Engine", SAE Paper, 2003-01-3218
12. Aizawa, T., Kosaka, H., Matsui, Y., "2-D Imaging of Soot Formation Process in a Transient Spray Flame by Laser-induced Fluorescence and Incandescence Techniques", SAE Paper, 2002-01-2669
13. Iida, N., "Combustion Analysis of Methanol-Fueled Active Thermo-Atmosphere Combustion (ATAC)

Engine Using a Spectroscopic Observation", SAE Paper, 940684

14. Yamasaki, Y. and Iida, N., "Auto-Ignition and Combustion Analysis of DME Air Homogeneous Charge Compression Ignition Engine by Radical Luminescence Measurement", 14th ISAF, 2002-VT-11, (2002)
15. Gaydon, A. G., "The Spectroscopy of Frames", London Chapman and Hall, 2nd edition, pp.127-143, 282-295, 352-353, (1974)
16. Sato, S. and Iida, N. "Analysis of DME Homogeneous Charge Compression Ignition Combustion", SAE Paper, 2003-01-1825
17. Curran, H. J., Pitz, W. J., Westbrook, C. K., Dagaut, P., Boettner, J-C. and Cathonnet, M., "A Wide Range Modeling Study of Dimethyl Ether Oxidation", International Journal Chemical Kinetics, 30-3, (2000), 229-241
18. Sakanashi, H., Yamauchi, M., Inada, Y., Choi, N. and Tezaki, A., "Laser-Induced Fluorescence Measurements of Formaldehyde in Homogeneous Charge Compression Ignition (HCCI) of Dimethyl Ether", Transactions of the Japan Society of Mechanical Engineers B, Vol.70, No.690, pp.503-510, (2004), (in Japanese)
19. Amano, T., Morimoto, S. and Kawabata, Y., "Modeling of the Effect of Air/Fuel Ratio and Temperature Distribution on HCCI Engines", SAE Paper, 2001-01-1024

Contact.

Norimasa Iida [*1], Yudai Yamasaki [*2] and Susumu Sato [*3]

Department of System Design Engineering Keio University, 3-14-1 Hiyoshi, Kohoku-ku Yokohama 223-8522 Japan

1. e-mail: iida@sd.keio.ac.jp
2. e-mail: yudai_y@fiv.t.u-tokyo.ac.jp
3. e-mail: y09797@educ.cc.keio.ac.jp

HP: http://www.iida.sd.keio.ac.jp/index2.html

Which Fuels For Low CO_2 Engines?
© P. Duret, X. Montagne (Eds.) and Editions Technip, Paris, 2004, pp. 45 – 58
27 rue Ginoux, 75015 Paris

THE EFFECT OF FUEL PROPERTIES ON HCCI ENGINE COMBUSTION CHARACTERISTICS AND PERFORMANCES

Gen Shibata[1], Koji Oyama[1], Tomonori Urushihara[2], Tsuyoshi Nakano[2]

1. Central Technical Research Laboratory, Nippon Oil Corporation
2. Nissan Research Center, Nissan Motor Co., Ltd

ABSTRACT – *A supercharged 4-cylinder engine was introduced to evaluate how fuel properties affect engine combustion and performance in homogeneous charge compression ignition (HCCI) operation. In this study, choosing from 12 hydrocarbon constituents, model fuels were mixed to have the same distillation but different research octane numbers (RON=70, 80, 92). For each fuel, RON distribution against distillation is controlled to maintain the same in-cylinder history of octane number in the vapor phase during the air-fuel compression process.*

*To confirm the appropriateness of model fuels and test procedures, regular gasoline((w)=winter blend) (RON=90) was also included. From the combustion analysis it was clear that the low temperature heat release depends on fuel characteristics. RON92 fuel has a small low temperature heat release and a high temperature heat release combusts slowly. RON92 fuel is considered to be a good anti-knocking fuel for HCCI, but at higher engine speeds, the high temperature heat release is excessively late resulting in frequent misfire. Although the higher octane number can yield better thermal efficiency, the engine could not be run at high speed. This trend was also evident with regular gasoline**(w)**.*

Comparing the HCCI engine test results of regular gasoline((s)=summer blend) and a mixture of only paraffins yielding the same octane number, the heat release patterns differ This means that the fuel has a possibility to improve engine performance, and the distillation, vaporization characteristics and fuel components also affect HCCI combustion.

The effect of fuel chemical components on low and high temperature heat release was investigated. To simplify the engine tests, 6 fuels, a base blend, a mixture of base blend+6.5% of n-hexane (n-paraffin), a mixture of base blend+6.5% of 2methylpentane (iso-paraffin), a mixture of base blend+6.5% of 4methyl-1pentene (olefin), a mixture of base blend+6.5% of cyclohexane (naphthene), and a mixture of base blend+6.5% of toluene (aromatic) were used. N-hexane, 2methylpentane, 4methyl-1pentene, and cyclohexane are all C_6 molecules but have different chemical structures. Toluene was used instead of benzene. The test results show that paraffin blended fuels have a large low temperature heat release and the high temperature heat releases start earlier than olefin, naphthene, and aromatic blend fuels. Adding only 6.5% of chemical blend substantially changes the HCCI heat release and HCCI engine performance.

PRF90.5, a fuel mixture consisting only of paraffins (iso-octane and n-heptane),exhibits earlier low temperature and high temperature heat releases. Fuel components must be one of the most important factors for HCCI combustion.

INTRODUCTION

HCCI is being considered as an alternative combustion process in the internal combustion engine. As a state-of-the-art internal combustion system, HCCI has the potential for nearly zero NOx and soot emissions, with diesel like engine efficiency.

HCCI engine combustion has a low temperature heat release **(LTHR)** and a high temperature heat release **(HTHR)**, and both heat releases occur within certain temperature ranges. Specifically, the LTHR occurs in the range 750-850K and the HTHR at around 900-1050K [1-3]. LTHR is one of the most important phenomena for HCCI engine operation. The heating value of the LTHR is fuel dependent, and consequently so is the crank angle phasing of the HTHR [4].

Both SI engine knock and HCCI combustion are autoignition processes in which combustion is driven by a combination of fuel characteristics and the time history of temperature and concentration.

Knowing that the knocking behavior of a fuel can be different for different engine conditions, and roughly equating the knocking phenomenon with HCCI combustion, leads one to assume that the fuel octane number and its octane sensitivity would be correlated with HCCI operating characteristics in an engine.

A recent study by Kalghatgi [5-6] suggests that the anti-knock quality of a fuel is related to RON, MON, and operating condition as shown in equation (1).

$$OI = RON - KS \quad (1)$$

OI:	Octane Index
RON:	Research Octane Number
S:	RON – MON, Octane Sensitivity
K:	Varies by engine and operating conditions

Traditionally, the octane index has been defined by OI=(RON+MON)/2, which is the special case of K=1/2 in the Kalghatgi definition.

Foster applied equation (1) to his HCCI test results and showed that in some cases OI has merit as an HCCI fuel index [7].

The K factor depends on engine speed and load, but is independent of fuel. This means that under different operating conditions, a certain fuel with given RON and MON will exhibit different knocking behavior.

Kalghatgi developed OI by using various fuels and changing operating parameters [8]. OI research should endeavor to evaluate the HCCI combustion through an understanding of the two different combustion stages leading to an index that is engine independent.

Kalghatgi's OI seems to be a simple HCCI index but the fuel components and distillation are also important. The fuel vaporization fraction in the cylinder during the compression process is different between the fuel chemicals [9]. Low boiling point hydrocarbons, such as n-pentane, are mostly vaporized prior to ignition, but high boiling point hydrocarbons, such as m-xylene, still exist as liquid. For fuels with the same octane number but different distillation, the vaporized fuel octane number histories during combustion are different. Therefore, not only OI but also the fuel vaporization characteristics (distillation) and fuel chemical components must be included as parameters of an HCCI combustion index. Thus, an HCCI fuel index can not be a simple index, but must depend on at least three or more parameters.

The chemical components effect on HCCI combustion is also investigated [10]. To simplify the engine tests, three fuels, a base blend, a mixture of base blend+6.5% of n-heptane and a mixture of base blend+6.5% of toluene, were used. N-heptane blended fuel accelerates the start of LTHR and enhances the LTHR heating value. The start of LTHR with toluene blended fuel is the same as the base blend but the LTHR heating value is reduced to 90% of the base blend. This is because toluene has no LTHR and works merely as a "LTHR diluent"

Investigation of the effect of the fuel octane number and fuel type on engine performance is the focus of this paper.

EXPERIMENTAL DESCRIPTION

ENGINE BENCH SETUP

The engine bench was in an environmental control chamber, where the temperature, humidity and atmospheric pressure were maintained at 20°C, 60% and 101.3kPa, respectively, to insure constant intake air conditions.

The specifications of the HCCI engine are given in Table 1. The schematic of the research engine bench is shown in Figure 1. The engine employed in the experiment is a research, supercharged, 4-cylinder engine with high compression pistons (compression ratio=15), 4-valves per cylinder, and centrally located spark plugs. The only heating of the intake air was that accompanying the pressure rise in the supercharger (from an engine development perspective, it may be more practical to use boost pressure and not try to control the temperature). The engine can be run either under SI combustion or HCCI combustion depending on the load and engine speed conditions (during HCCI operation, the spark signal is inhibited).

The engine is equipped with a pressure transducer for each cylinder and 400 cycles of crank angle resolved pressure data are transmitted to the PC. The quantity of injected fuel can be controlled for each injector to yield the same IMEP for each cylinder. For all test conditions in this paper, the water jacket temperature and engine oil temperature were kept at 80°C.

FUELS

Choosing from 12 hydrocarbon constituents, model fuels were mixed to have the same distillation but different octane numbers (RON=70, 80, 92).

Table 1. HCCI Engine Description

Engine type	4 cylinder MPI
Compression ratio	15
Bore	86mm
Stroke	86mm
Displacement	1998cc
Exhaust valve open	53°CA BBDC
Exhaust valve close	7°CA ATDC
Intake valve open	1°CA ATDC
Intake valve close	19°CA ABDC

The boiling point-research octane number (**RON**) map for the hydrocarbon constituents used in this study is shown in Figure 2. Five boiling point areas were defined, and for each area, two or three hydrocarbons, one with a low RON number, and another one or two having a high RON number, were chosen, as shown in Table 2.

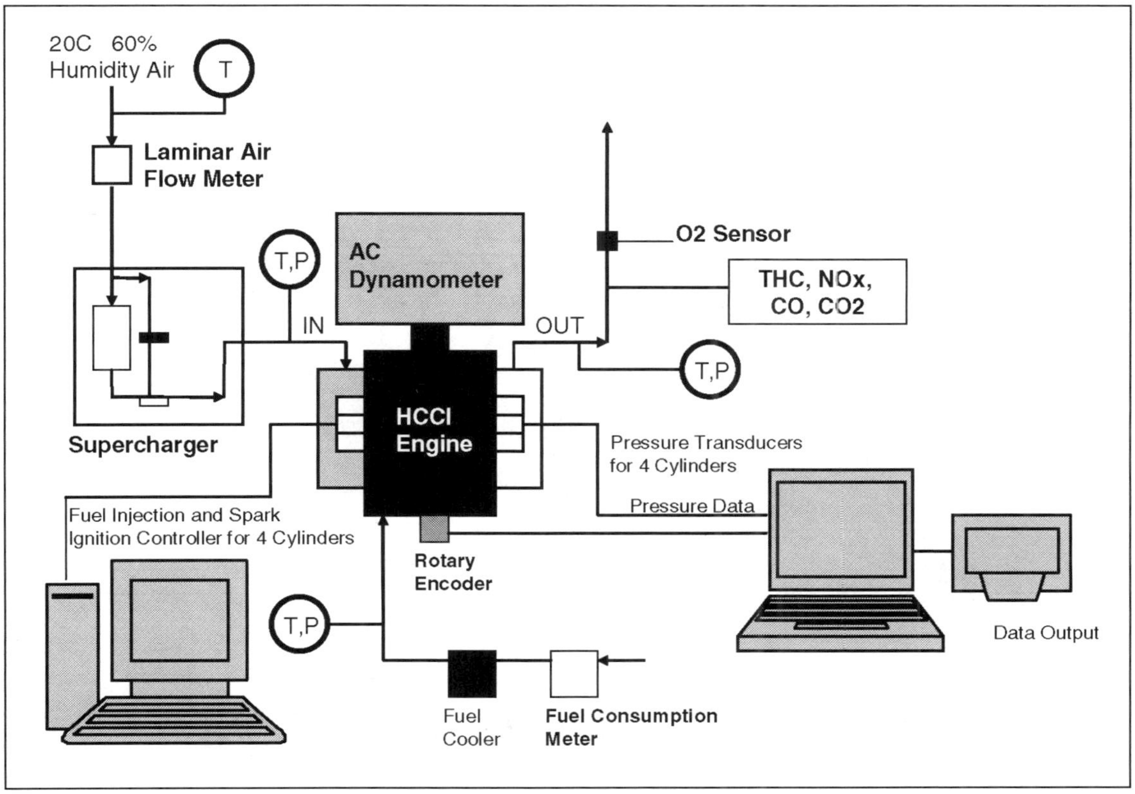

Fig. 1 Experimental Engine Setup Schematic

Table 2. Fuel Components for RON70, RON80 and RON92 Fuels

	Hydrocarbon	Boiling Point °C	RON
Area A (31-35°C)	iso-Pentane	31	86.0
	n-Pentane	35	48.5
	Isoprene	35	132.0
Area B (53-60°C)	4Methyl1pentene	53	104.0
	2Methylpentane	60	67.3
Area C (70-82°C)	n-Hexane	70	26.8
	Cyclohexane	81	85.9
	Cyclohexene	82	92.3
Area D (98-100°C)	n-Heptane	98	0.0
	2.2.4Trimethylpentane	100	100.0
Area E (139-140°C)	m-Xylene	139	127.3
	C9 iso-paraffin	140	65.7

For example, in Area E, RON66 through RON127 fuels, with a boiling temperature of 139-140°C, can be obtained by changing the mixture ratio of m-xylene and C_9 iso-paraffin. As shown in Figure 2, RON70 through RON92 is the common range for areas A to E.

For this range, fuels having the same distillation but different RON values can be obtained by mixing hydrocarbons with 20% by volume from each of areas A-E, where the ratio of components within each area is controlled to yield the desired RON value. The model fuel properties are shown in Table 3. These RON70, 80, and 92 fuels have the same distillation and the same distribution against distillation. To confirm the appropriateness of model fuels and procedures, regular gasoline((w)=winter blend) (RON90) was also included. The regular gasoline(w) is modified to be used for HCCI operation. These 4 fuels (RON70, RON80, RON92 and regular gasoline(w)) are used to investigate the relation between RON, LTHR and the start crank angle of the HTHR by means of an octane number test.

Furthermore, 6 more fuels (Base, MC3, MC4, MC5, MC6, MC10), shown in Table 4, were used to investigate how the LTHR timing depends on fuel chemistry (chemical components test).

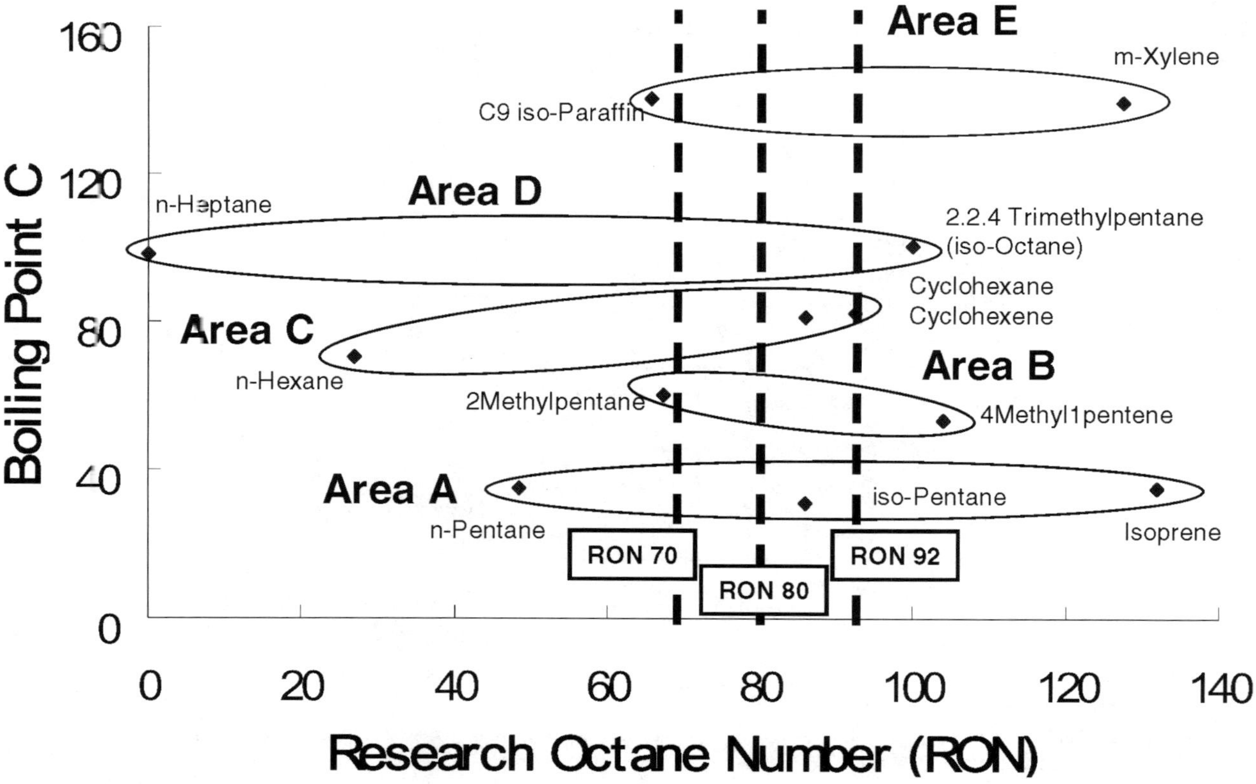

Fig. 2 The Boiling Point-Research Octane Number Map of the Fuel Components

Table 3. Description of the Six Test Fuels for Octane Number Test and Regular(s)-PRF90.5 Test

		RON70	RON80	RON92	Regular Gas(w)	Regular Gas(s)	PRF 90.5
Octane Number RON		70.0	80.0	92.0	90.0	90.5	90.5
Cetane Number		15.0	6.0	4.5	3.5	-	-
Density	g/cm3	0.6965	0.7076	0.724	0.7306	0.7290	0.6976
Reid Vapour Pressure	kPa	45.0	43.5	44.0	73.5	67.0	-
Distillation °C	0%	42.0	45.5	44.0	29.0	32.0	-
	10%	59.0	61.5	60.5	45.5	50.5	-
	30%	68.0	69.5	69.0	62.0	69.5	-
	50%	80.5	81.5	81.5	83.5	92.5	-
	70%	97.5	98.5	99.0	112.5	119.5	-
	90%	129.0	129.0	128.5	139.0	158.0	-
	95%	134.0	133.5	133.0	150.5	171.5	-
	EP	143.0	138.5	141.5	171.5	191.0	-
Fuel Composition vol%					vol%	vol%	vol%
Area A	iso-Pentane	11.5	0	0	n-paraffin	n-paraffin	iso-Octane
	n-Pentane	8.5	15.7	12.7	17.94	12.15	90.0
	Isoprene	0	4.3	7.3	iso-paraffin	iso-paraffin	n-Heptane
Area B	4Methyl1pentene	1.5	6.9	12.4	36.71	38.14	10.0
	2Methylpentane	18.5	13.1	7.6	olefin	olefin	
Area C	n-Hexane	5.3	2.0	0	9.26	17.11	
	Cyclohexane	14.7	18.0	0	naphthene	naphthene	
	Cyclohexene	0	0	20.0	3.37	7.62	
Area D	n-Heptane	6.0	4.0	1.6	aromatic	aromatic	
	2.2.4Trimethylpentane (iso-Octane)	14.0	16.0	18.4	31.89	24.98	
Area E	m-Xylene	1.4	4.7	7.9			
	C9 iso-paraffin	18.6	15.3	12.1			

Table 4. Description of the Six Test Fuels for Chemical Components Test

		BASE	MC3	MC4	MC5	MC6	MC10
Octane Number	RON	87.0	88.5	86.0	83.3	87.0	88.8
	MON	72.8	72.5	72.5	71.0	72.3	73.3
Density	g/cm3	0.7282	0.7246	0.724	0.7239	0.7315	0.7374
Reid Vapour Pressure	kPa	37.5	39.0	38.5	37.5	36.5	36.0
Lower Heating Value	J/g	43496	43538	43567	43566	43489	43254
Distillation °C	0%	52.5	51.0	53.0	52.0	51.5	52.0
	10%	66.5	66.0	66.0	67.0	67.0	67.5
	30%	73.0	71.5	72.0	73.0	73.5	75.0
	50%	82.0	80.0	80.0	81.0	82.0	84.5
	70%	95.5	94.0	93.5	94.0	94.0	97.5
	90%	113.0	113.0	112.0	113.0	112.0	113.0
	95%	124.0	126.0	124.0	125.5	125.0	122.5
	EP	145.0	145.0	145.5	146.5	144.0	144.0
Remarks		-	BASE+6.5% 4Methyl-1pentene (C6 olefin)	BASE+6.5% 2Methyl-pentane (C6 iso-paraffin)	BASE+6.5% n-Hexane (C6 n-paraffin)	BASE+6.5% Cyclohexane (C6 naphthene)	BASE+6.5% Toluene (C7 aromatic)

DEFINITION OF HEAT RELEASE DATA

Figure 3 and Figure 4 show the definitions used in the analysis of heat release in this paper. After the average of 400 cycles of pressure data was converted to heat release data, the minimum points A and B were taken to represent the start and end points of LTHR. The area C is the heating value of LTHR and the area D is the heating value of HTHR. The crank angle resolution is 0.25 CA deg. The HTHR start and end points were chosen as the x-axis intercepts.

OCTANE NUMBER TEST

(Test Fuels: RON70, RON80, RON92, Regular Gasoline((w)=winter blend, modified for HCCI operation))

Although many investigations have been conducted regarding the effect of octane number on HCCI combustion [7], in many instances distillation and RON distribution against distillation were allowed to vary between the fuels. An high speed in-cylinder gas sampling schematic is shown in Figure 5 and 6 [9]. The sample air-fuel mixture was taken at the spark plug position with a high-speed electric sampling valve for 27°CA duration just before the start of combustion.

Figure 7 is the fuel vaporization fraction for the engine cold and warmed-up conditions. This data shows that fuel volatility in the cylinder depends on the fuel even for a warmed-up engine.

For example 2-methylbutane is mostly vaporized prior to ignition, but for xylenes half of the fuel is still liquid in the cylinder.

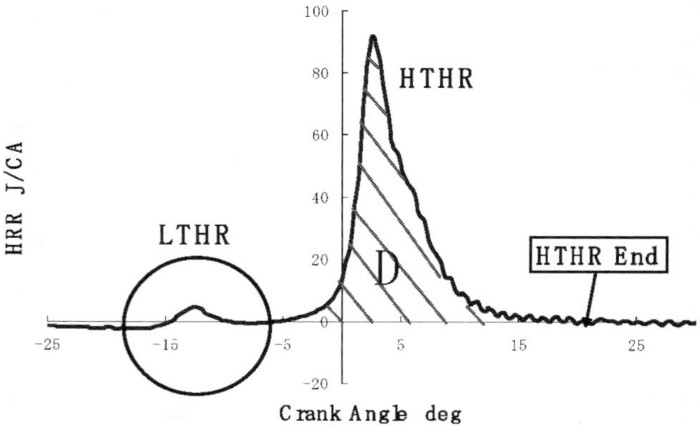

Fig 3 The Definition of Heat Release Data

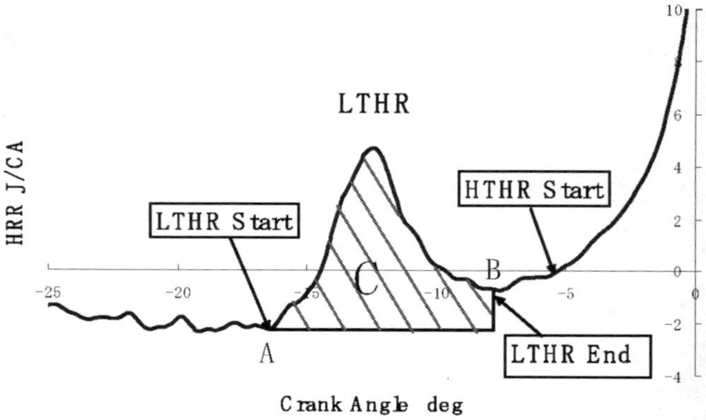

Fig. 4 The Definition of Heat Release Data

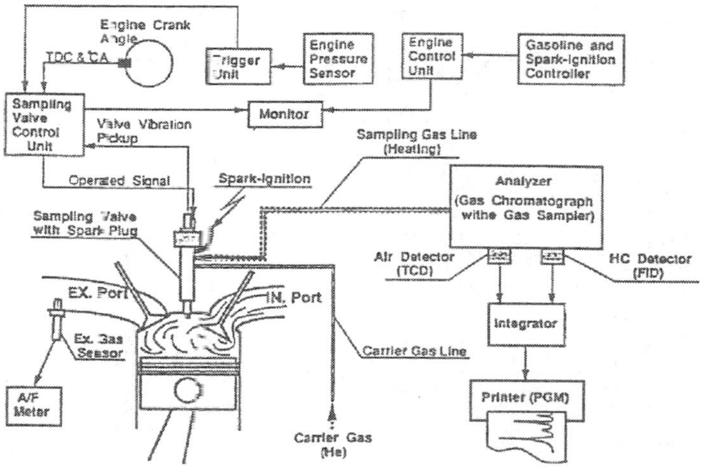

Fig. 5 High Speed In-cylinder Gas Sampling Schematic

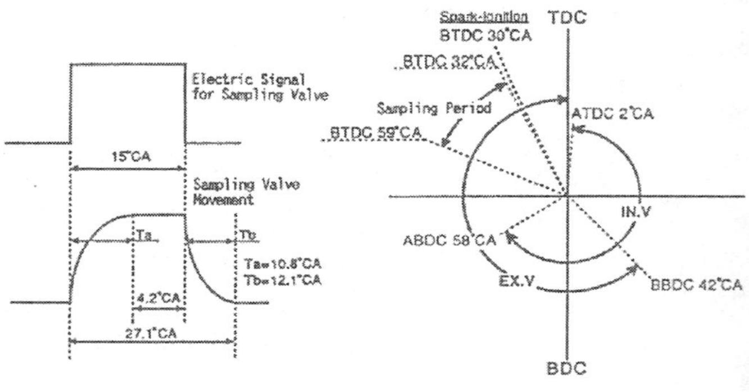

Fig. 6 Gas Sampling Valve timing Chart

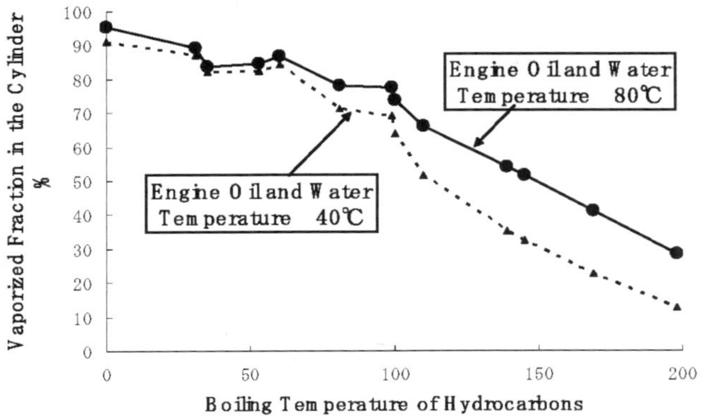

Fig. 7 Vaporization Fraction of Hydrocarbons in the Cylinder[9]

For the octane number test, while varying the octane number the distillation of the fuels must be kept the same in order not to control the octane number of the vaporized fraction. Thus, RON distribution against distillation is important. If the RON distribution against distillation is different, the vaporized fuel RON history during combustion changes. Model fuels RON70, RON80 and RON92 were selected to keep the same distillation and same RON distribution against distillation. These three model fuels have the same distillation (see Table 3) curves. Therefore, for the RON70, which has a 42-143°C distillation range, RON is 70 even within each of the boiling point areas A through E (see Figure 2). Regular gasoline(w) (RON90) was also included to confirm the appropriateness of model fuels and procedures.

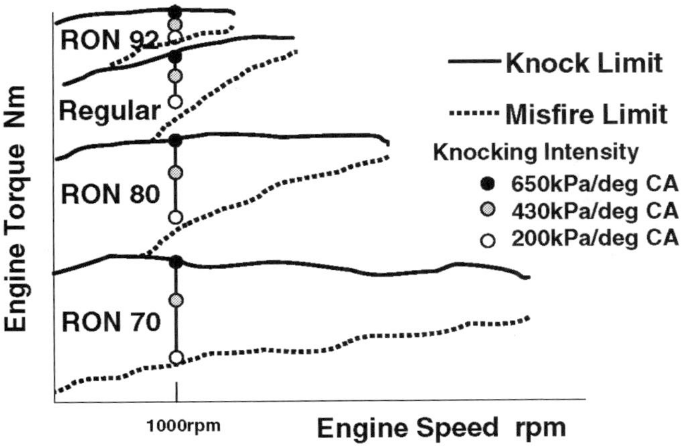

Fig. 8 Engine Speed-Torque Map

The different RON fuels have different LTHR profiles and the octane number test was conducted to investigate the fuel HCCI combustion interactions, especially the effect of combustion on HCCI engine characteristics.

The engine speed-torque map is shown in Figure 8. For each fuel, the allowable operating range is a wedge-shaped region [11]. The upper torque line of this region represents the knocking limit (*knocking intensity=650kPa/deg CA) and the bottom line represents the frequency misfire limit. Within this region the engine exhibits stable HCCI combustion.

The extent of these wedge-shaped regions depends on the fuel. RON92 fuel has a small region – the maximum engine speed is lower than in the case of RON70, but its maximum torque is higher. To compare these fuels, the engine tests were performed under the same knocking intensity. The test conditions are:

Fuels: RON70, RON80, RON92, Regular(w)

Engine Speed:	1000 rpm
Intake Manifold Pressure:	+26.7 kPa (gage)
Intake Manifold Temp.:	47-50°C
Knocking Intensity:	430 kPa/deg CA
Water & Oil Temperature:	80°C

Atmospheric Conditions at the Engine Bench

Temperature:	20°C (controlled)
Humidity:	60% (controlled)
Pressure:	101.3kPa (controlled)

Knocking Intensity (KI) is defined as the 400 cycle average of the maximum rate of pressure increase, $(dP/d\theta)_{max}$. To keep the same KI at 1000rpm, the engine IMEP (or torque) is different depending on the fuel, and the injected fuel quantity is therefore altered accordingly (see Figure 8).

Figure 9 shows the in-cylinder pressure data for 400 cycle averages. As the octane number is increased, the autoignition is delayed.

Figure 10 and Table 5 show the heat release data of these 4 fuels, calculated from the Figure 9 data. LTHR and HTHR regions are clearly distinguishable. The start crank angle of the LTHR is between 16.5-18.5 BTDC CA for the 4 fuels (see Figure 3 and Figure 4 for the definition of the start crank angle of LTHR and HTHR). This means that the trigger for LTHR is essentially the piston-compressed air temperature, although the start crank angle of the LTHR varies slightly depending on the fuel.

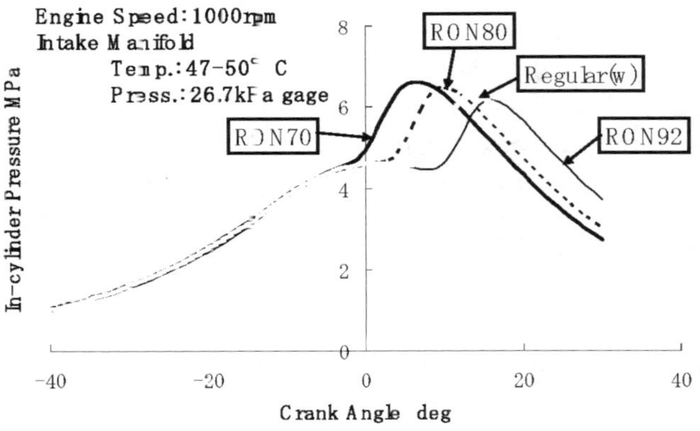

Fig. 9 Pressure Data of RON70, RON80, Regular(w) and RON92

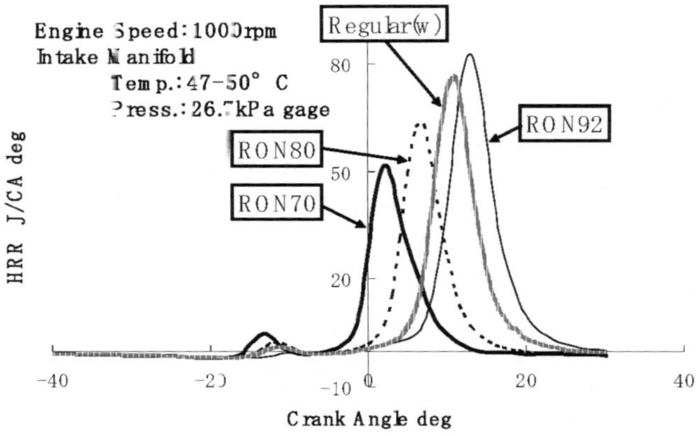

Fig. 10 Heat Release Data of RON70, RON80 Regular(w) and RON92

It was obvious that RON70 had the larger LTHR, and the duration of the LTHR was also shorter. Furthermore, concerning the duration from LTHR start to HTHR finish, the period for RON70 is shorter than in the case of RON92.

In the case of RON70 (see Figure 10) the HTHR occurs earlier than for RON92. This means that with RON70 there is a margin for high-speed operation, while the engine speed range with RON92 is restricted to low speeds. Of course, if the chemical reaction time is shortened, such as by heating of the intake air, the engine could be extended to higher speeds even with RON92.

Figure 11 shows the investigation of the temperature history of the LTHR and HTHR, calculated from the pressure (Figure 9), air-fuel quantity and engine volume data. The temperature at the start of HTHR is 989K for RON70 and 925K for RON92.

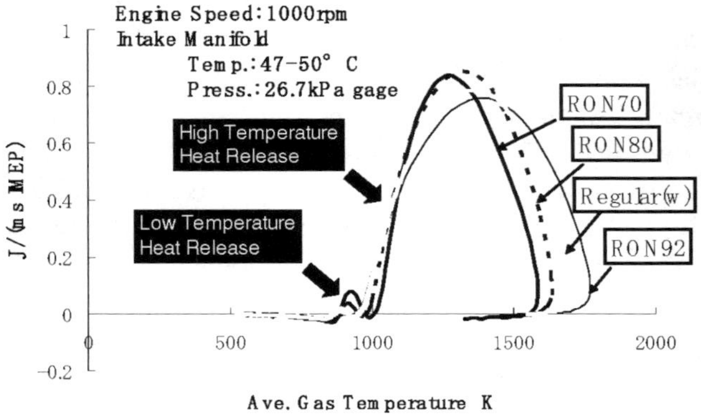

Fig. 11 Temperature History of HCCI Combustion

Table 5. Low Temperature and High Temperature Heat Release Data

		Unit	RON70	RON80	Regular Gas(w)	RON92
Low Temperature Heat Release	Heating Value	J	31.0	24.2	17.3	10.5
	Start Crank Angle	BTDC	18.5	18.0	17.5	16.5
High Temperature Heat Release	Heating Value	J	324	397.5	517.3	600.3
	Start Crank Angle	BTDC	5.0	2.0	2.5	2.0
Intake Air Temperature		°C	47.0	47.1	50.2	47.8
MEP		kPa	373.4	452.5	586.2	654.7
Thermal Efficiency		%	36.11	38.25	40.08	41.40
Air/Fuel Ratio			42.5	33.8	30	28.7

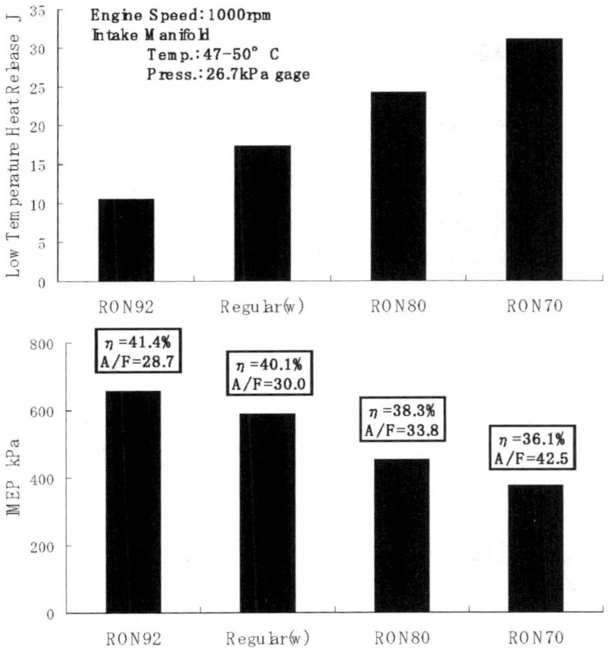

Fig. 12 Low Temperature Heat Release and Engine Performance at Constant Knocking Intensity

Figure 12 shows the effect of the heating value of LTHR on engine performance. At the same knocking level (KI=430 kPa/deg CA), the air-fuel ratio of RON70 is the most lean and both the IMEP and thermal efficiency are the lowest. The immutable fact is that the LTHR heating and the chemical reactions between the LTHR and HTHR decide the HCCI combustion. A fuel with a large heating value of LTHR can be described as an "Easy Knock, High Speed & Low Torque" fuel, and a fuel with a small heating value of LTHR can be described as a "Difficult Knock, Low Speed & High Torque" fuel. Details are shown in Table 6.

From this analysis, the following may be concluded.

Lower research octane number fuels have larger LTHR.

The larger LTHR enhances the temperature of the gas mixture and the HTHR combustion starts sooner.

As HTHR combustion becomes "too fast", HCCI knocking results and the engine can not be run at a fuel rich condition. Therefore, the thermal efficiency goes down.

The octane index by Kalghatgi has been suggested as an HCCI fuel index. In the next section, some experimental results were processed to obtain an indication of LTHR influence on HCCI combustion by the octane index.

CORRELATION OF HCCI OPERATION TO OCTANE SENSITIVITY

As mentioned in the introduction, Kalghatgi [5-6] introduced a new definition for the octane index as OI=RON – KS (equation (1)), where K is a parameter that depends on the engine and its operating conditions. Kalghatgi correlated his octane index, OI, to the knock limited spark advance, which was defined as the ignition timing when knock intensity (KI) reached a predetermined magnitude. Foster [7] applied this equation to pursue a correlation between the start of HCCI combustion and the fuel octane sensitivity. He attempted a correlation of the fuel's OI to the start of HTHR combustion, defined by CA10 (timing of 10% mass burn in HTHR combustion).

Table 6. The Relation Between Fuel Types and Engine Performances

Fuel Type	Engine Torque	Air Fuel Ratio Range	Engine Speed Range	Anti-Knocking Performance	Thermal Efficiency
Large LTHR Fuel	Small	Lean Side	Wide	Poor	Low
Small LTHR Fuel	Large	Rich Side	Narrow	Good	High

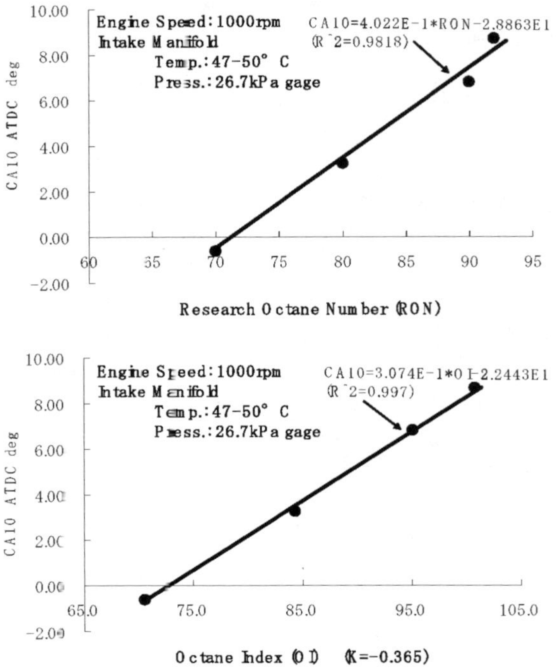

Fig. 13 Start of Combustion vs RON and Octane Index by Kalghatgi[5]

Table 7. Results from Octane Index Analysis (K=-0.365)

	RON	MON	S	OI	CA10
RON70	70	68.5	1.5	70.5	-0.61
RON80	80	68	12	84.4	3.23
Regular(w)	90	76	14	95.1	6.79
RON92	92	67.8	24.2	100.8	8.68

As our octane number test was done under the same knock intensity, these results are applicable to this equation. Figure 13 shows the start of HTHR, shown as CA10, versus RON and OI for an engine condition with supercharged air. Tabulated values of the fuel RON, MON, octane sensitivity, the computed OI and the crank angle for 10% energy release, CA10, are given in Table 7. For our test results, K is -0.365 and it was obvious that RON had a much stronger influence on the octane number test results than MON. But once the engine and test conditions are changed, K and OI also change. HCCI combustion may be defined as a "slow and mild knocking phenomenon" of SI engines. Some engine test results give us hints as to the meaning of RON and MON.

REGULAR GASOLINE((S)=SUMMER BLEND) AND PRF90.5

To investigate the fuel potential for changing LTHR, regular gasoline(s) (RON90.5) and PRF90.5 was used. Regular gasoline(s) is a mixture of more than 200 chemical components including paraffins, olefins, naphthenes, and aromatics. PRF90.5 is a mixture of 9.5% of n-heptane and 90.5% of iso-octane. N-heptane is a C_7 normal paraffin and its boiling point is 98°C. Iso-octane is a C_8 iso-paraffin and its boiling point is 100°C. There are no olefins, naphthenes, or aromatics in PRF90.5 fuel and the distillation and vaporization characteristics are completely different between regular gasoline(s) and PRF90.5. Engine test conditions were as follows:

Fuels: Regular Gasoline(s), PRF90.5	
Engine Speed:	1000 rpm
Intake Manifold Pressure:	+50.7kPa (gage)
Intake Manifold Temp.:	53-54°C
Torque:	86.7Nm
Water & Oil Temperature:	80 °C

Atmospheric Conditions at the Engine Bench

Temperature:	20°C (controlled)
Humidity:	60% (controlled)
Pressure:	101.3 kPa (controlled)

Figure 14 shows that the heat release rates are different between regular gasoline(s) and PRF90.5. The start timing of LTHR of PRF90.5 is earlier than that of regular gasoline(s). Figure 15 shows the temperature history of LTHR and HTHR. The HTHR starting temperature of regular gasoline(s) and PRF90.5 is the same, 910K. Regular gasoline(s) reaches 910K later in the cycle than than PRF90.5, because of the smaller heating value of LTHR.

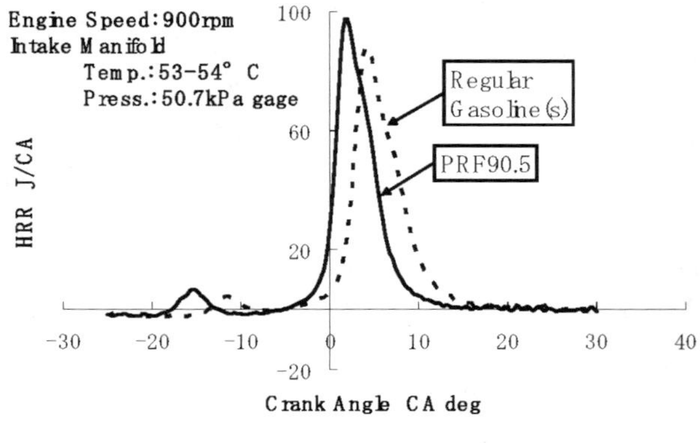

Fig. 14 HRR of PRF90.5 and Regular Gasolie(s)

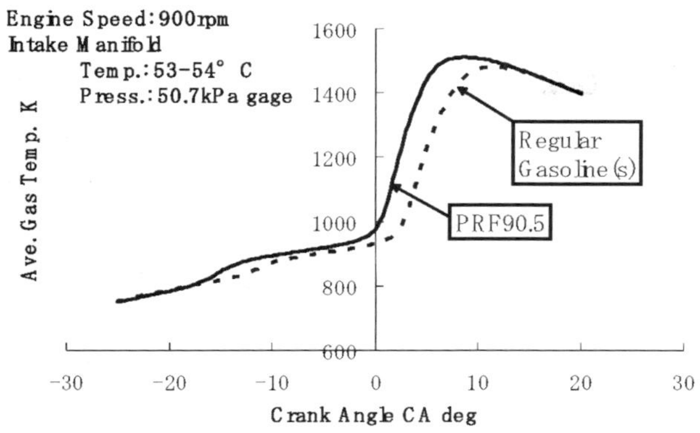

Fig. 15 Temperature History of PRF90.5 and Regular Gasoline(s)

HTHR combustion period of PRF90.5 is also shorter than that of regular gasoline(s). These HRR differences must be caused by the fuel distillation, vaporization characteristics and fuel components. Oil refineries change gasoline composition between winter and summer to maximize the SI engine emission and driveability performances. It is clearly important to also understand these effects on HCCI combustion.

CHEMICAL COMPONENTS TEST

(Test Fuels: Base, MC3, MC4, MC5, MC6, MC10)

In this section, the effect of the fuel chemical components (types) on LTHR is investigated. To simplify the effect of the chemicals, six fuels, (Base, MC3, MC4, MC5, MC6, and MC10) were used. MC3 is a mixture of 6.5% of 4methyl-1pentene (Olefin, C_6H_{12}) and 93.5% of Base. MC4 is a mixture of 6.5% of 2methylpentane (iso-paraffin, C_6H_{14}) and 93.5% of Base. MC5 is a mixture of 6.5% of n-hexane (n-paraffin, C_6H_{14}) and 93.5% of Base. MC6 is a mixture of 6.5% of cyclo-hexane (naphthene, C_6H_{14}) and 93.5% of Base. MC10 is a mixture of 6.5% of toluene (aromatic, C_7H_8) and 93.5% of Base. The C_6 aromatic, benzene, is considered to cause cancer and thus toluene was used instead of benzene. The additives are all C_6 chemicals, except toluene, but their chemical structures are different.

Table 4 shows these six fuel properties. These six fuel mixtures have nearly the same distillations, but the RON is different because of the added chemicals in the blend.

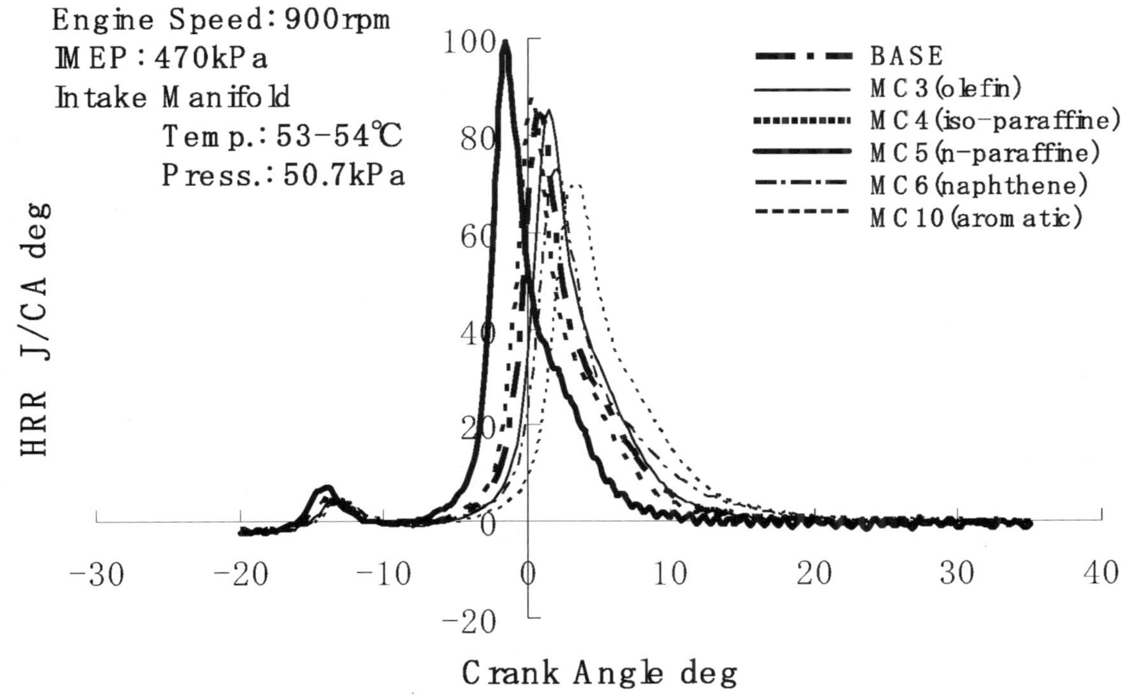

Fig. 16 Heat Release Data of Chemical Components Test

Table 8. Low Temperature and High Temperature Heat Release Data

		Unit	BASE	MC3	MC4	MC5	MC6	MC10
Low Temperature Heat Release	Heating Value	J	25.9	25.1	26.5	31.6	24.2	23.4
	Start Crank Angle	BTDC	17	18	17.75	18.5	16.75	16.75
High Temperature Heat Release	Heating Value	J	418.9	425	416	410.2	405	428.5
	Start Crank Angle	BTDC	5.75	5.25	6.75	8.5	6.5	4.5
Intake Air Temperature		°C	49.0	49.3	47.5	48.0	48.3	48.2
Air/Fuel Ratio (Exhaust O2 Sensor)			37.5	37.5	37.4	37.1	37.7	37.4

Engine test conditions were as follows:

Fuels: Base, MC3, MC4, MC5, MC6, MC10
Engine Speed: 900 rpm
Intake Manifold Pressure: +50.7kPa (gage)
Intake Manifold Temp.: 53-54°C
IMEP: 470 kPa
Water & Oil Temperature: 80°C

Atmospheric Conditions at the Engine Bench
Temperature: 20°C (controlled)
Humidity 60% (controlled)
Pressure: 101.3 kPa (controlled)

Figure 16 shows the rate of heat release data for these 5 fuels. The order corresponding to LTHR heating value is MC5, MC4, Base, MC3, MC6 and MC10, and the start of HTHR conforms to the same order (see Table 8).

The n-paraffin (n-hexane) in MC5 and the iso-paraffin (iso-paraffin) in MC4 have a large amount of LTHR. Base has 25.9J LTHR and LTHR heating value of MC3 (25.1J) is larger than 93.5% of Base (24.2J). This means that the olefin (4methyl-1pentene) in MC3 has a medium amount of LTHR. The LTHR heating value of MC6 is 24.2J and that of MC10 is 23.4J. This shows that the naphthene (cyclo-hexane) in MC6 and the aromatic (toluene) in MC10 have small or no LTHR. These fuel components, such as cyclo-hexane or toluene, have a function to dilute the mixture and decrease the LTHR proportional to the mixture ratio of cyclo-hexane or toluene.

In the comparison of regular gasoline(s) and PRF90.5, LTHR and HTHR were faster for PRF90.5. The total amount of olefins, naphthenes, and aromatics in the regular gasoline(s) is almost 50 volume-%, which delays the LTHR and HTHR combustion, while PRF90.5 consists only of paraffins.

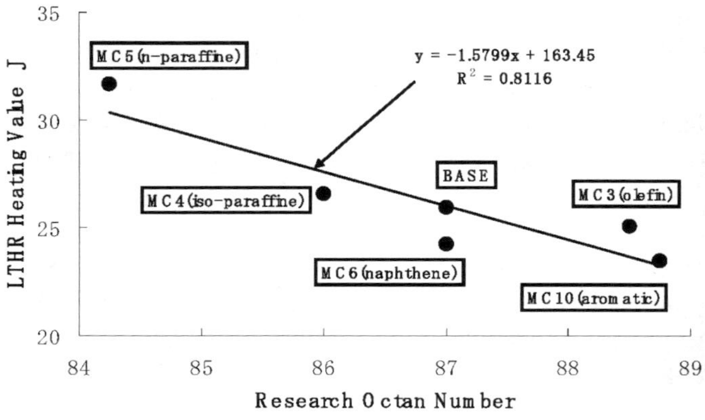

Fig. 17 Relation between Research Octane Number and Low Temperature Heat Release

Figure 17 shows the relation between the LTHR heating value and the research octane number (RON). In this test, IMEP is constant at 470kPa but the maximum rate of pressure increase and the excess air ratios are different. This is the reason why there is not a stronger relation between RON and the LTHR heating value (R^2=0.812). If the maximum pressure increase rate or the excess air ratio is the same, there is a strong relation between RON and the LTHR heating value, as shown by the octane number test.

SUMMARY

It is clear that fuels change the engine HCCI performance parameters, such as engine speed range, torque, air/fuel ratio range, thermal efficiency, and specific fuel consumption. To develop and test HCCI engines, knowing the effects of fuel characteristics on HCCI combustion is very important.

The magnitude of LTHR determines engine performance. A large LTHR fuel enhances the mixture temperature and initiates HTHR earlier than a small LTHR fuel. Therefore, a small LTHR fuel, such as RON92, can not be used at high speed, although its thermal efficiency is high and the engine can be run at fuel rich condition. The possible engine speed range and air/fuel ratio are strongly dependent on LTHR, which is depends on the fuels.

From the results of the regular gasoline(s) and PRF90.5 test, the fuel has a big potential to change the heat release and engine performance even at the same research octane number. There are several fuel factors that change the LTHR starting crank angle and its magnitude. Not only octane number but also the distillation and vaporization characteristics must be important factors. Those properties depend on the fuel composition which was addressed by the chemical components test.

N-hexane, a n-paraffins, exhibited large LTHR and caused the LTHR to begin at earlier crank angles. The order corresponding to LTHR heating value for C_6 hydrocarbons is n-hexane (n-paraffin), 2-methyl-pentane (iso-paraffin), 4methyl-1pentene (olefin), cyclo-hexane (naphthene) and toluene (C_7 aromatic), and the start of HTHR also conforms to this order. Cyclo-hexane and toluene have no LTHR and those chemicals function as diluents with regard to LTHR.

OI in equation (1), suggested by Kalghatgi, is one simple HCCI index. In our test case, K in equation (1) was -0.365 and RON had a much stronger influence on the octane number test results than MON. It is said that the K factor depends on engine speed and load but is independent of fuel. An HCCI combustion index that does not include distillation and vaporization characteristics is only partially satisfactory. HCCI index can not be a simple index and must depend on three or more parameters

CONCLUSIONS

The following conclusions are drawn from test results using a supercharged 4-cylinder engine with no intake air heating.

1. A fuel with a large heating value of LTHR is an "Easy Knock, High Speed & Low Torque" fuel, and a fuel with a small heating value of LTHR is a "Difficult Knock, Low Speed & High Torque" fuel. These tendencies were consistent for the fuels with the same distillation and same RON distribution against distillation.

2. Fuels having the same research octane numbers but different distillations showed different rates of heat release. This means that not only octane number but also fuel composition, distillation and vaporization characteristics are important factors for HCCI combustion.

3. A small change in chemical composition can change the HCCI combustion characteristics, including the amount and the phasing of LTHR that significantly affects the HTHR. The order corresponding to LTHR heating value for C_6 hydrocarbons is n-hexane (n-paraffin), 2methylpentane (iso-paraffin), 4methyl-1pentene (olefin), cyclo-hexane (naphthene) and toluene (C_7 aromatic), and start of HTHR conforms to the same order. Cyclo-hexane and toluene have no LTHR and those chemicals function as diluents with regard to LTHR.

4. For a future demand of compatibility of HCCI-SI engine operations, a more versatile HCCI fuel index must be obtained.

Contact. for further information please contact gen shibata, email: gen.shibata@eneos.co.jp

REFERENCES

1. Zheng, J., Yang, W., Miller, D.L., and Cernansky, N.P., "Prediction of Pre-ignition Reactivity and Ignition Delay for HCCI Using a Reduced Chemical Kinetic Model", SAE Technical Paper, No. 2001-01-1025, 2001
2. Epping, K., Aceves, S., Bechtold, R., and Dec, J., "The Potential of HCCI Combustion for High Efficiency and Low Emissions", SAE Technical Paper, No. 2002-01-1923, 2002
3. Tanaka, S., Ayala, F., and Keck., J.C., "A Reduced Chemical Kinetic Model for HCCI Combustion of Primary Reference Fuels in a Rapid Compression Machine", Combustion and Flame 133, Page 467-481, 2003

4. Tanaka, S., Ayala, F., Keck, J.C., and Heywood, J.B., "Two-stage Ignition in HCCI Combustion and HCCI Control by Fuels and Additives", Combustion and Flame 132, Page 219-239, 2003
5. Kalghatgi, G.T., "Fuel Anti-Knock Quality-Part I. Engine Studies", SAE Technical Paper, No. 2001-01-3584, 2001
6. Kalghatgi, G.T., "Fuel Anti-Knock Quality-Part II. Vehicle Studies-How Relevant is Motor Octane Number (MON) in Modern Engines?", SAE Technical Paper, No. 2001-01-3585, 2001
7. Aroonsrisopon, T., Sohm, V., Werner, P., Foster, D., Morikawa, T., and Iida, M., "An Investigation into the Effect of Fuel Composition on HCCI Combustion Characteristics", SAE Technical Paper, No. 2002-01-2830, 2002
8. Kalghatgi, G.T., "The Available and Required Autoignition Quality of Gasoline like fuels in HCCI engines at High Temperatures", SAE Technical paper, No. 2004-01-1969
9. Shibata, G., Omata, T., Isoda, T., Hosono, K., Nakamura, K., Tsuneishi, J., and Kawano, H., "The Development of Driveabirity Index and the Effects of Gasoline Volatility on Engine Performance", SAE Technical Paper, No. 952521, 1995
10. Shibata, G., Oyama, K., Urushihara, T., and Nakano, T., "The Effect of Fuel Properties on Low and High Temperature Heat Release and Resulting Performance of an HCCI Engine", SAE Technical Paper, No.2004-01-0553
11. Hiraya, K., Hasegawa, K., Urushihara, T., Iiyama, A., and Itoh, T., "A Study on Gasoline Fueled Compression Ignition Engine-A Trial of Operation Region Expansion-", SAE Technical Paper, No. 2002-01-0416, 2002

Which Fuels For Low CO_2 Engines?
© P. Duret, X. Montagne (Eds.) and Editions Technip, Paris, 2004, pp. 59 – 67
27 rue Ginoux, 75015 Paris

HCCI: FUEL AND ENGINE INTERACTION

Tom W. Ryan III, Darius Mehta, T. J. Callahan

Southwest Research Institute® San Antonio, Texas USA

ABSTRACT – *A study was performed to examine the effects of engine operating parameters on HCCI combustion using multiple fuels. The parameters studied include compression ratio, inlet air temperature and pressure, and EGR level. Extensive experiments were performed using Diesel fuel, gasoline, blends of gasoline and diesel fuel, and Fischer Tropsch Naphtha. Experiments were performed on a single-cylinder variable compression ratio research engine. The independent variables included compression ratio, EGR level, intake manifold conditions, and the fuel composition.*

Detailed heat release rate analysis was used to examine the effects of various characteristics of the heat release process on the engine efficiency. The results of the heat release rate analysis and cycle simulation modeling results were used to develop simple regression models and rules for controlling the heat release rate to produce the maximum efficiency. These results are used to suggest an optimum fuel formulation using a blend of gasoline and Fischer-Tropsch Naphtha.

INTRODUCTION

There are basically three modes of reacting fuels in reciprocating engines. Diesel engines depend upon diffusion combustion. In a modern diesel engine fuel is injected at a controlled rate to accomplish diffusion burning rates equal to the injection rate. Current gasoline engines depend upon flame propagation through a homogeneous fuel-air mixture. Combustion control in these engines is focused on control of the air fuel ratio to stoichiometric conditions, with periodic rich excursions to accomplish NO_x reduction in the 3-way catalyst. A new mode of combustion becoming popular in reciprocating engines is called Homogeneous Charge Compression Ignition, or HCCI. HCCI is defined as a method of reacting fuel in a reciprocating engine in which the fuel and air are premixed and in which the reactions are initiated through compression heating.

The majority of the recent work in HCCI has focused on overcoming two major issues associated with HCCI operation: mixture preparation and start of reaction control. In addition, much of the work is focused on the use of gasoline and diesel fuel. In the use of these two existing fuels there are two general approaches that are being pursued: one addressing the high ignition temperatures required with gasoline [1-5]*; and one focused on overcoming the mixture preparation and low ignition temperature problems associated with the use of diesel fuel [6-10]. A significant amount of work is also being done on the use of alternative fuels, mainly natural gas [11,12] and DME [13-15], which in a fashion represent very volatile replacement for gasoline and diesel fuel, respectively. There is also a growing opinion that there are opportunities to develop more ideal fuel formulations for use in HCCI engines [14,16], analogous to the historic optimization of gasoline and diesel fuels for application in their respective engines.

Southwest Research Institute has developed a very specific definition of HCCI, in which reaction is initiated as indicated above, and includes a limit on NO_x of 15 ppm and a smoke limit of zero BSN. The majority of the recent work has focused on overcoming two major issues associated with HCCI operation: mixture preparation and start of reaction control. This paper is the fifth in a series covering the study of HCCI reaction in a variable-compression ratio research engine. The first paper

* Numbers in Parenthesis Designate Entries in the Reference List

[17] focused on examination of the ignition, or start of reaction, process, and developing a method for prediction of the start of reaction in HCCI engines. The second paper [18] covered some aspects of the control of the start of reaction. The third paper introduced the concept of development of fuels specifically designed for use in HCCI engines. [16] The fourth paper was a summary of the combined effects of selected engine variables on the start of reaction of diesel fuel and a variety of other test fuel formulations [19]. This paper presents the results of an effort to use detailed heat release analysis and cycle simulation analysis to arrive at a fuel formulation that is more ideal in terms of boiling point distribution and ignition characteristics.

OBJECTIVE

The objectives of the work reported in this paper were to develop a better understanding of the heat release process in HCCI engines and to arrive at some rules for arriving at the optimum heat release rate. The ultimate goal was to arrive at a near optimum fuel formulation for use in HCCI engines.

EXPERIMENTAL APPARATUS AND PROCEDURES

TEST ENGINE

The test engine set-up, discussed in detail previously [17], and summarized in [19], was based on the use of a specially designed variable compression ratio engine. The description of the test set-up is repeated here for completeness.

The HCCI engine development at SwRI has been performed in a single cylinder research engine that incorporates several features that simplified the development process. The most useful of these features included variable compression ratio, an Exhaust Gas Recirculation (EGR) system, an electrically heated intake system, and a simulated turbocharger system. These design features allowed for examination of the effects of compression ratio, intake manifold pressure and temperature, and EGR on the start of reaction (SOR). The piston was a two piece design, consisting of a removable crown. This made it very simple to change the combustion chamber design. The work reported in this paper was accomplished using a simple open dish piston bowl design with low intake swirl (0.7) and a squish area of 51 percent. All tests reported in this paper were performed using a quiescent open chamber design. The specifications of the test engine are included in Table 1.

Table 1. Engine Specifications

Bore	96.8 mm
Stroke	95.3 mm
Rod Length	166.5 mm
Displacement	702 cc
Compression Ratio	7.5:1 to 16.5:1
Swirl Ratio	Less Than 0.7
Combustion Chamber	Shallow Dish

The initial effort focused on creation of a homogeneous mixture of fuel vapor and air in the intake runner. This required very high air and surface temperatures to accommodate the high boiling point of diesel fuel. The initial experiments demonstrated, however, that the fuel did not have to be totally vaporized, it is sufficient to create a homogeneous mixture of small fuel droplets and air. This finding led to the development of a port fuel injection scheme, in which an air assist fuel injector was located in the intake port and targeted at the intake valve. The fuel injector was an air assist injector developed at SwRI to produce a very finely atomized spray. The engine was also equipped with a traditional, mechanical pump-line-nozzle, direct-injection system. This system was used for baseline measurements and for engine start-up and temperature stabilization.

The intake air to the engine was supplied using an external compressor and air conditioning system that made it possible to control the intake manifold pressure and temperature. The EGR rate was controlled by adjusting the engine backpressure to create the appropriate pressure differential. EGR rate was monitored using simultaneous CO_2 measurements in both the intake manifold and the exhaust manifolds.

The engine was equipped with a SwRI developed engine control system that allowed for control of the injection timing and quantity. A wide range oxygen sensor in the exhaust was used to set and control the equivalence ratio. The equivalence ratio was also computed based on the gaseous emissions measurements. Exhaust particulate (PM) measurements were also made during the initial experiments, but it was observed that the PM soot levels were always near zero when operating in

HCCI mode. It was also observed that the Bosch Smoke Number (BSN) provided a similar indication of the successful achievement of HCCI operation. BSN was used as the indicator of the PM levels as well as an indicator of HCCI operation.

Engine airflow was measured using a rotameter and the fuel flow rate was measured gravimetrically. The gaseous emissions were measured using infrared instruments for CO and CO2, a chemilumenescent instrument for NO/NO2, and a ionization detector for unburned hydrocarbons. Particulate emissions (PM) measurements were made at a limited number of test conditions using a mini-dilution tunnel. The engine was also equipped with a water-cooled pressure transducer in-cylinder. The pressure data was digitized every 0.6 CA degrees, and used to compute the apparent heat release rate and the cumulative heat release.

TEST FUELS

The test fuels included diesel fuel, gasoline, blends of gasoline and diesel fuel, and Fischer-Tropsch naphtha. The properties of the diesel fuel, the gasoline, and Fischer-Tropsch naphtha are listed in Table 2. The gasoline and diesel fuel blends included 20, 40, 60, and 80 percent gasoline in diesel fuel, in addition to the neat gasoline and diesel fuel.

Table 2. Test Fuel Specifications

PROPERTY	FT NAPHATHA	DIESEL FUEL	GASOLINE
Heat of Combustion, MJ/kg			
Net	44.3	42.5	32.9*
Gross	47.7	45.3	45.4*
Sulfur, mass %	0.0	0.039	.023
Specific Gravity	0.7095	0.8485	.7436
IBP, °C	61	187	29
50%	141	263	105
95%	186	328	188
FBP	194	339	201
Carbon, mass %	83.98	86.83	86.74*
Hydrogen mass %	15.92	13.24	13.22*
Cetane Number (IQT)	51	43	13

* Estimated Values

TEST PROCEDURE

As indicated above, the primary engine independent variables included compression ratio, EGR rate, intake manifold temperature and pressure, and equivalence ratio. A majority of the tests were performed at 1200 rpm, but engine speed was also considered as one of the engine independent variables. After warm-up, the engine speed was selected, the compression ratio was fixed, and equivalence ratio, EGR rate, and intake manifold conditions varied to define the region of HCCI operation. At each stable HCCI test condition, the gaseous emissions, BSN, fuel consumption, and engine heat release data were recorded. The PM emissions were measured at selected test conditions, but it was observed early in the program that the PM dry soot emissions were always zero during HCCI operation, and that BSN values of zero provided a excellent indication of true HCCI operation.

The compression ratio was varied from 8 to 16.5:1. The EGR levels ranged from 0 to 50 percent. Air-fuel ratio was varied from richer than stoichiometric (12.5:1 for diesel fuel) to very lean (75:1 for diesel fuel). The intake temperature depended upon the properties of the test fuel, but the entire range was from 28 to 220°C. The intake manifold pressure was varied over the range from 1 to 2.1 Bar. Engine speed was varied from 600 to 2000 rpm, with the vast majority of testing done at 1200 rpm, but with limited testing at each 100 rpm increment from 600 to 2000 rpm.

All of the data reported in this paper was acquired at 1200 rpm, 14:1 compression ratio, and naturally aspirated conditions. The exception was gasoline because it would not react under these conditions. Gasoline was tested at 16:1 compression ratio.

CYCLE SIMULATION

The cycle simulations were performed using ALAMO_ENGINE, a code developed at a Southwest Research Institute (SwRI) [20]. Alamo_Engine is a zero-dimensional cycle simulation code. The air flow through the engine is computed using a valve flow model. The model includes a detailed thermodynamic analysis for the compression and expansion. It also has a detailed equilibrium and associated NOx model, and it does account for turbocharging using a simple model for calculation of the intake and exhaust manifold conditions based on an assumed turbocharger efficiency.

EXPERIMENTAL RESULTS AND DISCUSSION

As described previously [19], the combustion, or perhaps better phrased, the reactions in an HCCI engine generally occur in two phases. This is demonstrated in Figure 1, which shows a typical heat release rate diagram for a typical US diesel fuel. Also shown in the figure are various characteristics that have been identified as important in relationship to HCCI operation and engine efficiency. These characteristics include the start of the pre-reaction (or cool flame reactions), the start of the main reaction, and the interval between the start of the pre and the main reactions, called the phasing.

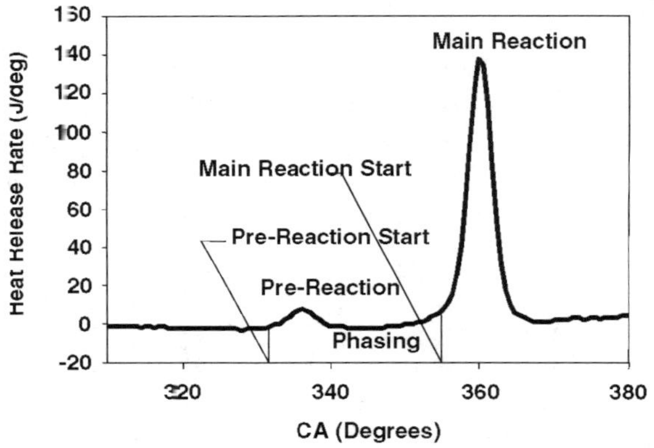

Fig. 1 Heat Release Rate Diagram for Diesel Fuel at 1200-rpm, Zero EGR, and 8:1 CR

ENGINE EFFICIENCY

The controlling factors in engine efficiency are the combustion efficiency and the timing of the heat release event. As indicated above, control of the start of reaction is a very critical issue in the development of HCCI engines. As indicated in Reference 19, the start of the pre-reaction, the magnitude of the of the pre-reaction, and phasing are all important in defining the overall timing of the heat release.

Figure 2 shows the heat release rate diagram for a test using the 20% blend of gasoline in diesel fuel. It is clear that the pre-reaction is very early and the resulting main reaction completed well before TDC. The resulting indicated thermal efficiency for this run was 19%. This same heat release rate diagram was shifted in 5 degree increments and used in the Alamo_Engine cycle simulation to predict the effect on the indicated thermal efficiency. The results of these analyses are presented in Figure 3, where the indicated thermal efficiency is plotted versus the start of the main reaction. As can be seen, the maximum thermal efficiency occurs when the peak of the main heat release rate is 12 degrees after TDC. It should be noted that the entire heat release event, including the pre-reaction was shifted. For this particular run the pre- reaction represented 4 percent of the total heat release, so the centroid of the heat release is displaced from the centroid of the main heat release event. It is clear from the figure that the location of the heat release has a significant effect on the thermal efficiency of the engine, as expected, based on previous experience of the authors [21]. Additional cycle simulation results also indicated that the shape of the diagram, in terms of the magnitude of the pre-reaction and the phasing also play significant roles in defining the engine thermal efficiency. These observations led to a more thorough examination of the heat release rate in the HCCI engine.

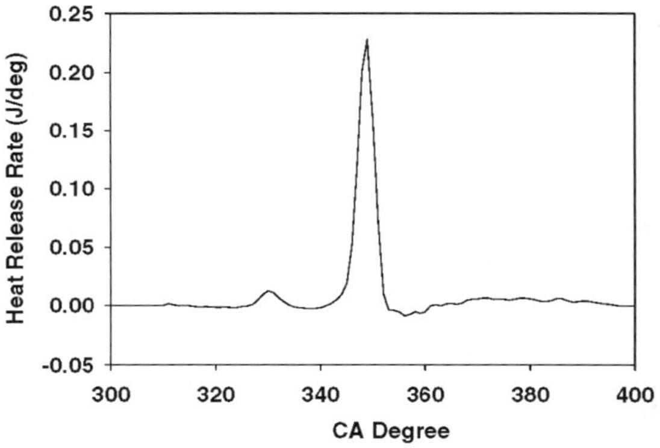

Fig. 2 Heat Release Rate Diagram for 20% Gasoline in Diesel Fuel

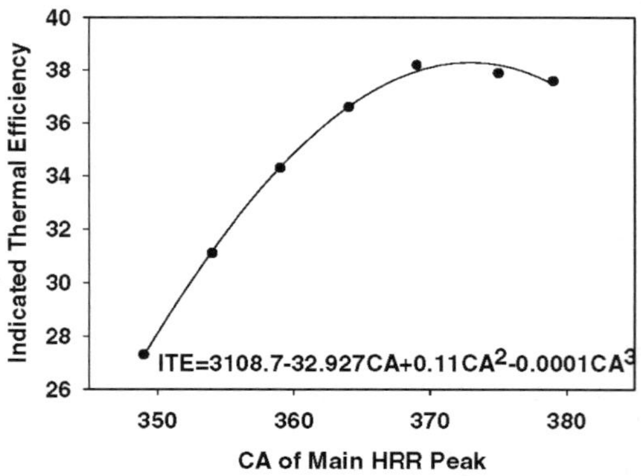

Fig. 3 Indicated Thermal Efficiency versus the Timing of the Main Heat Release

PRE-REACTION DELAY

The start of the pre-reaction was demonstrated previously [19] to be dependent primarily on temperature, with initiation of the pre-reaction occurring when the compression temperature reaches a critical level. The relationships between the pre-reaction delay times (PRD) and the compression temperature at TDC are shown in Figure 4 for gasoline, diesel fuel, and blends of gasoline and diesel fuel. The pre-reaction delay time is defined as the time in milliseconds from 46° BTDC to the start of the pre-reaction. The significance of 46°, as described in Reference 19, is based on the fact that this is the time in the compression stroke where significant temperature increases begin to occur. Also shown in the figure are the relationships for the pre-reaction delay time as functions of the TDC compression temperature.

As can be seen, the data are bound by the results for the gasoline, with relatively long delay times, and diesel fuel with the shortest delay times. These data reflect the difference in the propensities of the two fuels to autoignite, with gasoline more resistant to autoignition than diesel fuel. The interesting observation from this data is the fact that as little as 20% gasoline in the diesel fuel produced delay characteristics in the blend that are more like gasoline than diesel fuel. Increasing the concentration of gasoline concentration above 20% had no additional effect. On average, the delay time of the diesel fuel was 30% shorter than that of gasoline. 20% gasoline in diesel fuel reduced this difference by 83%. If diesel fuel is considered in this case to be the reactive component (lower ignition temperature, shorter delay time) and gasoline the non-reactive component, it is clear that the non-reactive component dominates the initiation of the pre-reaction. Previous work [22] with Fischer-Tropsch Naphtha (FTN) and natural gas, indicated a somewhat opposite effect, where the more reactive FTN dominated the pre-reaction delay, with natural gas having almost no effect.

If the Fischer Tropsch Naphtha (FTN) data from the current experiment are added to the results in Figure 4, as shown in Figure 5, the differences between the gasoline and the diesel fuel appear to be less significant. As can also be seen in Figure 5, the ignition temperatures for the FTN are much lower than for any of the other fuels, indicating a higher propensity to autoignite, or a fuel that is even more reactive than the diesel fuel. This is indicated by the fact that the regression line for the FTN falls below that of the diesel fuel. As can be seen in Figures 4 and 5 the coefficients of the regression equations are very similar, indicating that it is perhaps possible to grossly lump all of the data together to develop a generalized expression for the pre-reaction delay time, as shown in Figure 6.

These results indicate that the initiation of the pre-reaction can be affected significantly by the addition of relatively small quantities of blending material. In the cases described above, the reactivity of the non-reactive component varied greatly, from methane in [22], to the gasoline and FTN used in these experiments. It should also be noted that the truly non-reactive components have very little pre-reaction. Natural gas does not exhibit this phenomena and gasoline demonstrates very little pre-reaction heat release. On the other hand, the very reactive components tend to exhibit a large pre-reaction heat release. These characteristics are discussed in the next section.

PRE-REACTION RATIO

It has been observed that there appears to be a relationship between the reactivity, or tendency to autoignite at low temperature, and the amount of energy liberated in the pre-reaction. A new term has been defined to provide a measure of this tendency, or Pre-Reaction Ratio (PRR), defined as the fraction of the total heat release that occurs in the pre-reaction.

The PRR for the gasoline-diesel fuel blends are presented in Figure 7. There are three observations that can be made from this data. First, diesel fuel, as the reactive component, has a significantly larger pre-reaction than the less reactive gasoline. Second, the PRR increases in all cases, as the PRD becomes longer. This apparently indicates that the longer residence time prepares more mixture for the pre-reaction. Third, as occurred with the pre-reaction delay, the addition of as little as 20% gasoline made the diesel fuel behave more like gasoline than diesel fuel, as indicated by the fact that all of the blend data were clustered about the gasoline line. These differences are, however, small in comparison to the

PRR observed with the FTN, as shown in Figure 8. As can be seen, the PRR's for the FTN were factors of 4 to 6 larger than those for the diesel fuel and gasoline. The FTN has both a shorter delay, or PRD, at a given temperature, as well as a larger heat release, or PRR, in the pre-reaction than any of the other fuels.

PHASING

As defined above, phasing is a measure of the time from the start of the pre-reaction to the start of the main reaction. It was demonstrated in Reference 19 that the phasing is related to the magnitude of the pre-reaction. The relationships for the gasoline-diesel fuel blends are shown in Figure 9. As can be seen, the phasing decreases as the pre-reaction increases. It can also be seen that the more reactive diesel fuel has a significantly longer phasing than gasoline. In addition, as with the other combustion parameters, the addition of as little as 20% gasoline makes the diesel fuel behave more like gasoline than diesel fuel. The diesel fuel phasing is on average 94% longer than gasoline. The use of 20% gasoline accomplishes 75% of this difference, with the addition of more gasoline having little added effect.

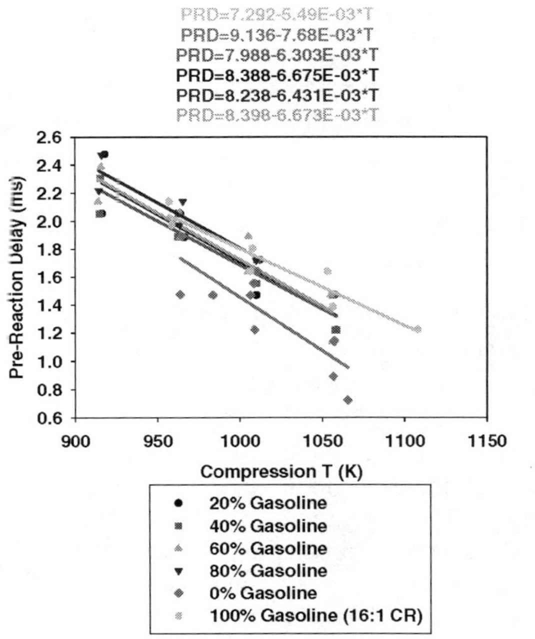

Fig. 4 Pre-Reaction Delay versus TDC Compression Temperature for the Gasoline and Diesel Fuel Blends

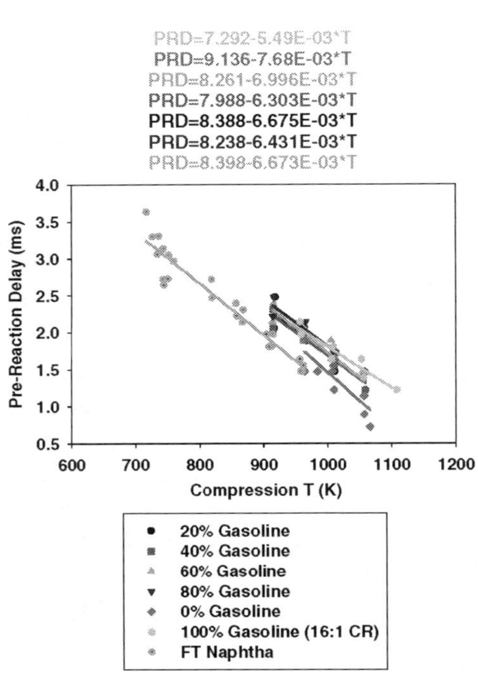

Fig. 5 Pre-Reaction Delay versus TDC Compression Temperature for the Gasoline and Diesel Fuel Blends, and FTN

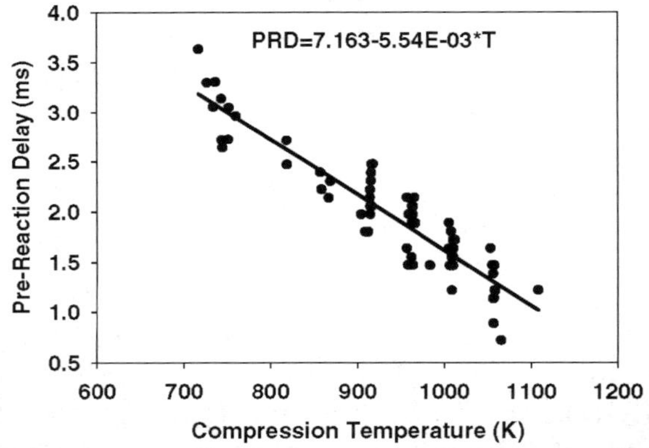

Fig. 6 Pre-Reaction Delay versus TDC Compression Temperature for all Fuels

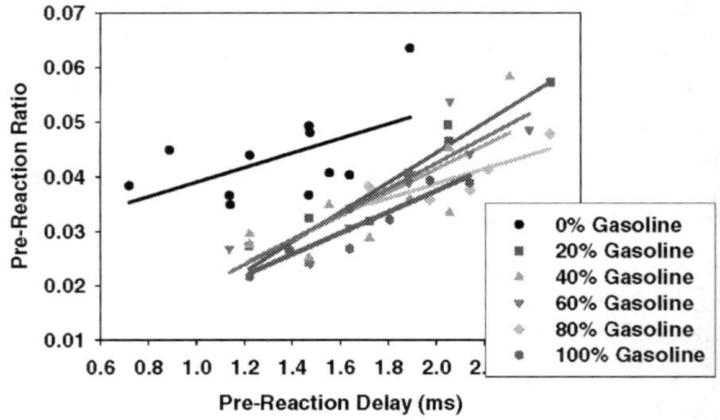

Fig. 7 PRR versus PRD for Gasoline and Diesel Fuel Blends

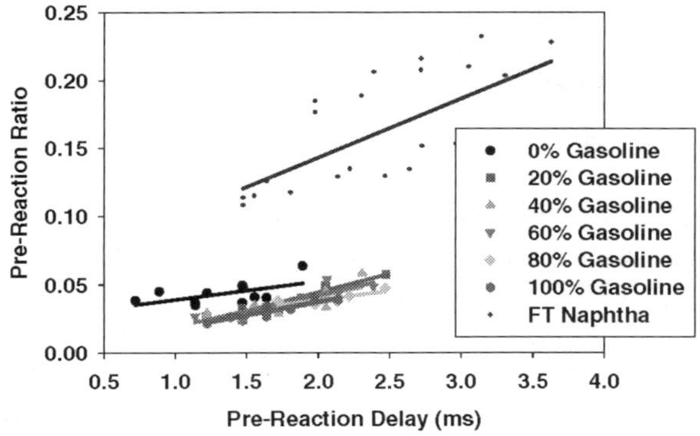

Fig. 8 PRR versus PRD for Gasoline and Diesel Fuel Blends and FTN

The FTN data is added to the plot in Figure 10. It is clear that he pre-reaction heat release is much larger than either gasoline or diesel fuel, by a factor of 4. It is also clear the phasing is much shorter, 33% as compared to gasoline and 65% as compared to diesel fuel.

SUMMARY AND PREDICTION

The results of these experiments indicate that various reactive and non-reactive fuel components can be blended to produce what might be more optimum formulations of HCCI engine fuels. For instance, it appears that FTN could be blended with gasoline to arrive at a near optimum fuel. In this case the FTN is the reactive component and the gasoline is the non-reactive component. In the case of gasoline and diesel fuel, both of which have pre-reaction heat release, the non-reactive component tends to dominate the ignition process and the timing of the main heat release. This was not the case with FTN and natural gas as reported in [19], where the reactive component (FTN) dominated the ignition process, with the non-reactive methane dominating the effective phasing. It should be noted, however, that in the case of methane, there was no evidence of pre-reaction, and the actual phasing for the FTN component was not affected by the presence of methane.

It is felt that in a 20-80 blend of FTN and gasoline, the gasoline will inhibit the start of the pre-reaction and slow the initiation of the reactive component. This is desirable because recall that the pre-reaction delay (PRD) is too short with FTN, resulting in very early reaction. It is hypothesized that the large pre-reaction heat release with the FTN on the other hand will provide sufficient energy to achieve reaction of the gasoline, with the appropriate phasing.

Alamo_Engine was used to calculate the compression temperature for the test engine used in this program, but operating as high boost (3 bar) and moderate EGR (10%) with a 25:1 fresh air-fuel ratio, conditions that represent a heavy-duty full load condition. The ALAMO-ENGINE simulation provided compression temperature and the fueling requirements to achieve the load, but operating in conventional diesel mode. The results of the simulation are presented in Table 3.

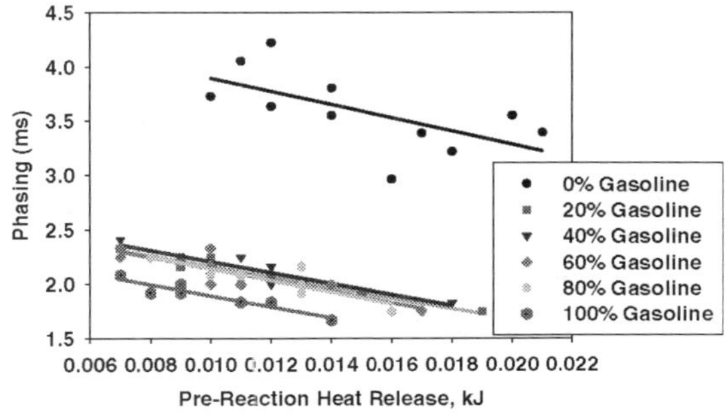

Fig. 9 Phasing versus Pre-Reaction Heat Release Gasoline-Diesel Fuel Blends

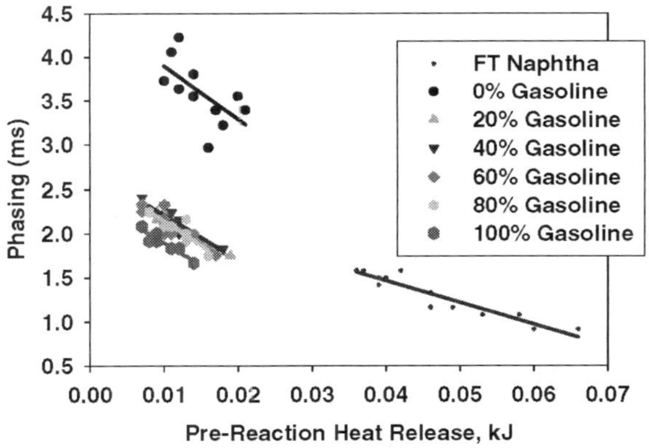

Fig. 10 Phasing versus Pre-Reaction Heat Release Gasoline-Diesel Fuel Blends and FTN

Table 3. Cycle Simulation Results for Diesel Engine Baseline

Parameter	Value
MAP (kPa)	300
Fuel Flow Rate (kg/s)	0.00114
Fresh Air Flow Rate (kg/s)	0.0292
Torque (N-m)	108.7
Indicated Power (kW)	21.76
Indicated Thermal Efficiency	44.2
BSFC ((g/kW-hr)	195.1

The computed compression temperature at TDC was 860K. This temperature was used in the equations for FTN and gasoline to predict the PRD for both, or 2.24 and 2.57 ms, respectively. Also, as shown above, the Pre-Reaction Ratio (PRR) is related to the Pre-Reaction Delay (PRD). The corresponding PRR's for FTN and Gasoline are 0.154 and 0.0484, respectively, based on linear regression of the data presented in Figure 8. These ratios were use to predict the magnitudes of the pre-reaction heat release. The pre-reaction heat releases where then used in the linear regressions presented in Figure 10 to predict the phasing for FTN and gasoline, or 1.79 and 1.61 ms, respectively.

If the FTN functions as it did with methane, the PRD will be dominated by the FTN, and the PRD for the blend would be that of the FTN or 2.24 ms. If, on the other hand, the gasoline inhibits the reaction of the FTN as it did with the diesel fuel, then the PRD for the blend would be 83% of the difference between the two components, then the PRD for the blend would be 2.51 ms. So, it is predicted that the 20/80 blend of FTN in gasoline would have a PRD between 2.24 and 2.51 ms. At the conditions used in the cycle simulation analysis above, these delay times correspond to starts of reaction at 22° and 19° BTDC, respectively.

The pre-reaction heat release for FTN was unchanged as methane was added to increase load, as reported in Reference 19. In the experiments reported here, the presence of gasoline inhibited the pre-reaction of the diesel fuel, with 20% gasoline having the effect of reducing the PRR by approximately 30%. If the FTN dominates the start of the pre-reaction, the PRD would be 2.24 ms (22° BTDC) and the PRR would be 0.154 (.0268 kJ). On the other hand, if the gasoline functions as it does with diesel fuel, the PRR for the blend would be reduced by 30%, resulting in a PRR of 0.108 (0.0188 kJ). Thus it is predicted that the PRR for the 20/80 blend would be in the range from 0.154 to 0.108.

The logic in predicting the phasing is a bit more complex because of the interactions between the fuels, in terms of the potential effect of the gasoline on the pre-reaction heat release of the blend, as well as the interaction of the fuels in terms of the effects of gasoline on phasing. If it is assumed that the FTN dominates, then the PRR of the blend will be 0.154 and the predicted phasing will be 1.79 ms (19.3°). On the other hand, if the gasoline behaves as it did with diesel fuel, the PRR of the blend will be 0.108, and the phasing is predicted to be 1.66 ms (17.9°)

The results of this exercise indicate that if a 20/80 blend of FTN and gasoline are run at 1800 rpm and 3 bar boost, the start of the pre-reaction will occur in the range from 22° BTDC to 19° BTDC and the main reaction will start in the corresponding range from 2.7 ° BTDC to 1.1° BTDC. If it is assumed that the gasoline has the expected role, the following combustion characteristics are predicted:

Start of Pre-Reaction =19 ° BTDC
Start of Main Reaction = 1.1 ° BTDC
PRR = 0.108

Assuming that the fuel energy input is the same as the baseline diesel engine run, and the combustion efficiency is 95%, then the proportions of the heat release between the pre and the main reactions are:

Pre-Reaction Heat Release = 0.334 kJ
Main Reaction Heat Release = 2.76 kJ

If it is assumed that the heat release rate diagram for this HCCI condition can be approximated as shown in Figure 11, the resulting predicted thermal efficiency is 40%, based on cycle simulation analysis.

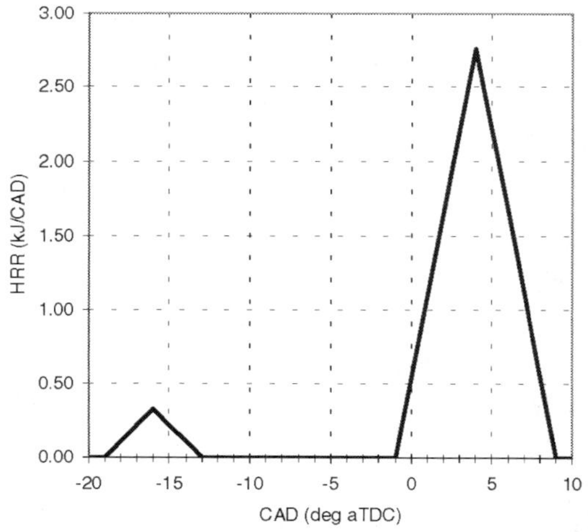

Fig. 11 Predicted Heat Release Rate Diagram for 20/80 Blend of FTN and Gasoline

Various blends of FTN and gasoline, including the proposed 20/80 will be tested and the results reported at a later date. In these experiments the composition of the gasoline will be determined and an effort will be made to identify the effects of the blend composition on the performance in the HCCI engine.

Acknowledgements. The work was financially supported as an Internal Research Program of Southwest research Institute. The authors would like to thank Mr. Bill Gray for his management of the laboratory engine experiments and Mr. Steve Schneider for performing the engine experiments.

REFERENCES

1. Zheng, J., D. Miller, N.P. Cernansky, D. Liu, and M. Zhang, "The Effects of Active Species in Internal EGR on Preignition Reactivity and on Reducing UHC and CO Emissions in Homogeneous Charge Engines" SAE Paper 2003-01-1831, June 2003.
2. Osborne, R.J., G. Li, S.M. Sapaford, T.H. Lake, and M.R. Heikal, "Evaluation of HCCI for Future Gasoline Powertrains" SAE Paper 2003-01-0750, March 2003.
3. Koopmans, L., H. Stroms, S. Lundgren, O. Backlund, and I. Denbratt, "Demonstating a SI-HCCI-SI Mode Change on a Volvo 5-Cylinder Electronic Valve Control Engine" SAE Paper 2003-01-0753, March 2003.
4. Chen, R., N. Milovanovic, J. Turner, and D. Blundell, "The Thermal Effect of Internal Exhaust Gas Recirculation on Controlled Auto Ignition" SAE Paper 2003-01-0751, March 2003.
5. Urushihara, T., K. Hiraya, A. Kakuhou, and T. Itoh, "Expansion of HCCI Operating Region by the Combination of Direct Fuel Injection, Negative Valve Overlap and Internal Fuel Reformation" SAE Paper 2003-01-0749, March 2003.
6. Hasegawa, R., and H. Yanagihara, "HCCI Combustion in DI Diesel Engine", SAE Paper 2003-01-0745, March 2003.
7. Peng, Z., H. Zhao, and N. Ladommatos, "Effects of Air/Fuel Ratios and EGR Rates on HCCI Combustion of N-Heptane, a Diesel Type Fuel" SAE Paper 2003-01-0747, March 2003.
8. Shimazaki, N., T. Tsurushima and T. Nishimura, "Dual Mode Combustion Concept with Premixed Diesel Combustion by Direct Injection Near Top Dead Center" SAE Paper 2003-01-0742, March, 2003.
9. Su, W., T. Lin, and Y. Pei, "A Compund technology for HCCI Combustion in a Diesel Engine Based on the Multi-Pulse Injection and the Bump Combustion Chamber" SAE Paper 2003-01-0741, March, 2003.
10. Hasegawa, R., I. Sakata, T. Koyama, H. Yanagihara, "Numerical Analysis of Ignition Control in HCCI Engine" SAE Paper 2003-01-1817, May, 2003.
11. Olsson, J-O, P. Tunestal, B. Johansson, S. Fiveland, R. Agama, M. Willi, D. Assanis, "Compression Ratio Influence on Maximum Load of a Natural Gas Fueled HCCI Engine" SAE Paper 2002-01-0111, March, 2002.
12. Hiltner, J., S. Fiveland, R. Agama, M. Willi, "System Efficiency Issues for Natural Gas Fueled HCCI Engines in Heavy-Duty Stationary Applications" SAE Paper 2002-01-0417, March, 2002.
13. Kim, H., S. Cho, K. Min, "Reduced Chemical Kinetic Model of DME for HCCI Combusiton", SAE Paper 2003-012-1822, May, 2003.
14. Ogawa, H., N. Miyamoto, N. Kaneko, H. Ando, "Combustion Control and Operating Range Expansion with Direct Injection of Reaction Suppressors in a Premixed DME HCCI Engine" SAE Paper 2003-01-0746, March, 2003.
15. Shudo, T., Y. Ono, and T. Takahashi, "Ignition Control by DME-Reformed Gas in HCCI Combustion of DME" SAE Paper 2003-01-1824, May, 2003.
16. Ryan, T. W., and A. Matheaus, "Fuel Requirements for HCCI Engine Operation" SAE Paper 2003-01-1813, May, 2003.
17. Ryan, T.W. and T. Callahan, "Homogeneous Charge Compression Ignition of Diesel Fuel" SAE Paper 961160, May, 1996.
18. Gray, A.W. and T.W. Ryan, "Homogeneous Charge Compression Ignition of Diesel Fuel" SAE Paper 971676, May, 1997.
19. Ryan, T.W., T.J. Callahan, and D. Mehta, "HCCI in a Variable Compression Ratio Engine-Effects of Engine Variables" SAE Paper 2004-01-1971, 2004.
20. Dodge, L.G.et. al. "A PC-Based Model for Predicting NOx Reductions in Diesel Engines" SAE Paper 962060, October 1996.
21. Callahan, T.J., D.M. Yost, and T.W. Ryan III, "Acquisition and Interpretation of Diesel Engine Heat Release", SAE Paper 852068, Oct. 1985.
22. Stanglmaier, R.H., T. Ryan, and J. Souder, "HCCI Operation of a Dual-Fuel Natural Gas Engine for Improved Fuel Efficiency and Ultra-Low NOx Emissions at Low-to-Moderate Engine Loads" SAE Paper 2001-01-1897, May 2001

Which Fuels For Low CO_2 Engines?
© P. Duret, X. Montagne (Eds.) and Editions Technip, Paris, 2004, pp. 69 – 78
27 rue Ginoux, 75015 Paris

EFFECT OF FUEL PROPERTIES ON CAI/HCCI COMBUSTION AND EMISSION IN A 4-STROKE GASOLINE ENGINE

Kayiu Man[1], Hua Zhao[1], Tom Ma[1], Aaron J. Oakley[2]

1. Department of Mechanical Engineering, Brunel University, UK
2. Ford Motor Company, UK

ABSTRACT – *This paper presents results from an ongoing experimental programme researching the in-cylinder conditions necessary to obtain homogeneous CAI (or HCCI) combustion in a 4-stroke engine. The fuels under investigation include Unleaded Gasoline, Primary Reference fuels, Methanol and their mixtures. This work concentrates on establishing the CAI/HCCI operating range with regard to Air/Fuel ratio and Exhaust Gas Recirculation and their effect on the ignition timing, combustion rate and variability, Indicated thermal efficiency, and engine-out emissions. Detailed maps are presented, defining how each of the measured variables changes over the entire CAI/HCCI region.*

INTRODUCTION

CAI combustion involves the compression-ignition of a premixed combustible charge and requires the auto-ignition of the cylinder contents combined with a means of controlling the subsequent heat release rate. The charge products must be raised to a temperature that will allow it to auto-ignite under compression yet the heat release rate must be limited as not to incur destructive engine knock.

The development of CAI has come some way since the days of Onishi et. al. [1] and Noguchi et. al. [2] who first realized the benefits and applied CAI successfully to a 2-stroke gasoline engine. Today's researchers are concerned with 4-stroke engines to eliminate problems associated with the gas exchange process of 2-Stroke engines that result in higher emissions.

Najt and Foster [3] applied successfully HCCI to a 4-Stroke gasoline engine with elevated initial charge temperatures and demonstrated the reduction in fuel consumption and emissions.

Since then, many techniques have been adopted to increase the operating range of CAI/HCCI, including the use of alternative fuels, intake charge heating and higher compression ratios.

Christensen et al [4] used iso-octane, ethanol and natural gas as the fuels under investigation. It was found that although a higher compression ratio (21:1) was used, it was still necessary to implement intake charge heating for ethanol (80°C) and natural gas (120°C). It was later demonstrated [5] that EGR helps reduce HC emissions and sufficient pre-heating would decrease CO emissions.

Shudo and Ono [6] used concentrations of dimethl ether (DME) and methanol reformed gas (MRG) to achieve HCCI. DME and MRG can both be produced from methanol using endothermic reactions and thus reduces the need to carry two separate fuel blends. It was found that with optimum compression ratio, DME and MRG in HCCI combustion results in higher thermal efficiency and a comparable power output as compared with SI combustion fueled with MRG.

A. Oakley et. al. [7, 8] used an elevated inlet charge temperature (320°C) to investigate the dilution effects on CAI of hydrocarbon and alcohol fuels. It was demonstrated that alcohol fuels have a much higher tolerance to air and EGR dilution when compared to hydrocarbon fuels and consequently no misfire region was observed thus extending the CAI region. Furthermore, it was also observed that Methanol exhibited the lowest emissions when compared to hydrocarbon fuels.

It is known that alcohol fuels are superior in CAI combustion but the commercial use of alcohol fuels is currently limited to certain markets (e.g. South

America) for reasons of climate and cost. The aim of this paper is compare the hydrocarbon and alcohol fuels and their effect on emissions when running on CAI.

The fuels under investigation in this paper are primarily reference fuels, Gasoline, Methanol and Ethanol based on the results obtained by Oakley [9].

ENGINE SETUP

The engine used for experimental work is that of a Ricardo E6, single cylinder research engine with variable compression ratio that can be varied from 4.5 to 20 even during engine operation. It uses overhead poppet valves and has a bore of 76mm and a stroke of 111mm. The combustion chamber is cylindrical in shape. The speed of the engine is controlled manually using a swinging field AC dynamometer.

CONTROL SYSTEMS – The engine is modified to allow exhaust gas recirculation (EGR) via an external pipe from the exhaust manifold back into the inlet manifold. A gate valve controls the amount of exhaust gas entering the inlet manifold. The exhaust is admitted back into the inlet manifold approximately one meter upstream from the inlet port allowing a homogeneous air/EGR mixture to be established and also to reheat the EGR through an air heater. Since all tests are carried out with WOT, a reduced inlet manifold pressure can not be used to admit external EGR and thus a gate valve in the exhaust line is used to control the back pressure during the exhaust stroke. EGR is achieved through another gate valve approximately 1 metre upstream of the inlet port. This allows the EGR/air mixture to become homogeneous before entering the cylinder and also to reheat the EGR through the air heater.

A 3kW air heater is closed-looped controlled to allow a fixed inlet charge temperature of 320°C, held accurately within ±1°C. The inlet charge temperature is measured at the inlet port to allow accurate control of the EGR/air entering the chamber irrespective of EGR rate. Coolant temperature is also closed-looped controlled, set at 80±0.2°C to ensure that the head temperature remains constant irrespective of engine load. Fuel is delivered in the inlet port at a pressure of 2.7 bar using a Bosch port injector. A purpose built electronic control unit controls the injection.

MEASUREMENT SYSTEMS – A water cooled pressure transducer linked to a PC is used for heat release analysis. The data acquisition system uses a program based on Labview® and developed at Brunel University to convert pressure traces in-order to calculate parameters such as IMEP. Knock analysis may also be measured via a band pass filter and the internal clock of the computer as opposed to using the crank decoder.

Emissions are measured using a NOx (Signal 4000VM) and HC analyzer (Signal 3000HM FID). Sample lines are heated to minimize errors associated with water condensation and species absorption. O2, CO2 and CO are measured using a Richard Oliver K650 analyzer in the inlet and exhaust lines.

A/F ratio and EGR rates are calculated analytically using the inlet and HC readings via an in-house program written in Excel.

EXPERIMENTAL PROCEDURE

The engine parameters were set as follows:

Engine Speed:	1500rpm
Airflow:	WOT
Inlet Charge Temperature:	320±1°C
Coolant Temperature:	80±0.2°C
Oil Temperature:	55±1°C
Compression Ratio:	11.5
Exhaust Backpressure:	0.15 bar gauge
Injection Timing:	70°CA BTDC Compression Stroke

These conditions were obtained by running the engine in a warm-up procedure. This was accomplished by running the engine in SI mode and switching to CAI mode with both modes running for 1 hour consecutively. The transition between SI mode to CAI mode was proceeded by switching the heater on and the fuel flow rate gradually reduced over time so that SI combustion is able to occur without the onset of knock. Auto-ignition begins to occur approximately when the inlet charge temperature reaches 270°C. At this point, the ignition (spark) is switched off and CAI regime takes over.

Testing is carried out once stability of the engine is ascertained. The fuel flow rate is adjusted independently through the injector at WOT to obtain different A/F ratios. EGR rates is gradually increased in increments of approximately 5% for each fuel setting, from zero to the maximum permissible for combustion to occur.

RESULTS AND DISCUSSION

The satisfactory CAI operating region of each fuel was first determined for a range of A/F ratios and EGR rates, as shown in Figure 1. The CAI boundaries were limited by knock, misfire and partial burn.

KNOCK REGION – An increase in fuelling rate (decrease in lambda) leads to an increase in combustion rates, thus causing rapid in-cylinder pressure rises. This in-turn leads to the knocking effect and is physically damaging to the engine.

MISFIRE – Increasing the EGR rate introduces higher rates of CO_2 and H_2O in the intake charge, which causes the occasional failure of ignition.

PARTIAL BURN REGION – As the fuelling rate is reduced (increase in lambda), the combustion temperature in this region becomes too low to fully oxidize a significant proportion of the fuel. Thus a high concentration of CO and uHC can be observed.

Figure 1 shows the CAI operating region established for the three primary reference fuels (PRF) of 80 RON, 95 RON, and 100 RON, 95RON unleaded gasoline, Methanol, and Ethanol. It can be seen that, Methanol posses the largest obtainable CAI region followed by Ethanol, Gasoline and PRFs respectively. This is thought to be due to the ignition timing of the fuels under the imposed test conditions. The ignition timing is dependant on the EGR rates for the PRF whilst for the alcohol fuels; this is not found to be the case, as to be shown later on.

The trends of the PRFs do not differ much in the size of CAI range although it can be seen that the higher the RON number, the amount of air dilution to resist knock is higher. There is a similar trend in the partial burn regions of PRFs. Furthermore, it can be seen that 80 RON is more tolerant to increases in EGR than that of 100 RON. The PRFs tested under CAI conditions can not operate under stoichiometric conditions and is less suited for CAI operation.

Gasoline, although possessing the same octane number as that of 95 RON PRF, can operate under stoichiometric conditions but is less resistant to knock. The partial burn and misfire limits are extended, thus allowing gasoline to be more viable for CAI operation.

The two alcohol fuels tested were Ethanol and Methanol. Both have the larger CAI operating range than any of the other fuels tested with Methanol being the superior of the two. Under the imposed test conditions, Methanol experienced no misfire even at EGR rates of over 70%, giving rise to the vast operating CAI region. Ethanol experienced a little misfire at stoichiometric conditions but at a far greater EGR rate than that of hydrocarbon fuels. The main drawback to alcohol fuels is that HC emissions and IMEP values were not ideal and well be discussed later.

The following discussion of results concentrates on the four fuels, 95 RON unleaded Gasoline, 95 RON PRF, Ethanol and Methanol.

IMEP – Figures 2-5 show the IMEP trends for the aforementioned four fuels. Gasoline shows the highest attainable IMEP at approximately 3.8 bar, with the PRF possessing a small reduction due to stoichiometric combustion not being possible. The alcohol fuels also has a reduced attainable maximum load compared to gasoline with a maximum of 3.6 and 3.4 bar IMEP for ethanol and methanol respectively. This trend is due to the advanced autoignition timing and faster subsequent heat release of the alcohols as to be shown below.

HC – The trends for the specific unburned HC emissions are given in Figures 6-9. The trends for the fuels are very similar, with minimum emissions obtained under the highest load operation, and always in the region of 5-10 g/kWh. The highest emissions can be seen when the air dilution is also highest and is due to the reduction in combustion temperature that prevents complete oxidation in this region. However, the hydrocarbon fuel trends with both air and EGR dilution are much more severe than those for the alcohol fuels, which in part accounts for the alcohols attaining a much wider acceptable operating region.

CO – CO emission trends are presented in Figures 10-13. 95 RON and gasoline fuels exhibit similar trends, although the rate of increase in CO emissions with air dilution is much higher for 95 RON. CO trends for the alcohol fuels are markedly different. The rate of increase of CO for alcohol fuels is much lower than that of the hydrocarbon fuels at lamda 5 and higher. As EGR rates increase beyond 60%, CO trends become highly dependant on the EGR rate, which is only moderate for the hydrocarbon fuels leading to an extended partial burn region for the alcohols.

NOx – NOx values for all fuels tested were extremely low and are considered negligible. A reduction of upto 95% when compared to normal SI operation was observed. Therefore, NOx emissions have been excluded in this paper.

IGNITION TIMING – Ignition timing for CAI combustion is defined as 10% of the mass fraction burnt (MFB) of the fuel. Trends of ignition timing can be seen in Figures 14-17. The trends for gasoline and PRF fuels are very similar and it can be seen that air dilution affects ignition timing at higher EGR rates. At lower EGR dilution, timing appears to be dependant mainly on EGR rate. However, trends differ significantly close to the knock region. PRFs ignition timing rapidly advances close to the knock boundary, whereby gasoline shows little or no dependence on air dilution. For alcohol fuels, it would appear that the EGR has no effect on the ignition timing over the entire map, although Ethanol shows a mere dependency. This trend gives reason to the extended CAI region of alcohol fuels as they exhibit high tolerance to EGR dilution and their ignition timings are much more advanced than that of hydrocarbon fuels.

COMBUSTION DURATION – Combustion duration trends shown in Figures 18-21, demonstrate that gasoline and 95 RON are very similar, with combustion durations increasing with air dilution at low EGR rates. Under close to stoichiometric conditions, durations increase rapidly with EGR rate and causes combustion instability under these conditions. Similar trends can be seen for alcohol fuels at low EGR rates, where combustion duration is mainly dependant on sir dilution. However, as EGR rates increases beyond approximately 40%, duration trends with air dilution eventually become entirely dependant on the EGR rate.

CONCLUSION

Alcohol fuels exhibit a larger CAI range in comparison to hydrocarbon fuels with Methanol possessing the largest operating range but with the drawback of a lower power output and higher HC emissions.

There is no direct correlation of RON and knocking combustion. Alcohol fuel with higher RON numbers appeared more ignitable and less resistant to knocking combustion in the CAI operation under the experimental conditions tested.

IMEP is lowest for alcohol fuels and is attributed to the advanced autoignition. PRFs exhibit a lower IMEP than that of gasoline as they are unable to operate under stoichiometric conditions.

Higher HC emissions are observed with hydrocarbon fuels in comparison with alcohol fuels. The rate of increase of CO is higher for hydrocarbon fuels and is less dependant on the EGR rate than that of alcohol fuels. This leads to the increase in CAI range and more specifically, the partial burn region for alcohol fuels.

REFERENCES

1. Onishi, S., Hong Jo, S., Shoda, K., Do Jo, P., and Kato, S., “Active Thermo-Atmosphere Combustion (ATAC) – A New Combustion Process for Internal Combustion Engines”, SAE Paper 790501, 1979
2. Noguchi, M., Tanaka, T., and Takeuchi, Y., “A Study on Gasoline Engine Combustion by Observation of Intermediate Reactive Products During Combustion”, SAE Paper 790840, 1979
3. Najt, P., and Foster, D., “Compression-Ignited Homogeneous Charge Combustion”, SAE 830264, 1983
4. Christensen, M., Johanssom, B., and Einwell, P., “Homogeneous Charge Compression Ignition (HCCI) Using Isooctane, Ethanol and Natural Gas – A Comparison with Spark Ignition Operation” SAE Paper 972874, 1994
5. Christensen, M., and Johanssom, B., “Influence of Mixture Quality on Homogeneous Charge Compression Ignition” SAE Paper 982454, 1998

6. Shudo, T., and Ono, Y., "HCCI Combustion of Hydrogen, Carbon Monoxide and Dimethyl Ether", SAE Paper 2002-01-0112, 2002
7. Oakley, A., Zhao, H., and Ladommatos, N., "Experimental Studies on Controlled Auto-Ignition (CAI) Combustion of Gasoline in a 4-Stroke Engine", SAE Paper 2001-01-1030, 2001
8. Oakley, A., Zhao, H., Ladommatos, N. and Ma, T., "Dilution Effects on the Controlled Auto-Ignition (CAI) Combustion of Hydrocarbon and Alcohol Fuels", SAE Paper 2001-01-3606, 2001
9. Oakley, A., "Experimental Investigations on Controlled Auto-Ignition Combustion in a Four-Stroke Gasoline Engine", 2001

Contact. K. Man kayiu.man@brunel.ac.uk

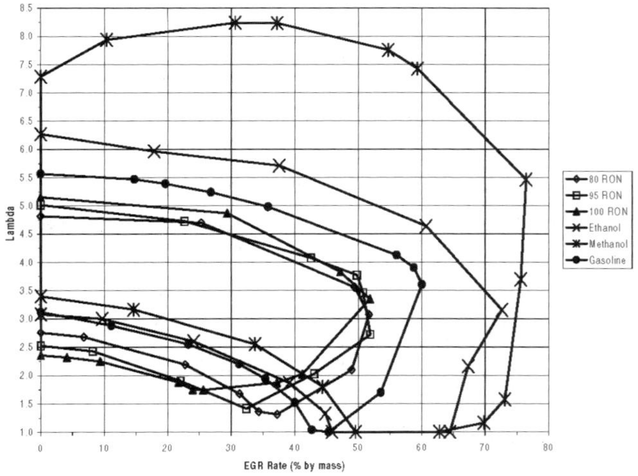

Fig. 1 CAI Boundary

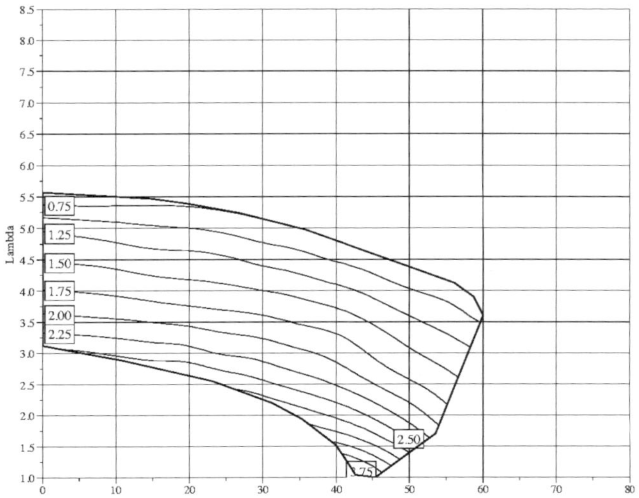

Fig. 2 IMEP for Gasoline CAI combustion (bar)

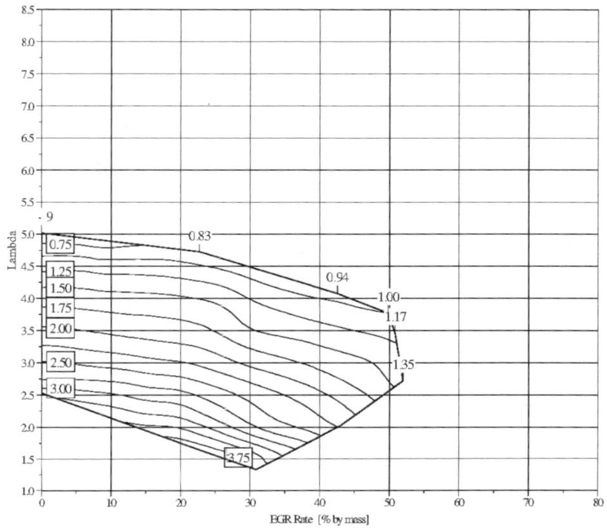

Fig. 3 IMEP for 95 RON PRF CAI combustion (bar)

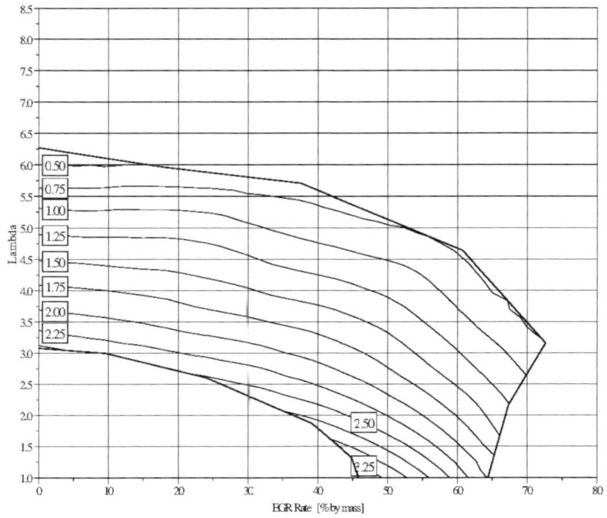

Fig. 4 IMEP for Ethanol CAI combustion (bar)

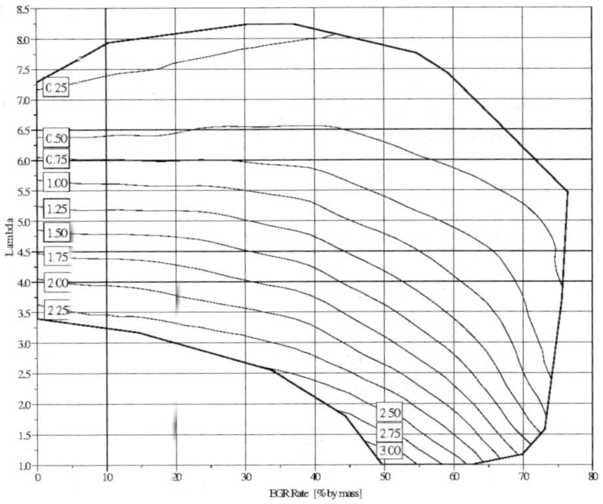

Fig 5 IMEP for Methanol CAI combustion (bar)

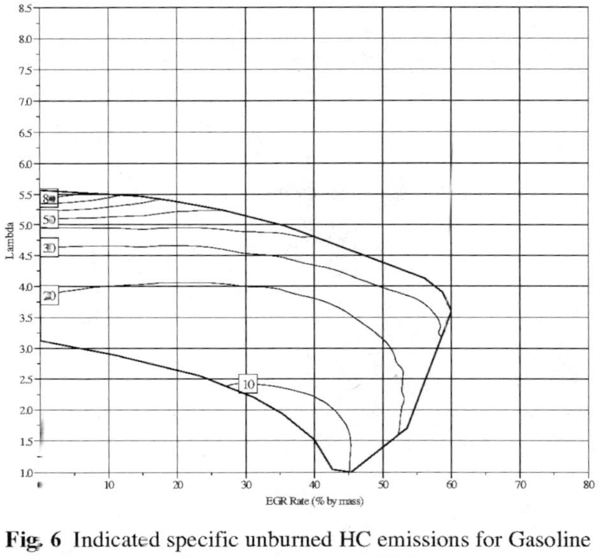

Fig. 6 Indicated specific unburned HC emissions for Gasoline CAI combustion (g/kWh)

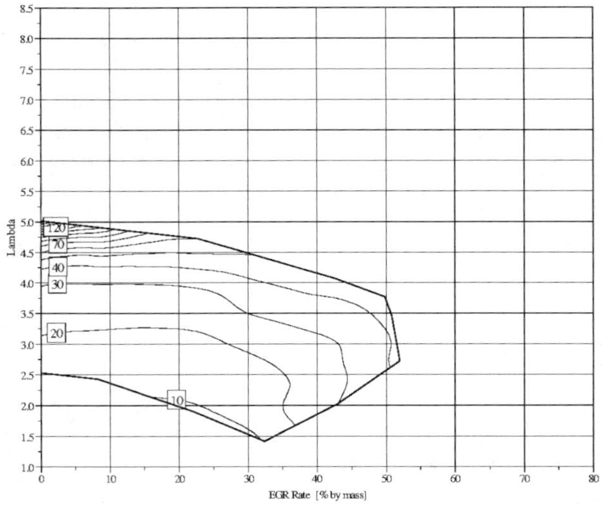

Fig. 7 Indicated specific unburned HC emissions for 95 RON PRF CAI combustion (g/kWh)

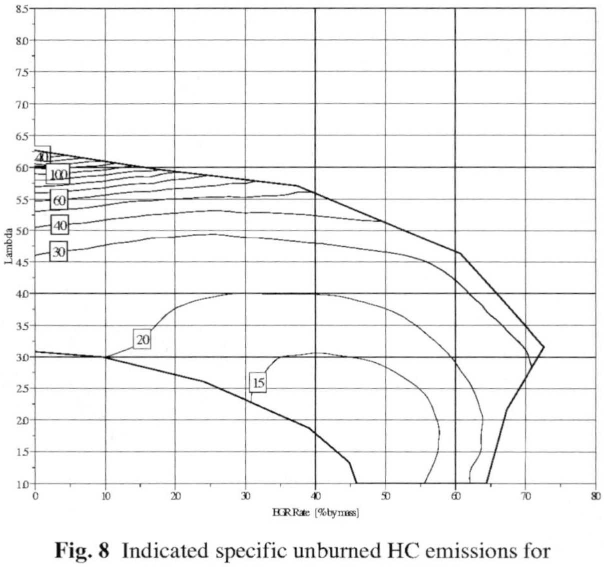

Fig. 8 Indicated specific unburned HC emissions for Ethanol CAI combustion (g/kWh)

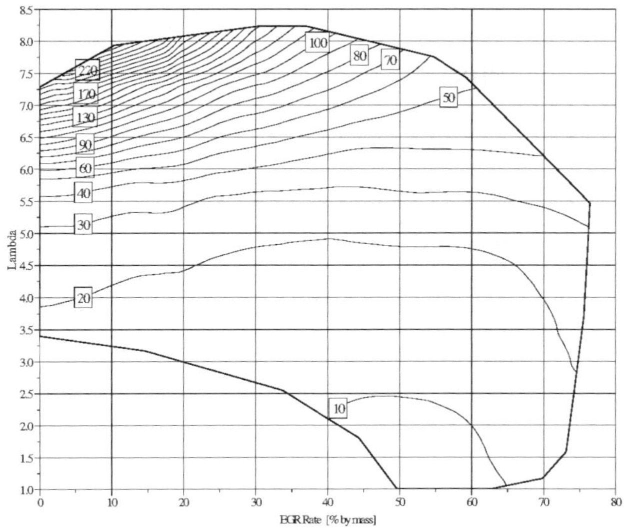

Fig. 9 Indicated specific unburned HC emissions for Methanol CAI combustion (g/kWh)

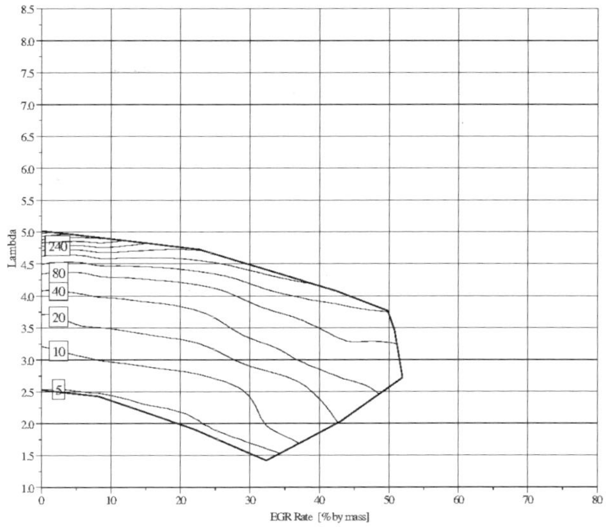

Fig. 11 Indicated specific unburned CO emissions for 95 RON PRF CAI combustion (g/kWh)

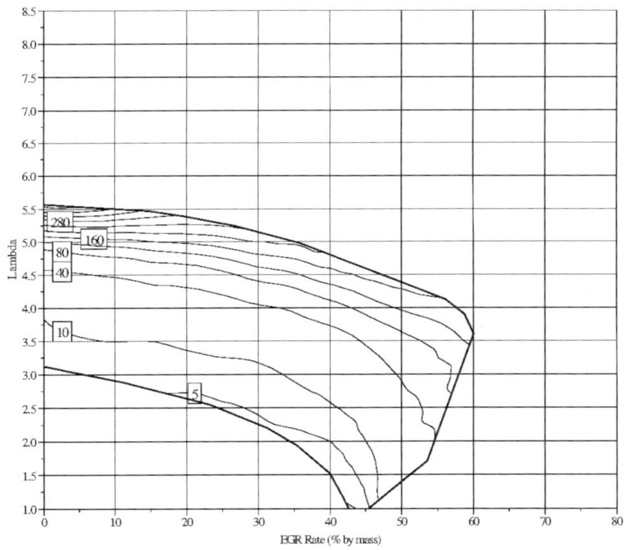

Fig. 10 Indicated specific unburned CO emissions for Gasoline CAI combustion (g/kWh)

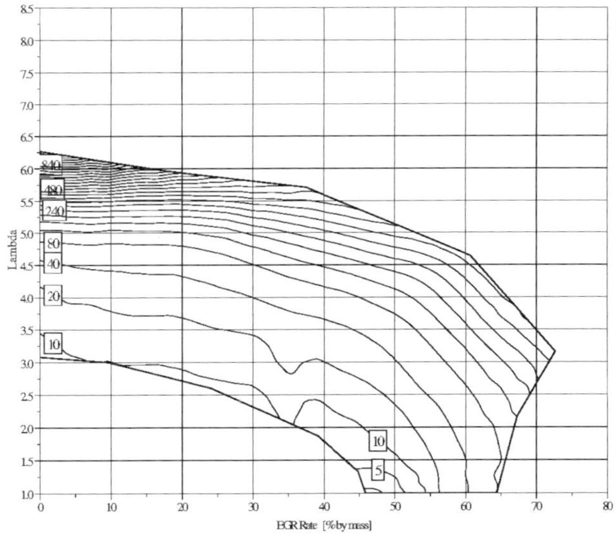

Fig. 12 Indicated specific unburned CO emissions for Ethanol CAI combustion (g/kWh)

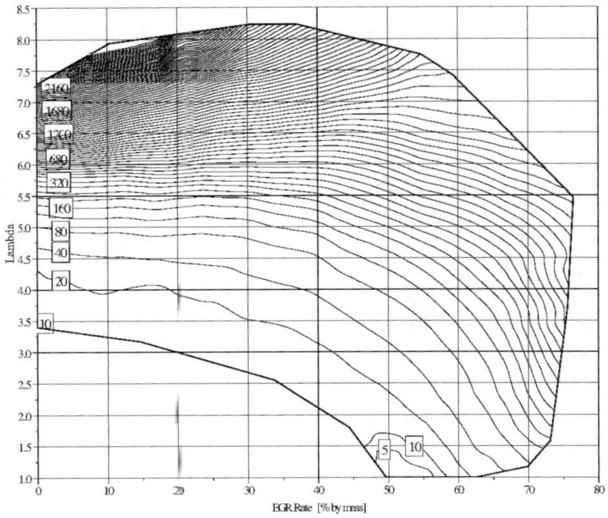

Fig. 13 Indicated specific unburned CO emissions for Methanol CAI combustion (g/kWh)

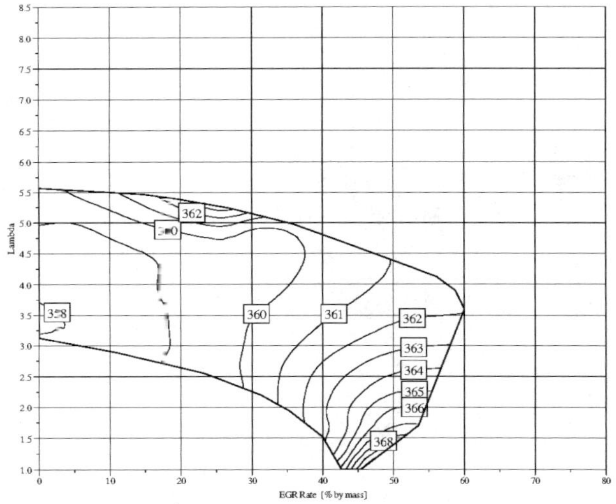

Fig. 14 Ignition timing (10% MFB) for Gasoline CAI combustion (°CA)

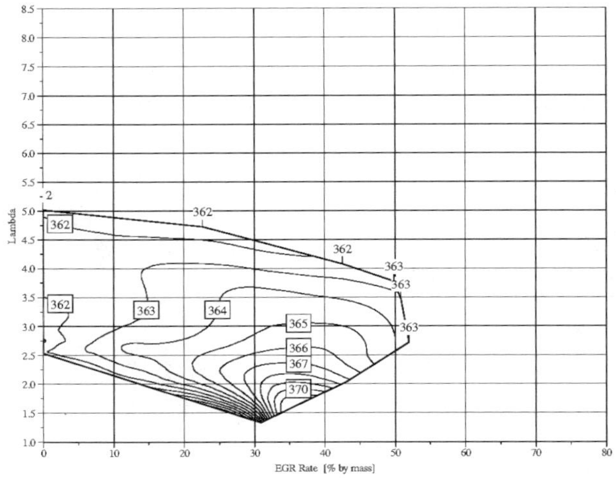

Fig. 15 Ignition timing (10% MFB) for 95 RON PRF CAI combustion (°CA)

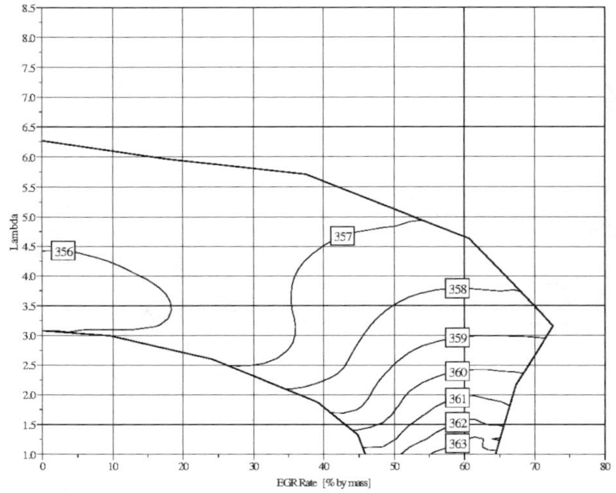

Fig. 16 Ignition timing (10% MFB) for Ethanol CAI combustion (°CA)

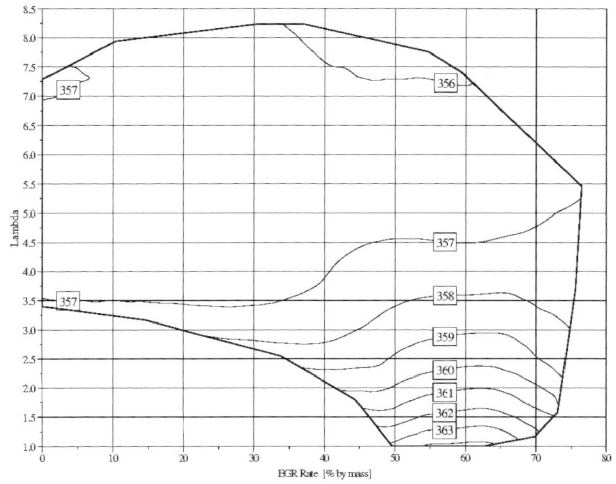

Fig. 17 Ignition timing (10% MFB) for methanol CAI combustion (°CA)

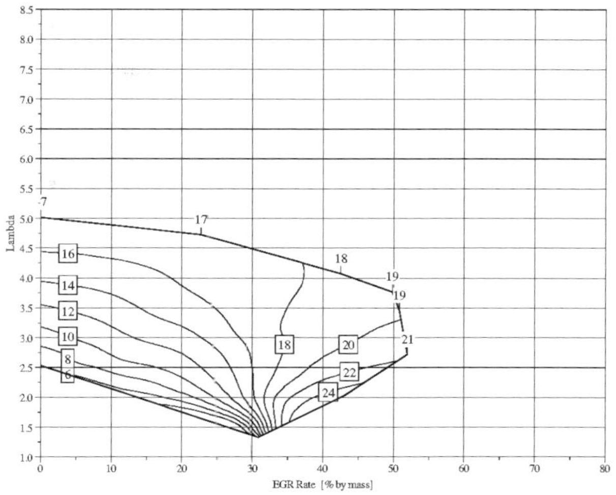

Fig. 19 Combustion duration (10-90% MFB) for 95 RON PRF CAI combustion (°CA)

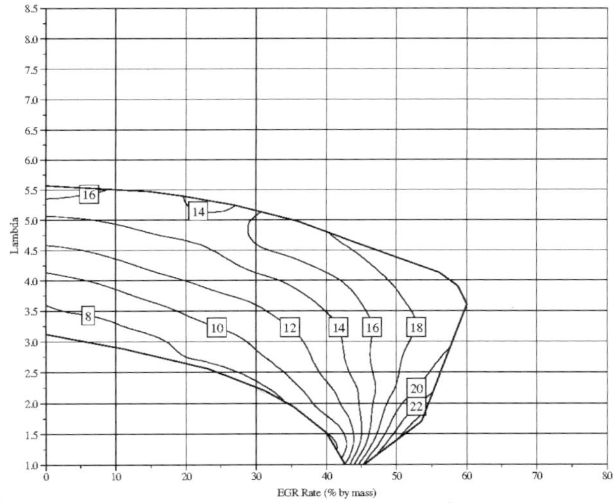

Fig. 18 Combustion duration (10-90% MFB) for gasoline CAI combustion (°CA)

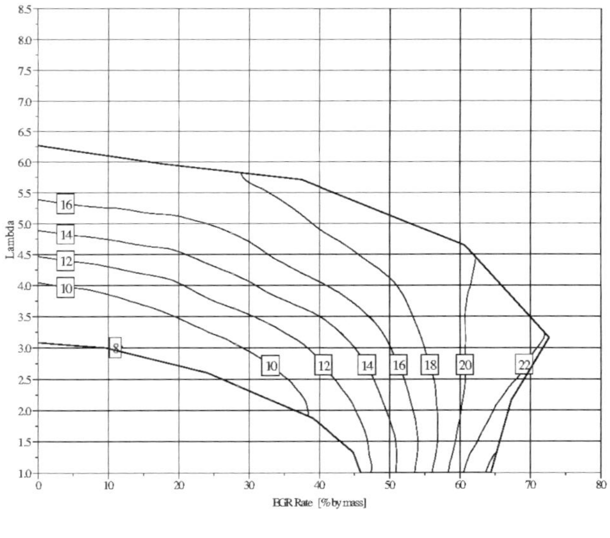

Fig. 20 Combustion duration (10-90% MFB) for ethanol CAI combustion (°CA)

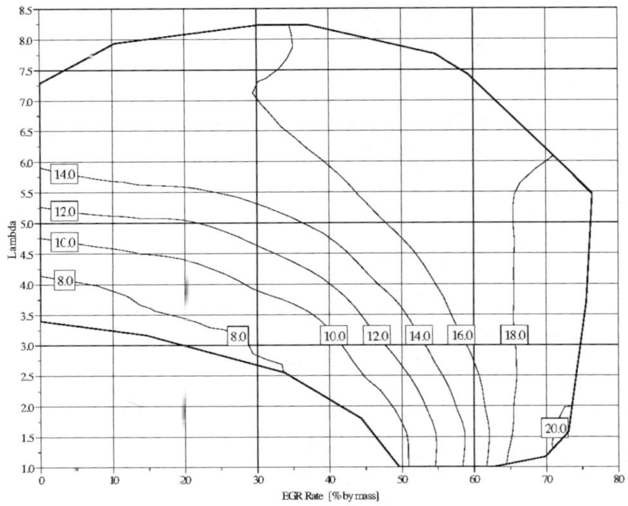

Fig. 21 Combustion duration (10-90% MFB) for methanol CAI combustion (°CA)

Which Fuels For Low CO_2 Engines?
© P. Duret, X. Montagne (Eds.) and Editions Technip, Paris, 2004, pp. 79 -- 87
27 rue Ginoux, 75015 Paris

LATEST DEVELOPMENTS IN HEAVY DUTY DIESEL HCCI

Kevin Duffy[1], Eric Fluga[1], Steve Faulkner[1], David Heaton[1], Charles Schleyer[2], Rafal Sobotowski[2]

1. Caterpillar 2. ExxonMobil

ABSTRACT – *Implementing a practical homogeneous charge compression ignition engine has numerous technical challenges. Among these are ensuring a proper mixture preparation, controlling combustion phasing and cylinder pressure rise rates, and expanding the operating range both to higher loads and idle/light loads. Methods to control combustion phasing include inlet manifold temperature and pressure, injection timing, valve timing and fuel properties. Caterpillar and ExxonMobil are working together to understand heavy duty HCCI using advanced engines, fuel combustion chemistry, and combustion models. Recent progress on part load HCCI operation in both single and multi-cylinder engines using diesel fuels will be given. Initial results from a single-cylinder engine have shown that HCCI technology can achieve very low PM and NOx emissions at light-medium load with a range of distillate fuel properties including US market fuel. Future work will address higher loads and other technical challenges.*

INTRODUCTION

Homogeneous charge compression ignition is a rapidly evolving technology that offers great potential for meeting future NOx and particulate emissions regulations while maintaining good fuel conversion efficiency. The interest in this combustion method is growing rapidly among OEMs, research institutes, government labs, universities and consulting companies as indicated by the large increase in papers at technical conferences worldwide. Different approaches to achieve low emissions have been developed by these groups and an excellent summary of the state-of-the-art up to 2003 is found in Zhou et al.[1]

Many studies have been done on fuel effects and desired fuel properties for HCCI engines (e.g., Ryan and Callahan [2], Kelley-Zion and Dec [3], Shibata et al.[4]). Each fuel brings with it unique attributes which impact storage, delivery, vaporization, and ultimately combustion characteristics. Major commercial fuel types are diesel and gasoline and new engine technology using these fuels can be introduced most readily. For the on-highway truck and off-highway machine markets, diesel is the dominant fuel. Distribution and storage networks are established and fuel injection equipment has been developed to work with diesel fuel. Therefore, it is of practical interest to develop an HCCI combustion method that uses diesel fuel.

The primary reason HCCI systems are being developed is because of their potentially ultra-low emissions capability that will be needed to meet future worldwide emissions regulations and the possible avoidance of costly aftertreatment systems. Figure 1 shows the near to mid term regulations for the U.S. and Europe.

HCCI systems will likely target the US on-highway 2010 and off-road Tier 4b regulations due to the extremely low NOx emissions levels, although forms of HCCI could be used for practically all upcoming regulations. NOx levels of 0.2-0.3 g/hphr translate into <50 ppm NO at all engine operating conditions (<10 ppm at most), and the only other known methods to achieve these levels involve the use of expensive and undeveloped NOx aftertreatment technology such as NOx adsorbers and SCR systems. If a true homogeneous mixture is achieved, rich regions in the cylinder are avoided and solid carbon levels are effectively zero, avoiding the need for particulate traps. Hydrocarbon and CO levels are also legislated and HCCI combustion has

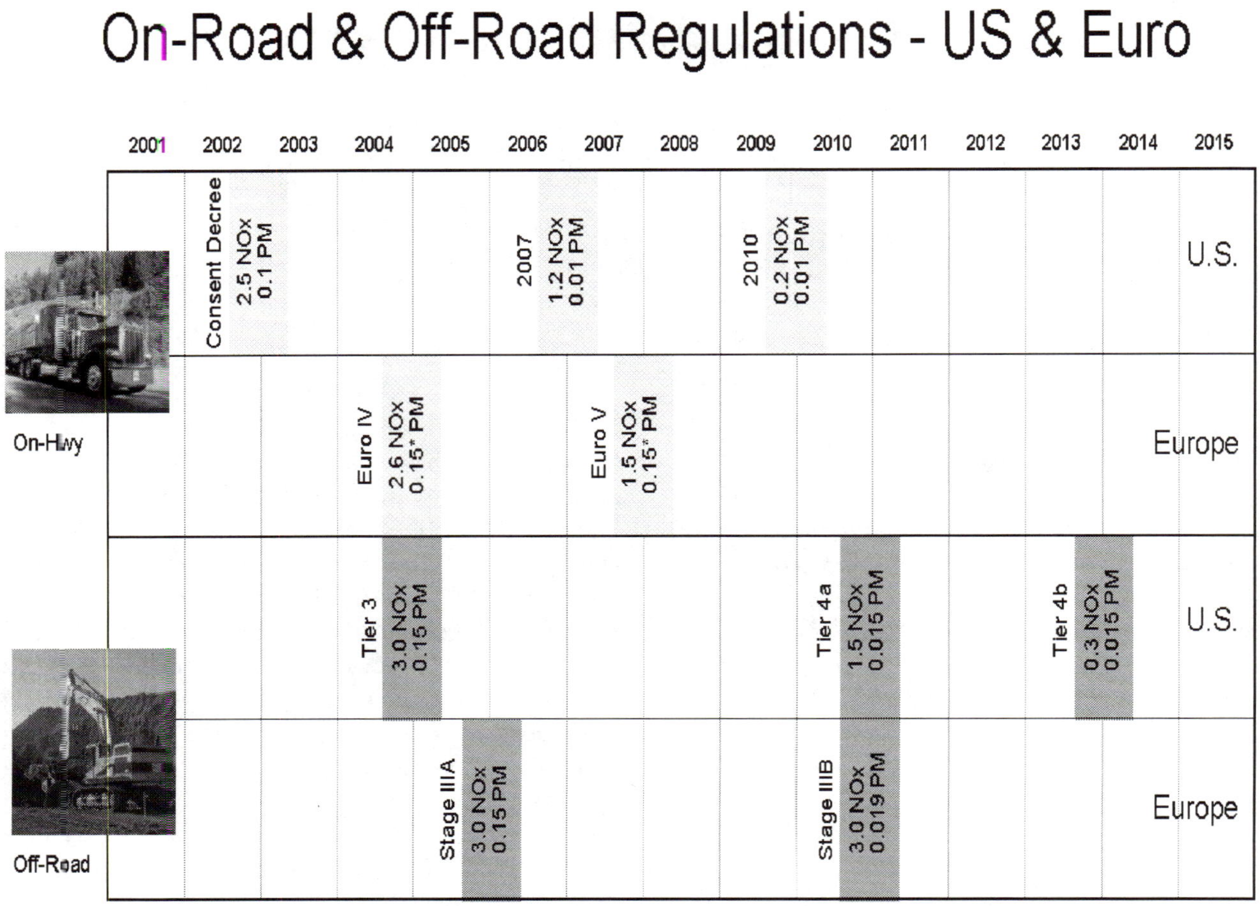

Fig. 1 U.S. and European Emissions Regulations

inherently high levels of these emissions (Dec and Sjoberg [5]), especially at light loads (low equivalence ratios) where temperatures are low limiting oxidation catalyst performance. So even if HCCI combustion methods are successful at eliminating the need for NOx and PM traps, oxidation catalysts will still be needed.

CHALLENGES

The primary challenge of most HCCI development activities is achieving these ultra low NOx and PM emissions over the entire power spectrum and legislated emissions cycles the engines must operate within. For certain applications such as passenger cars and light trucks, in some countries the emissions cycles only subject the vehicle to part load operation so an HCCI strategy that achieves low emissions up to ½ load and then uses more conventional methods at higher loads may be a perfectly acceptable solution. However, for on-highway trucks and off-road machines, the emissions cycles are such that the engines must produce ultra-low NOx and PM levels from light load up to full load. Therefore, the ideal HCCI solution is one which works at all engine operating conditions and this has proven to be the most difficult obstacle to overcome by most doing HCCI development. The primary reason for this is the rapid rate of combustion near top dead center as more fuel is injected to increase the power output of the engine. These high combustion rates can exceed the structural limits of the engine cylinder components (piston, rings, head, etc.) and often are accompanied by high NOx emissions and increased heat loss (Tsurushima et al.[6]).

Figure 2 shows a speed and load (BMEP) map typical of an on-highway truck engine. The red lines indicate load levels achieved by Nissan using their modulated kinetics (MK) HCCI-like combustion approach. While these BMEP levels are not extremely high, they cover a significant portion of the Japanese emissions cycle, and thus the MK

HCCI method is well suited for that particular engine/vehicle application. Shown by the solid black line is the full load curve for a Caterpillar on-highway C15 truck engine rated at 550 hp. The blue symbols are operating conditions achieved on a single cylinder engine variant of the C15 using a form of in-cylinder, diesel fueled HCCI. All conditions were run with less than 5 ppm NOx and less than 0.1 AVL smoke. The highest BMEP level of 1550 kPa is approximately ¾ load and represents one of the highest reported power levels achieved using diesel fuel in an HCCI engine.

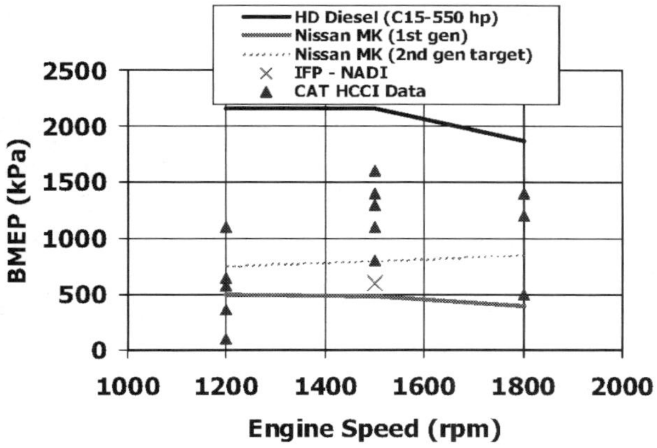

Fig. 2 Operating map of HCCI engine

Numerous other challenges associated with the practical implementation of in-cylinder diesel HCCI on truck and machine engines include proper fuel injection hardware to achieve a homogenous (or homogeneous-enough) air-fuel mixture, control measures to ensure proper combustion phasing and transient engine operation, light load HC/CO reduction, cold start with reduced compression ratios, and sensitivity to fuel property variation.

EXPERIMENTAL SETUP

The engine used in this study is a single cylinder Caterpillar 3401 engine with specifications given in Table 1. A modified hydraulically intensified fuel injector was used to provide a uniform spray distribution.

Intake and exhaust surge tanks were used to provide boost and backpressure levels that are representative of actual multi-cylinder turbocharger operation. No oxidation catalyst was used so the HC and CO levels reported are all engine-out values. A Kistler cylinder pressure transducer was used to record high speed data at 0.1 crank angle resolution.

Table 1. Caterpillar single cylinder engine specifications

Displacement	2.44L
Bore/Stroke	137.2 / 165.1 mm
Connecting Rod Length	261.6 mm
Valves per cylinder	4
Injection System	Intensified Hydraulic
Swirl Ratio	~0.4
Inlet Valve Open, Close	320.3, 568.7 deg
Exhaust Valve Open, Close	119.5, 380.0 deg

Heat release analysis was done using a standard 1st Law calculation (without heat transfer) available on the AVL Indimeter data acquisition system. Exhaust gas emissions of CO, HC, NOx and CO_2 were measured with a Horiba EXSA analyzer. An AVL smoke meter was used for soot measurement.

TEST MATRIX

Two sets of tests were conducted to evaluate engine variables injection timing, boost, and inlet manifold temperature and to evaluate fuel effects. The boost and backpressure levels were set to match (as close was known at the time) the actual inlet and exhaust manifold conditions on a multi-cylinder engine with turbocharging. The engine was run at three different speeds (1200, 1500, 1800 rpm) over a range of loads from approximately ¼ to ¾ load. Acceptable HCCI operation is defined as NOx emissions < 0.2 g/hphr, AVL smoke < 0.1, maximum cylinder pressure rise rate < 25 MPa/ms, and IMEPcov < 5%. The baseline fuel used is a conventional #2 Diesel with properties listed in Table 2 below. Additional fuels used to evaluate fuel property effects are described later.

Table 2. Baseline Fuel Properties for #2 Diesel

PARAMETER		TEST METHOD	VALUE
Density at 15 C, kg/m3		D4052	850.2
Cetane Number		D613	44.7
Distillation, C	10%	D86	214
	50%		256
	90%		315
SFC Aromatics, % m/m	Total	D5186	28.0
Sulfur, ppm		D5453	400
Kinematic Viscosity @ 40 C, cSt		D445	2.41

TEST RESULTS AND DISCUSSION

For in-cylinder diesel fuelled HCCI, large inlet manifold temperature increases to enhance fuel vaporization are not needed as in port injected Diesel HCCI. However, changes in the inlet manifold temperature do influence the compression temperature history and affect combustion phasing. Figure 3 shows the effect of changing inlet manifold temperature over a 20C range at a part load operating condition for a fixed injection timing and boost level. For 33C and 43C combustion is phased near TDC. NOx and smoke emissions are low. Increasing the inlet manifold temperature to 53C causes an advance of the cool flame combustion (first peak on the heat release curve at 15 BTDC) which, in turn, leads to an advance in the main heat release of approximately 5 degrees. A knocking type combustion occurs with increased cylinder pressure and rise rate. This results in an increase in NOx emissions but the fuel consumption improves by 7% since combustion is phased in a slightly more optimal location centered around TDC and the combustion efficiency improves with lower HC and CO. A further 5C increase in the inlet manifold temperature results in even more severe knocking and NOx emissions quickly rise to several hundred ppm.

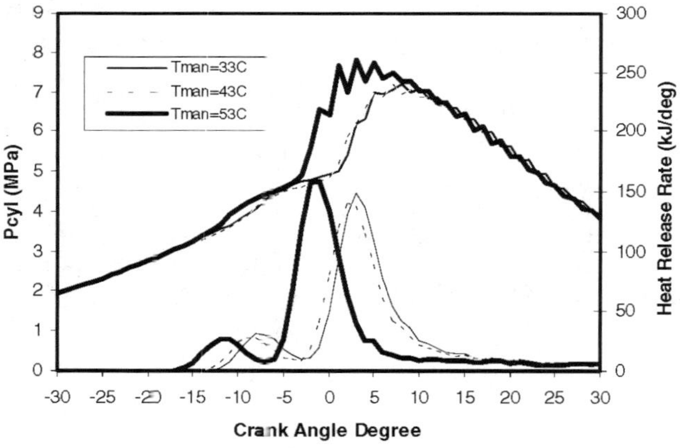

	33C	43C	53C
NOx (ppm)	2	2	8
Smoke (AVL)	0.2	0.16	0.07
HC (ppm)	1219	1014	715
ESFC change	0	-2%	-7%

Fig. 3 Effect of inlet manifold temperature at 1500 rpm, ¼ load

Injection timing is another critical parameter for in-cylinder Diesel HCCI combustion and emissions control. Figure 4 shows trends in emissions as timing is retarded from very early levels to closer to top dead center. In general, at very early timings, low in-cylinder density and spray angle effects lead to fuel overpenetration and impingement on cylinder liner surfaces which cause excessive HC, CO and smoke levels. In addition, this results in unacceptable fuel dilution of the oil. At these early timings, the NOx emissions are very low since ample time for fuel/air mixing leads to very lean, low temperature combustion. For the very late injection timings, the compression temperatures are much higher and the spray/piston bowl interactions become important. This results in significantly lower HC, CO and smoke levels. However, since the available time for fuel evaporation, mixing and the kinetic reactions to proceed before combustion occurs is shortened, the NOx emissions increase rapidly. An intermediate injection timing region is used where low NOx and smoke can be realized with moderate levels of HC and CO.

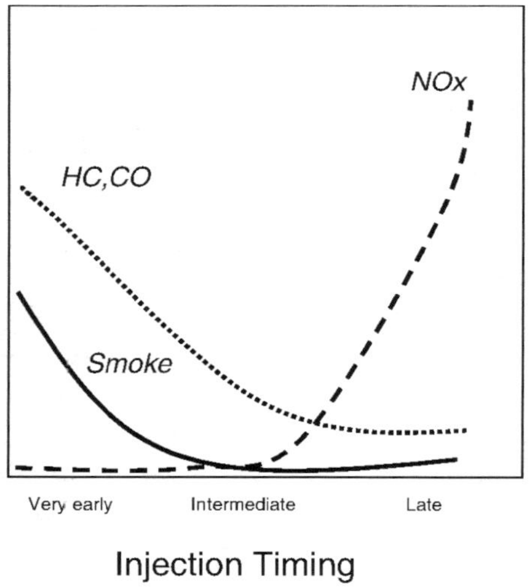

Fig. 4 Effect of injection timing on emissions

Injection timing also affects combustion phasing as shown in Figure 5. For an intermediate baseline timing the NOx and smoke emissions are extremely low and combustion is optimally phased at TDC. Heat release from cool flame chemistry is evident at approximately 9 degrees BTDC. A 5 degree retard of injection timing still results in low NOx and smoke while HC levels drop at constant CO. The cool flame heat release still begins at 9 degrees

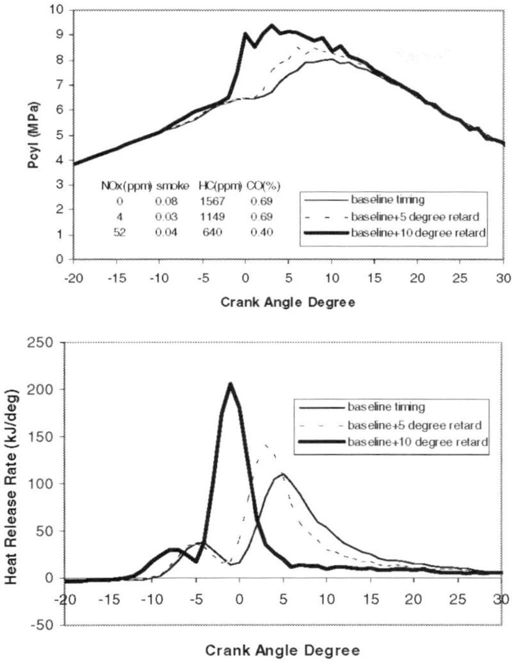

Fig. 5 Effect of injection timing on combustion phasing at 1500 rpm and ¼ load.

BTDC and the main combustion is phased slightly before TDC. Another 5 degree retard in timing leads to an earlier onset of cool flame heat release which in turn causes a more rapid main heat release at 5 degrees BTDC. This results in a significant tripling of the rate of cylinder pressure rise and a rapid increase in NOx emissions to over 50 ppm.

The effect of boost pressure on combustion phasing and performance/emissions is shown in Fig 6. The baseline boost level phases main combustion properly at TDC. This also results in 0 ppm NOx and very low smoke levels. Increasing the boost by 20 kPa causes a slight advance in the cool flame combustion phasing and subsequent advance in main combustion phasing by 3 degrees. While NOx and smoke emissions still remain very low, the advanced heat release phasing and increased pumping work results in a 5% increase in fuel consumption in spite of a combustion efficiency increase due to the HC/CO reduction. If the boost pressure is reduced by 30 kPa from the baseline, the cool flame combustion phasing does not change, but the magnitude of the cool flame heat release decreases. This is indicative of less active radical formation from the cool flame chemistry which in turn slows the transition to the main heat release phase. Main combustion is phased about 2 degrees retarded relative to the baseline case. NOx emissions remain low but are starting to increase, and due to reduced air-fuel ratio, the smoke emissions rise to 0.4 AVL. Combustion efficiency remains similar to the baseline, but the retarded heat release phasing results in a 6% increase in fuel consumption. These results show the high degree of sensitivity to boost level on combustion phasing, rate of pressure rise and performance and emissions levels.

A screening study was conducted to explore fuel property effects on HCCI engine performance and exhaust emissions, focusing on cetane number and aromatic content. The fuel matrix developed for this

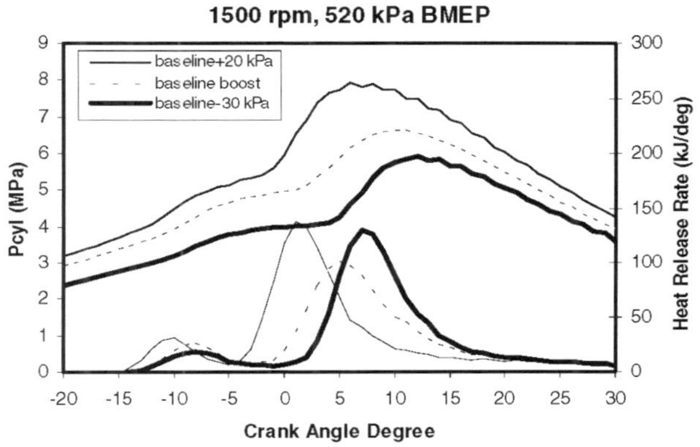

	BL+20kPa	Baseline	BL-30kPa
NOx (ppm)	1	0	13
Smoke (AVL)	0.01	0.06	0.4
HC (ppm)	890	1605	1514
BSFC (% change)	+5%	baseline	+6%

Fig. 6 Effect of boost level on performance and combustion

study is shown in Figure 7. Test fuel properties are presented in Table 3. Note engine performance for fuels in Tables 2 and 3 could not be compared because of small variation in engine operating conditions.

The impact of fuel cetane was evaluated using fuels F1 and F2 with cetane numbers of 46.7 and 55.4, respectively. The effect of aromatic content was investigated by comparing fuels F2 and F3 whose total aromatic content was 28.7 and 44.7%, respectively. To the extent possible, other fuel properties were held constant across the three fuels to isolate cetane number and aromatic content effects.

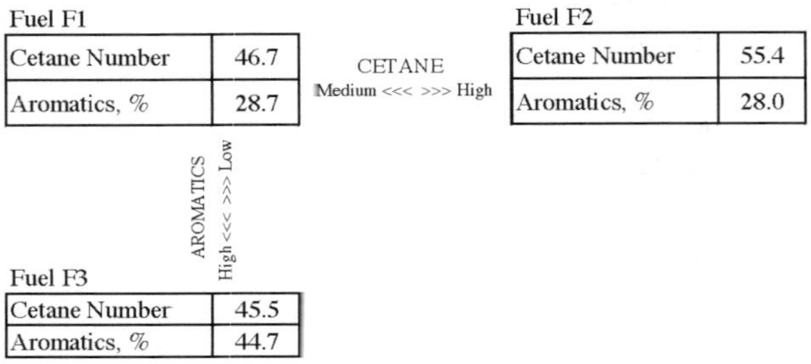

Fig. 7 Test fuel matrix

Table 3. Test Fuel Properties

PARAMETER		TEST METHOD	TEST FUEL		
			F1	F2	F3
Density, kg/m3		D4052	0.8433	0.8326	0.8503
Cetane Number		D613	46.7	55.4	45.5
Distillation, F	IBP	D86	176	158	187
	5%		200	189	206
	10%		207	204	211
	20%		217	227	218
	30%		227	246	228
	40%		239	260	237
	50%		253	271	249
	60%		267	282	262
	70%		281	293	281
	80%		296	307	304
	90%		313	325	324
	95%		328	341	335
	EP		335	350	343
SFC Aromatics, % m/m	1-ring	D5186	20.7	20.3	29.3
	2+ ring		8.0	7.7	15.4
	Total		28.7	28.0	44.7
Sulfur, ppm		D5453	191	44	208
Kinematic Viscosity @ 40 C, cSt		D445	2.42	2.64	2.22
Heat of Combustion, MJ/kg	Gross	D240	45.7	45.9	44.7
	Net		43.0	43.2	42.2

The fuels were tested at engine speeds of 1200, 1500 and 1800 rpm, and engine loads of 25 and 50%. Some measurements were taken at predetermined operating conditions (air/fuel ratio, intake manifold pressure, etc.). Additional tests were also performed with the objective of further optimizing engine performance and emissions.

The fuel effects study was focused on engine operating conditions characterized by NOx emissions < 0.2 g/hphr and AVL smoke numbers < 0.1. The former corresponds to US EPA 2010 NOx emission standard for heavy-duty engines, while the latter is roughly equivalent to the 2010 particulate emission requirement of 0.01 g/hphr. All fuels could meet these emission levels indicating that this HCCI combustion system was robust to a range of fuel properties.

As a rule, increasing cetane number advanced the start of combustion timing. This effect, illustrated in Figure 8 for fuels F1 and F2, is not desirable. In fact, it is counterproductive from the viewpoint of achieving higher load operation in HCCI engines.

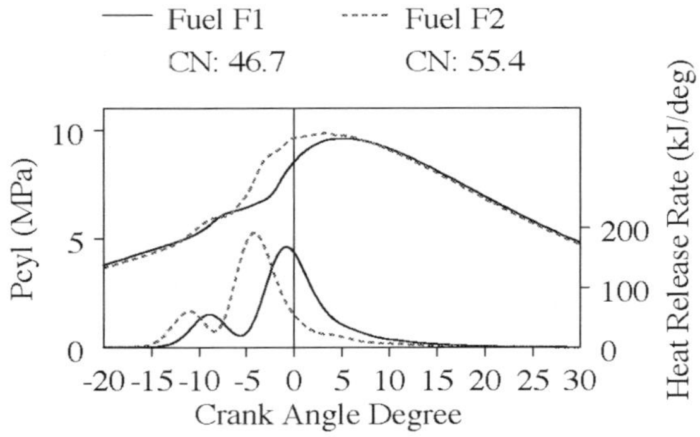

Fig. 8 Effect of cetane number on combustion phasing

Increasing cetane number makes it more difficult to achieve optimum combustion phasing and maximize thermal efficiency of the engine within the constraints of the cylinder pressure and rise rate limits. Further investigations covering high load operation are necessary to fully address this issue.

Figures 9 through 13 show NOx, AVL smoke, HC, CO and thermal efficiency of the test engine operated on fuels F1 and F2. Consistent with Figure 4, NOx emissions increased as fuel injection timing was retarded, while smoke, HC and CO emissions were reduced. Thermal efficiency also tended to improve with retarded injection timing, in line with lower HC and CO emissions. Overall, the effects of cetane number on engine performance and emissions were small and tended to disappear as injection timing was retarded at all engine operating

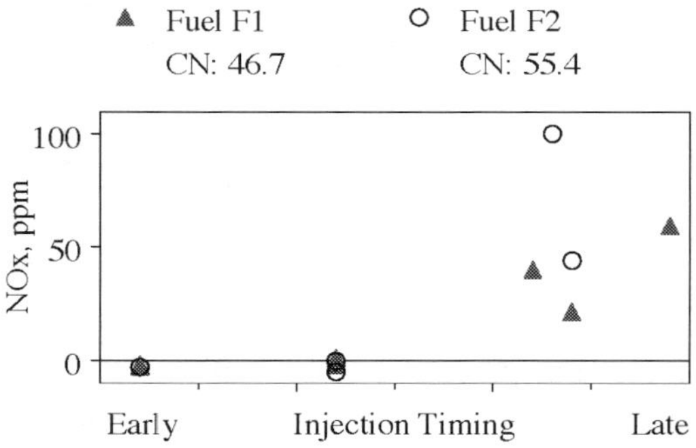

Fig. 9 Effect of injection timing and cetane number on NOx at 1200 rpm, 25% load

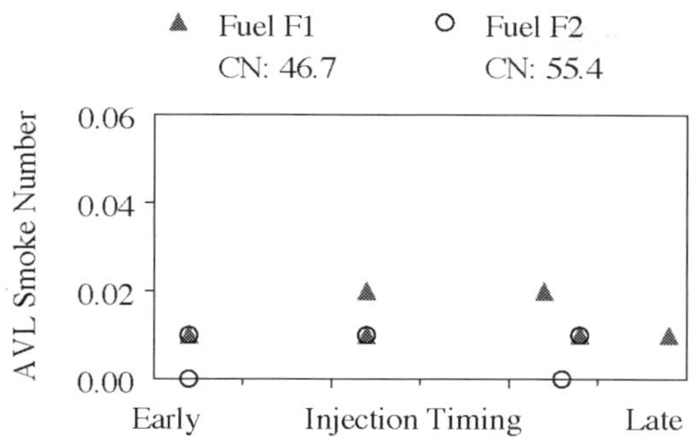

Fig. 10 Effect of injection timing and cetane number on smoke at 1200 rpm, 25% load

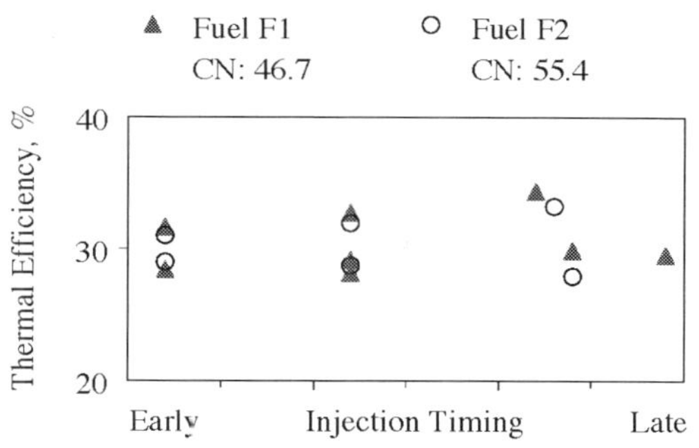

Fig. 13 Effect of injection timing and cetane number on thermal efficiency at 1200 rpm, 25% load

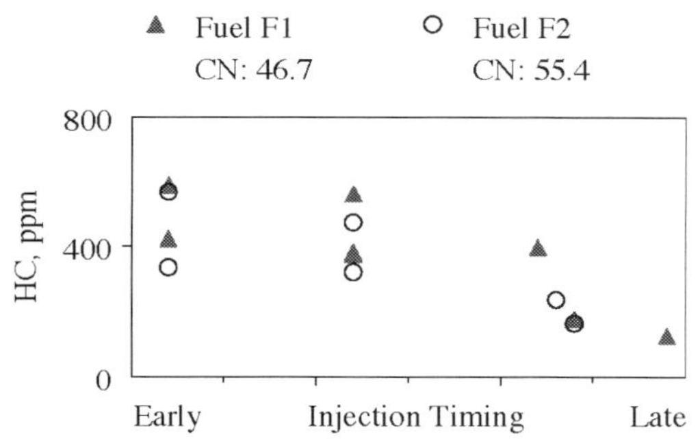

Fig. 11 Effect of injection timing and cetane number on HC at 1200 rpm, 25% load

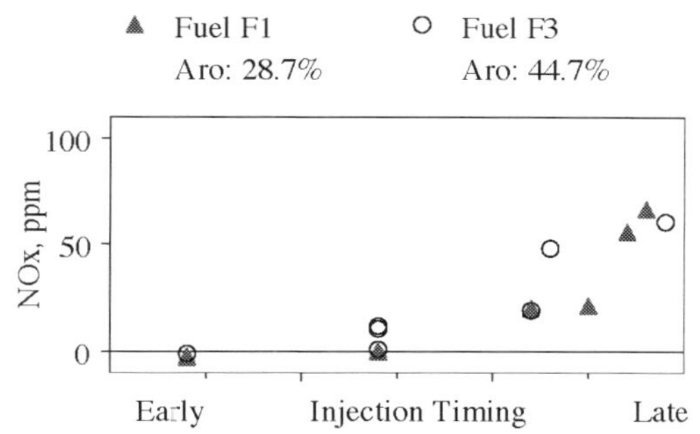

Fig. 14 Effect of injection timing and aromatic content on NOx at 1500 rpm, 25% load

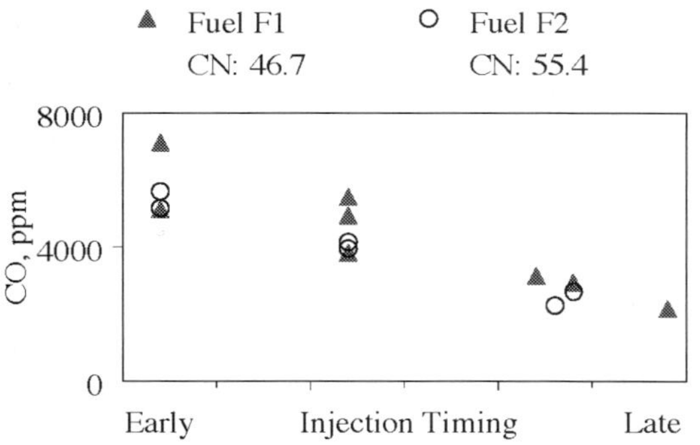

Fig. 12 Effect of injection timing and cetane number on CO at 1200 rpm, 25% load

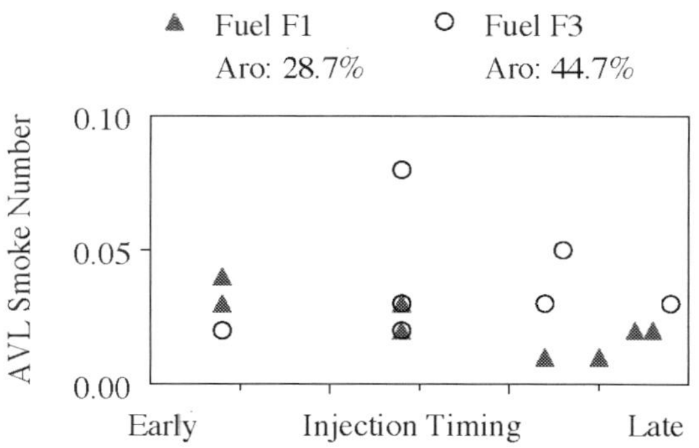

Fig. 15 Effect of injection timing and aromatic content on smoke at 1500 rpm, 25% load

conditions used in this study. Where its effect was detectable, cetane number increase seemed to improve CO, HC and smoke emissions at advanced fuel injection settings. These small effects of cetane number which were observed may be attributed to increased fuel reactivity and advanced start of combustion timing associated with cetane number increase.

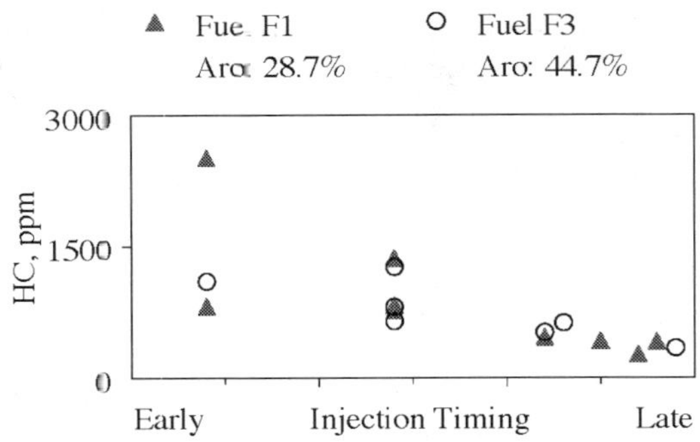

Fig. 16 Effect of injection timing and aromatic content on HC at 1500 rpm, 25% load

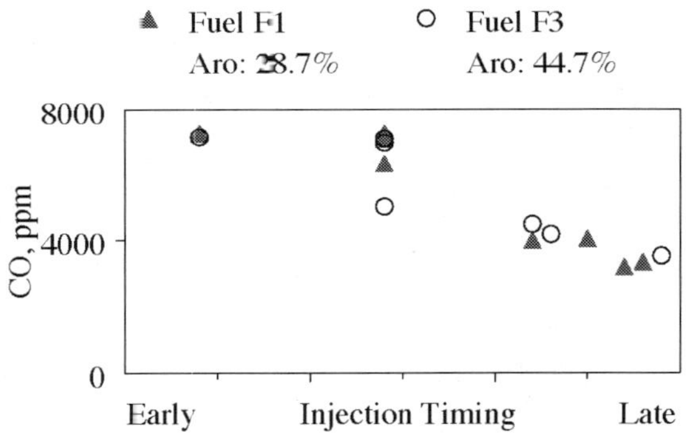

Fig. 17 Effect of injection timing and aromatic content on CO at 1500 rpm, 25% load

The effect of aromatic content of the fuel on exhaust emissions and thermal efficiency is shown in Figures 14 through 18 for the 1500 rpm, 25%load operating point. In general, the observed effects were small and followed no clear trends for the engine operating conditions used in this study. Neither did cylinder pressure and heat release data indicate distinct trends related to this parameter. These results suggest that this HCCI combustion system could be relatively insensitive to the aromatics content of diesel fuel.

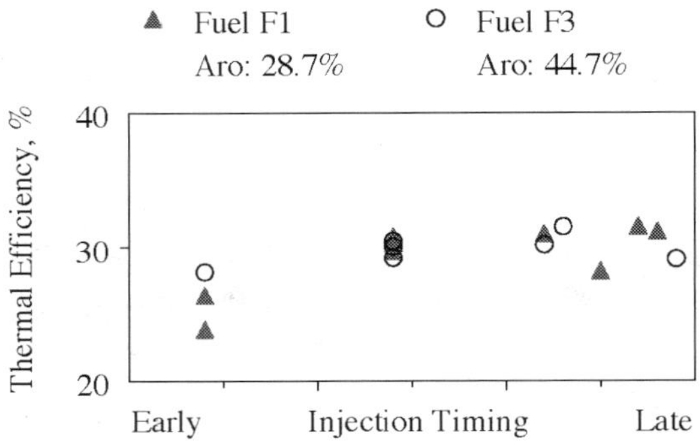

Fig. 18 Effect of injection timing and aromatic content on thermal efficiency at 1500 rpm, 25% load

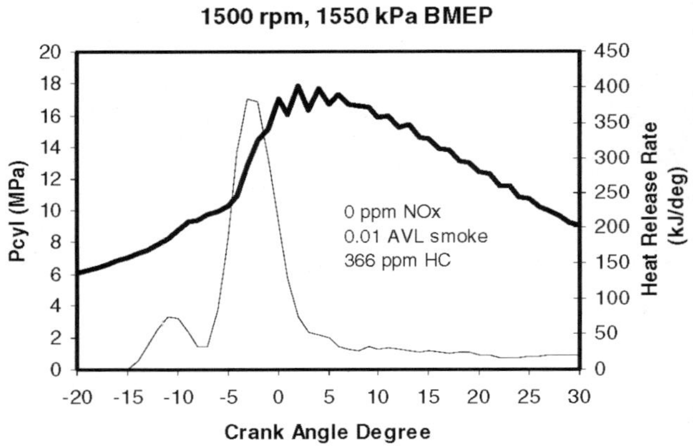

Fig. 19 HCCI operation at ¾ engine load with diesel fuel

As mentioned previously, the load levels achieved with this in-cylinder Diesel injection HCCI strategy are believed to be the highest in the world attained to date. Figure 19 shows the cylinder pressure and heat release rate measured at 1500 rpm and 1550 kPa BMEP (1750 kPa IMEP) on the single cylinder test engine. Diesel fuel characterized in Table 2 was used. In these tests, NOx and smoke emissions were negligible, while HC levels were reasonable for this load level. Combustion was phased slightly early of optimum, but the peak cylinder pressure was within current production limits for this engine and the rise rates were not excessive.

SUMMARY AND CONCLUSIONS

Significant progress has been made to increase the load capability of an HCCI engine using an in-cylinder, direct injection of diesel fuel method. Three quarter load (1750 kPa IMEP) has been attained on a single cylinder engine using various methods to control combustion phasing and minimize engine out emissions. These methods include inlet manifold temperature and pressure, injection timing, and fuel properties.

Increasing inlet manifold temperature by 20C can result in a 5 degree change in combustion phasing which, in turn, leads to a 7% decrease in fuel consumption. Optimal injection timing is critical to maintain low NOx and smoke while keeping HC and CO at moderate levels. Ten degree retard in injection timing can lead to severe differences in NOx, HC and CO levels and phase combustion by approximately 5 degrees leading to excessive cylinder pressure rise rates. Changes in boost level can also dramatically affect combustion phasing. Increasing boost phases combustion earlier and decreasing boost retards combustion phasing. The resulting impact on air-fuel ratio can lead to smoke increases if boost is too low.

Caterpillar and ExxonMobil are working together to understand heavy-duty HCCI. Initial results from a single-cylinder engine have shown that HCCI technology can achieve very low PM and NOx emissions at light-medium load with a range of distillate fuel properties including US market fuel. This study was a preliminary evaluation of fuel effects in an advanced HCCI engine. Over the engine conditions tested, cetane number and aromatic content have only a small effect on the performance and exhaust emissions of Caterpillar's HCCI combustion system. Increasing cetane number makes it more difficult to achieve optimum combustion phasing and maximize thermal efficiency of the engine within the constraints of the cylinder pressure and rise rate limits. Future work will address higher loads and other technical challenges.

REFERENCES

1. F. Zhou, T. Asmus, D. Assanis, J. Dec, J. Eng, and P. Najt (2003) Homogenous Charge Compression Ignition (HCCI) Engines: Key Research and Development Issues. SAE PT-94.
2. T. Ryan and T. Callahan (2002) Homogeneous Charge Compression Ignition of Diesel Fuel. SAE Paper 961160.
3. P. Kelley-Zion and J. Dec (2000) A Computational Study of the Effect of Fuel Type on Ignition in HCCI Engines. Proc. Comb. Inst., Vol. 28, part 1, pp. 1187-1194.
4. G. Shibata, K. Oyama, T. Urushihara and T. Nakano (2004) The Effects of Fuel Properties on Low and High Temperature Heat Release and Resulting Performance of an HCCI Engine. SAE Paper 2004-01-0553.
5. J. Dec and M. Sjoberg (2003) A Parametric Study of HCCI Combustion: Sources of Emissions at Low Loads and the Effects of GDI Fuel Injection. SAE Paper 2003-01-0752.
6. T. Tsurushima, E. Kunishima, Y. Asaumi, Y. Aoyagi and Y. Enomoto (2002) The Effect of Knock on Heat Loss in Homogeneous Charge Compression Ignition Engines. SAE Paper 2002-01-0108.
7. S. Kimura, O. Aoki, Y. Kitahara, and E. Aiyoshizawa (2001) Ultra-Clean Combustion Technology Combining a Low Temperature and Pre-mixed Combustion Concept for Meeting Future Emissions Standards. SAE Paper 2001-01-0200.
8. B. Gatellier, B. Walter and M. Miche (2001) New Diesel Combustion Process to Achieve Near Zero NOx and Particulates Emissions. IFP International Congress, Paris, France, pp. 43-52.

Which Fuels For Low CO_2 Engines?
© P. Duret, X. Montagne (Eds.) and Editions Technip, Paris, 2004, pp. 89 – 98
27 rue Ginoux, 75015 Paris

NEW HCCI / CAI COMBUSTION PROCESS DEVELOPMENT: METHODOLOGY FOR DETERMINATION OF RELEVANT FUEL PARAMETERS

Nicolas Jeuland, Xavier Montagne, Pierre Duret
IFP Powertrain Engineering

ABSTRACT – *To pass the simultaneous emission requirements for near zero pollutant and for low CO_2 level, which are the challenges of future powertrains, many research studies are currently carried out over the world on new engine combustion processes, such as the Controlled Auto-Ignition (CAI^{TM}) for gasoline engines and the Homogeneous Charge Compression Ignition (HCCI) for diesel engines. These new processes have the potential to reduce, by 10 to 100, NOx and particulate (PM) emissions in comparison with a typical gasoline or diesel engine. The main issue in the development of such combustion modes remains the limited range of engine speed and load that enable a correct engine running, with acceptable levels of NOx and noise emissions. One of the promising ways to increase the CAI combustion range lies in the research of more adapted fuels, optimized to enhance combustion initiation by modifying and controlling their auto-ignition characteristics and/or their physical and chemical properties.*

A new procedure is set-up to select the fuels, which allows a better characterization of the fuel impact on engine running in CAI, for various conditions of speed and load. The use of this methodology with a wide range of fuels, formulated according to their auto-ignition characteristics, volatility and chemical composition underlines the complexity of the action of fuel and points out the most influent fuel characteristics for a wide range of running conditions.

INTRODUCTION

Fuels and engines used in road transportation have to face two main challenges in a highly competitive economy:

- To reduce pollutant emission levels to such values that air quality in the cities complies with World Health Organization standards.
- To reduce the emissions of carbon dioxide (CO_2), regarded as the major greenhouse gas contributing to global warming and climate changes.

The stoichiometric gasoline four stroke engine with a 3-way catalyst has the ability to reach ultra-low emission levels for passenger cars such and to pass the Euro 4 requirements and even the future Euro 5 ones. On the other hand, its efficiency remains moderate when compared to Diesel engines. Different strategies to improve its efficiency were recently considered.

- Direct Fuel Injection (DI) technology could be adopted, with stratified charge and lean burn combustion systems on a limited operating range. This lowers fuel consumption through reduced pumping and reduced heat transfer losses, but with a major issue to reach ultra-low pollutant emission levels because of the need of a sophisticated and fuel sensitive DeNOx after–treatment device.
- Another approach would be to keep the conventional stochiometric 3-way catalyst engine for its low emission levels and to reduce fuel consumption by using new technologies such as variable valve actuation, down-sizing with supercharging…

Beside these two approaches, there is a strong industrial need for an alternative highly efficient lean burn combustion system, which could provide similar CO_2 advantages as stratified Diesel engines while solving the NOx emission problem of lean burn combustion, without depending on fuel sensitive and expensive after-treatment systems.

To fulfill these challenges, a lot of studies are currently carried out, especially on new combustion processes such as the Controlled Auto-Ignition (CAI™) for gasoline engines and the Homogeneous Charge Compression Ignition (HCCI) for Diesel engines. These new processes offer the potential to reduce strongly nitric oxide emissions (NOx) and particle emissions (PM) in comparison with a typical gasoline or diesel engine.

Their principle consists in preparing a highly diluted mixture of air, fuel and residual gases, in achieving its simultaneous ignition in the whole space of the combustion chamber and in closely controlling the combustion event for best performance in terms of efficiency and NOx/PM emissions (timing and heat release rate during the cycle). Thanks to the thermo-chemistry of the charge, the thermal NOx formation and the soot production are in principle much lower than in the diffusion flame typical of the conventional diesel or in the propagation flame typical of the conventional spark ignition engine. Figures 1 and 2 show an example of very the promising results that can be obtained [1, 2].

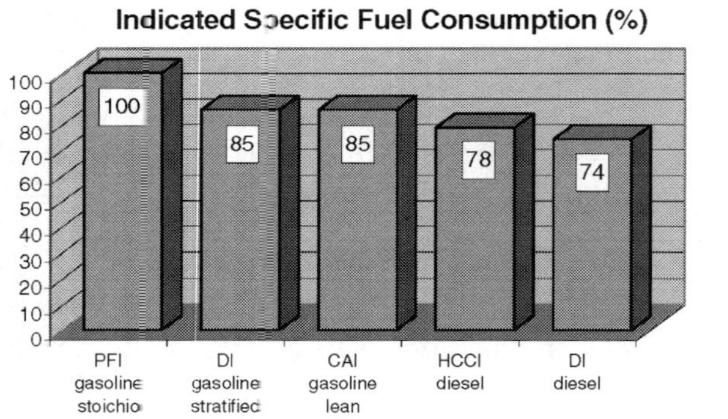

Fig. 1 Compared fuel consumption between various gasoline and diesel engine combustion processes (1500rpm – 3 bar IMEP)

For the gasoline engine application, the generic term of HCCI is sometimes also used. The reason why the terms "gasoline Controlled Auto-Ignition" (or gasoline CAI) are here preferred to "gasoline Homogeneous Charge Compression Ignition" is that in the approach described here, it is possible to auto-ignite the fresh charge in a 4-stroke gasoline engine without needing intake air heating nor increased compression ratio. It means that the same engine is also able, without engine design modification, to move from gasoline CAI at partial load to SI (Spark Ignition) at full load for maintained performance. Thanks to 15 years of 2-stroke CAI combustion background, the scientific methodology used was to transfer the 2-stroke combustion process to the 4-stroke engine, trying to reproduce the same in-cylinder conditions during the compression stroke. The first results of this innovative approach, based on the use of a high quantity of stratified internal EGR in a 4-stroke engine, as it can be achieved in 2-stroke engines [4-7], were published and clearly detailed in previous papers [8-10]. This "negative valve overlap" approach [11] enable to run successfully, probably for the first time, a 4-stroke engine in gasoline controlled Auto-Ignition. As this gasoline CAI is not homogeneous, the "H" letter of "HCCI" appears to be inappropriate. This is the reason why the name "gasoline CAI" is much preferred to "gasoline HCCI" [12].

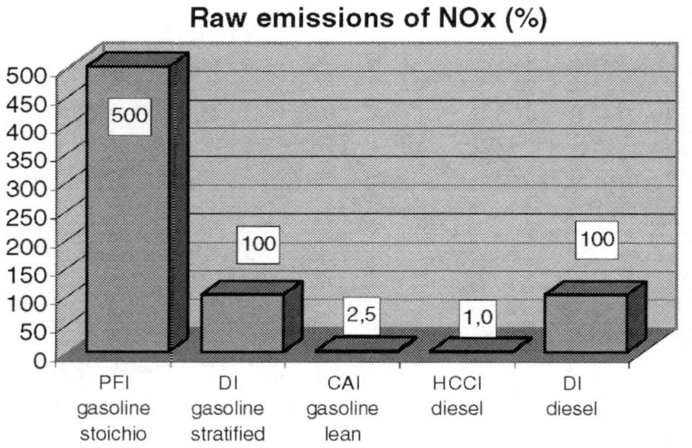

Fig. 2 Compared NOx emissions between various gasoline and diesel engine combustion processes (1500rpm – 3 bar IMEP)

DEVELOPMENT OF A FUEL CHARACTERIZATION METHODOLOGY

One of the key issues with the CAI combustion process, based on the use of a high rate of internal EGR, is the close control of the switch from the CAI mode to the classical SI mode. It is needed both at very low load (for combustion stability reasons) and at high load (to avoid knock). To solve this concern, several manufacturers developed variable valve trains able to change valve timing from cycle-to-cycle [13, 14]. These systems could help to get more flexible cycle to cycle combustion control and to extend the CAI application range, which is the other key issue of the concept. But for the time being, the devices remain rather complex, expensive and they induce a significant increase in the number of engine parameters to control.

Another parameter which could significantly affect the performance of new CAI & HCCI combustion processes is the fuel composition, as shown by Foster [15]. The use of specific fuels or additives could be an interesting solution to increase the benefits of the CAI combustion process, by extending its application range [16-18]. The IFP background and know-how already pointed out this significant influence of fuel composition and physical properties on CAI combustion (especially in terms of operating range, pollutant emissions and noise...). In previous work done on 2-stroke CAI engines, IFP showed that the benefits of such new combustion processes could be greatly enhanced with more appropriate fuel (and/or additives) formulations, which have to be studied and developed. Before starting such studies, it was necessary to define a test procedure enabling the evaluation of the fuel effect on the CAI.

METHOD DESCRIPTION

Basic principle

The method has already been described in previous publications [19, 20] then, only the most important points are summarized here.

The method is based on an index, called "CAI Number" or CAN. This index was initially developed for 2-strokes engines [9] and it has then been adapted to 4-strokes engines [19]. Its principle can be summarized as follows:

- Definition of engine running limitations (intake air pressure, combustion noise, NOx emissions, engine stability).
- Evaluation of the CAI running zone for each fuel with these parameters.
- Comparison of this CAI running zone with the one measured for a reference fuel.

This running zone defined by this method is illustrated in figure 3.

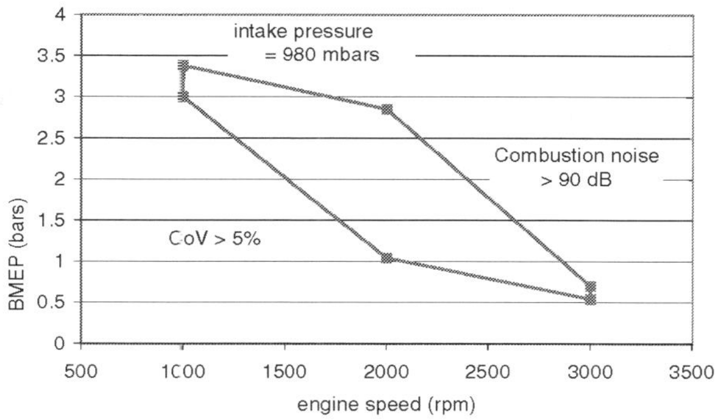

Fig. 3 CAI running zone definition

The CAI running zone is limited at low speed / low load by the engine stability: when the temperature is too low to ensure a correct auto-ignition of the air / fuel mix, the IMEP becomes highly unstable (misfiring). It is also limited at high load, when several cumulative phenomena tend to limit its operating range. The combustion noise can reach the 90dB limitation and in some cases, the maximum acceptable NOx level can be reached. Another high load limiting factor could be, with some special "low noise/low NOx/CAI" fuels, the air flow intake itself. As explained before, when the maximum intake pressure is reached, the maximum volumetric efficiency becomes limited by the "negative valve overlap" configuration used in this study to get CAI part load operation. Therefore, it must be pointed out that in the following results, the two first criteria (noise and NOx) are truly depending on the fuel effect while the third one (limited volumetric efficiency) could sometimes (with fuels adapted for high load CAI) limit the level of maximum load in CAI, independently of the fuel effect.

This method, tested with various fuel types, appeared to be able to discriminate between fuels, according to their impact on the CAI running zone. Nevertheless, this method also appeared to be too global to enable a correct interpretation of the combustion phenomena. Indeed, 2 fuels can have a similar index and still have radically different behaviors, as shown in figure 4.

In this example, the 2 best fuels have a radically different behavior, despite a similar CAN index:

fuel 1 tends to enhance CAI combustion at low speed / low load, while fuel 2 tends to increase the CAI combustion in both low load and high load conditions. Consequently, a unique calculated value does not seem to be sufficient to fully describe all the fuel behaviors observed with 4-stroke CAI running conditions and has to be completed with additional information.

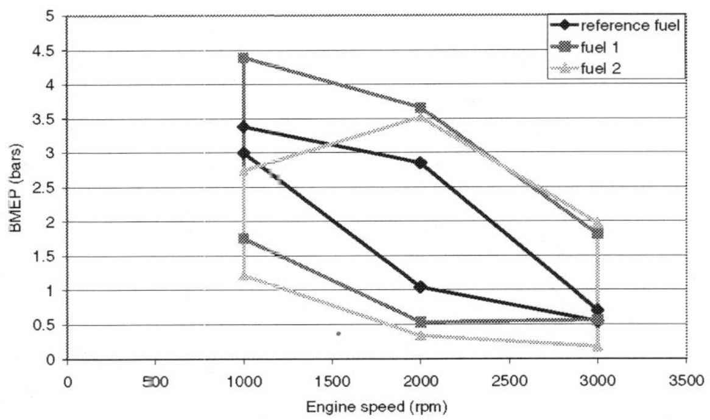

Fig. 4 Comparison of fuel effect on CAI running zone

Modification of the characterization procedure

A comprehensive index, providing information on the fuel effect on the whole CAI running zone has to describe the behavior at every speed and load running conditions. That for, the CAN index for CAI 4-stroke engines was divided into 4 parts to describe the effect of fuel on the 4 types of running conditions (Figure 5):

- Low speed / low load (zone 1),
- Low speed / high load (zone 2),
- High speed / high load (zone 3),
- High speed / low load (zone 4).

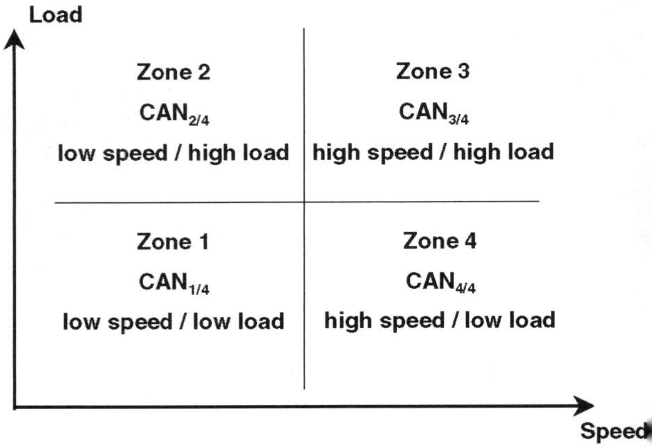

Fig. 5 Speed / load field division in 4 zones.

For each zone, one indicator can be determined. To calculate them, the CAI range obtained with a reference fuel (figure 6, area 1-2-3-4-5-6) was divided into 4 zones, defined according to the mean value of minimum and maximum BMEP obtained for each engine speed.

- 1000 rpm $\left(\text{point A} = \dfrac{(1)+(2)}{2}\right)$
- 2000 rpm $\left(\text{point B} = \dfrac{(3)+(6)}{2}\right)$
- 3000 rpm $\left(\text{point C} = \dfrac{(4)+(5)}{2}\right)$

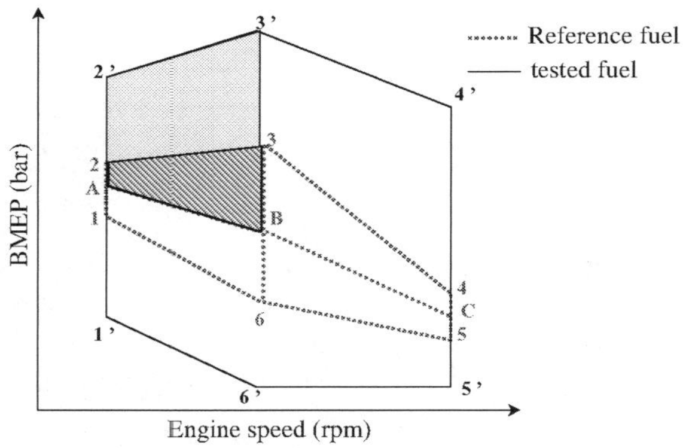

Fig. 6 Principle of CANs calculation

The 4 CANs are then calculated according to the following formulas:

$$CAN_{2/4} = \frac{\text{area}\,(A-2'-3'-B)}{\text{area}\,(A-2-3-B)}$$

$$CAN_{1/4} = \frac{\text{area } (A-1'-6'-B)}{\text{area } (A-1-6-B)}$$

$$CAN_{3/4} = \frac{\text{area } (B-3'-4'-C)}{\text{area } (B-3-4-C)}$$

$$CAN_{4/4} = \frac{\text{area } (B-6'-5'-C)}{\text{area } (B-6-5-C)}$$

This method is more complicated than the first characterization method, but it enables much more accurate analyses of the phenomena linked to CAI combustion.

IDENTIFICATION OF RELEVANT FUEL PARAMETERS FOR CAI RUNNING

The method described above has been applied on a set of fuels to obtain information on the most influent fuel parameters.

DESCRIPTION OF THE FUEL MATRIX

The tested fuels were divided in groups, according to their main characteristics.

- Reference fuel ("REF"),
- Reference Fuel + additives ("REF+"),
- Oxygenated compounds, pure ("O") or blended ("BO"),
- Gasoline-like fuels ("GLF"),
- Diesel-like fuels ("DLF").

Their compositions are detailed in table 1 and their main characteristics in table 2.

"REF"	Reference EN228 gasoline fuel (RON 97; MON 86)
"REF+$_1$"	Blend of reference fuel with 0.2% NEH (Nitrate Ethyl Hexyl, a commonly used procetane additive) 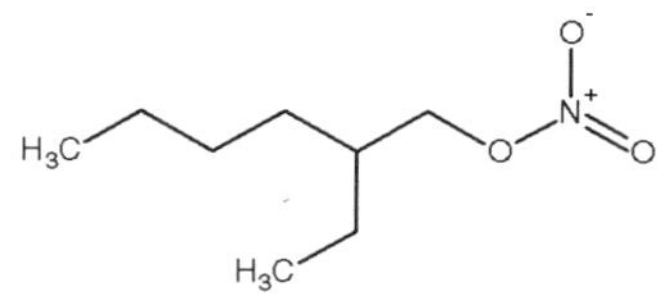
"REF+$_2$"	Blend of reference fuel with 0.2% of ditertiobutyl peroxide, another commercial procetane additive 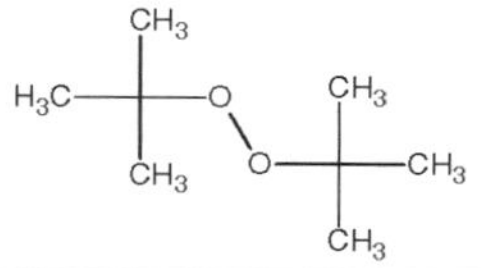
"DLF1"	Low sulfur diesel fuel (50ppm)
"DLF2"	Euro3 diesel fuel (350ppm sulfur)
"DLF3"	n-heptane
"DLF4"	Wide cut kerosene, called Jet B or NATO F45, characterized by its high volatility (initial boiling point around 80°C) and an intermediate cetane index (35)
"DLF5"	Fischer-Tropsch fuel, characterized by a very high cetane number
"GLF"	Blend of fuels DLF4 (Jet B) and REF (gasoline fuel), 50% each
"GLF2"	Blend of fuels REF (EN 228 gasoline) and DLF2 (EN590 diesel fuel)
"O1"	Ethanol
"O2"	Blend of ethanol and ethylal (50%vol each) 
"O3"	Blend of methanol and methylal (50% each). This fuel was included in the fuel matrix but could not be tested, due to an engine failure. 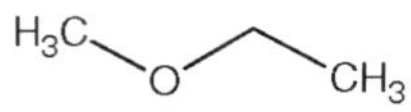
"BO"	Blend of fuels DLF4 (JetB) and O3 (methanol / methylal)

Table 1. Tested fuels composition

	RON *	Cetane number *	Density (g/l)	IBP (°C)	FBP (°C)
REF+1	70 - 80	< 0	740	< 50	150 - 200
REF+2	70 - 80	< 0	740	< 50	150 - 200
DLF1	< 0	50	840	> 200	> 300
DLF2	< 0	50	830	> 200	> 300
DLF3	0	56	680	50 - 100	100 - 150
DLF4	< 0	35	810	< 50	200 - 250
DLF5	< 0	93	760	150 - 200	> 300
GLF	< 50	10	770	< 50	200 - 250
GLF2	30 - 40	40	780	< 50	> 300
O1	50 - 60	-	820	50 - 100	50 - 100
O2	> 100	-	790	50 - 100	50 - 100

* : calculated values

Table 2. Main test fuels characteristics

IMPACT OF FUEL PARAMETERS ON CAI RUNNING

Cetane index

The figure 7 shows the impact of the cetane number on CAN1/4 and CAN2/4 (low speed running points).

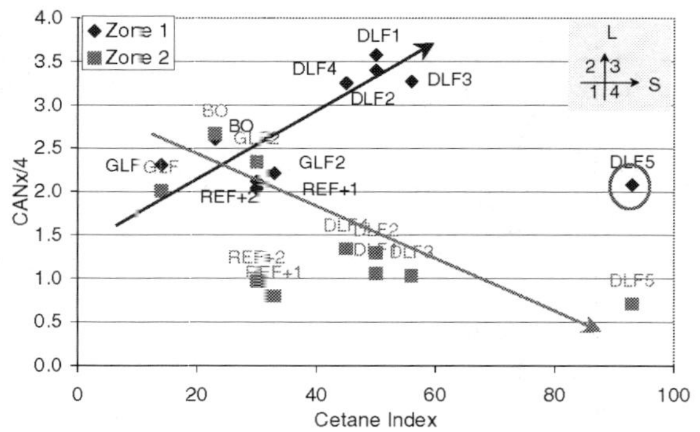

Fig. 7 Impact of fuel cetane index on CAI running zone

This figure shows that the impact of the cetane index is strongly different when considered at low load, (zone 1: increase in CAI range when the cetane index increases) or at high load (zone 2: decrease in CAI range when the cetane index increases).

- *At low load*, the engine temperature is low, and the main issue for CAI combustion is the start of the combustion. If the cetane index is high, the fuel auto-ignition is facilitated, even if the temperature remains low, and the engine stability is enhanced, inducing a wider CAI combustion zone.
- *At high load*, on the opposite, the CAI zone is limited by the noise level, which is linked to the start of the combustion. If the fuel cetane index is high, combustion begins earlier in the cycle, the heat release rate rises and noise increases.

This figure also shows that other physical parameters have to be considered. For instance, DLF5 (Fischer-Tropsch fuel), with a very high cetane index, should be the most efficient in enlarging the zone 1 (low load, low speed). This analysis shows that, on the contrary, this fuel is linked with one of the smallest $CAI_{1/4}$.

Density

As the engine used for this study was a direct injection gasoline engine, the density of the fuel could have an influence on the combustion, via processes such as air / fuel mixture quality.

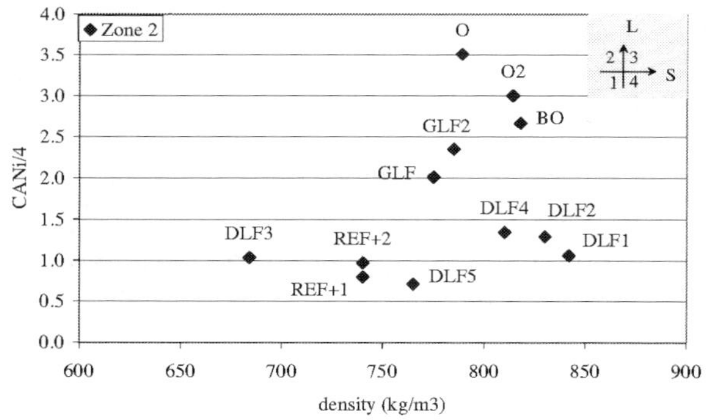

Fig. 8 Correlation $CAN_{2/4}$ - density of the fuel

At first look, no clear link appears between fuel density and CAN2/4. Nevertheless, it can be observed that the Diesel-like fuels have a peculiar behavior: while with other fuels, the CAI zone is linked to the density (higher density means wider CAI combustion zone), for Diesel-like fuels, no clear link is observed (figure 8).

In zone 3 (high speed, high load), this behavior is unchanged, as shown in the figure 9.

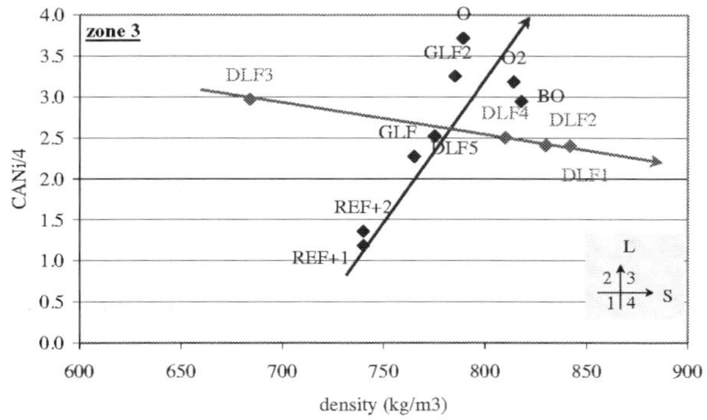

Fig 9 Correlation $CAN_{3/4}$ - density of the fuel

This analysis shows that:

- for low cetane fuels, a higher density tends to increase the CAI combustion zone,
- for high cetane fuels, the density is a parameter of secondary importance.

Latent heat of vaporization

The latent heat of vaporization (LHV) is the energy needed to vaporize 1kg of the product. A molecule with a high LHV traps an important quantity of energy during its vaporization, and thus decreases the temperature of the air/fuel mixture.

This parameter is strongly linked to the oxygen content of the molecule as oxygen induces the formation of intermolecular H-bonds (Van der Waals bonds), and consequently an important increase of the LHV. As a consequence, all the fuels have close LHV, except oxygenated fuels.

The impact of this parameter on CAI running is linked to the running point (low or high load), as shown in figures 10 and 11.

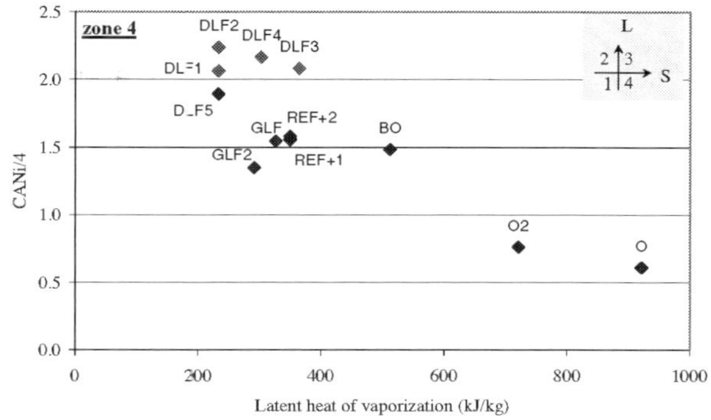

Fig. 10 Correlation between LHV and CAN4/4

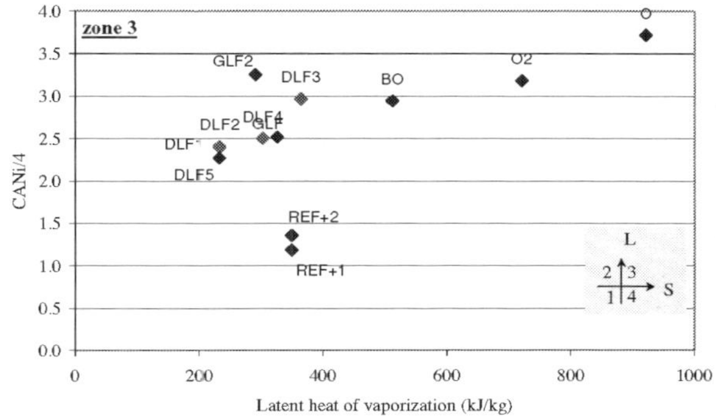

Fig. 11 Correlation between LHV and CAN3/4

The impact of this parameter on the CAI zone is different according to the running point.

- *At low load* (zone 4), an increase in LHV decreases the CAI zone. Indeed, the cooling effect induces a decrease in cylinder temperature and, as a consequence, some ignition difficulties.
- *At high load* (zone 3), on the contrary, the cooling effect tends to calm down the combustion and consequently to enlarge the CAI zone.

Oxygen content

Three fuels from the matrix were containing oxygenated compounds: "O" (pure ethanol), "O2" (blend ethanol / ethylal) and "BO" (blend JetB / Methanol / Methylal).

The figure 12 focuses on the results obtained for these fuels.

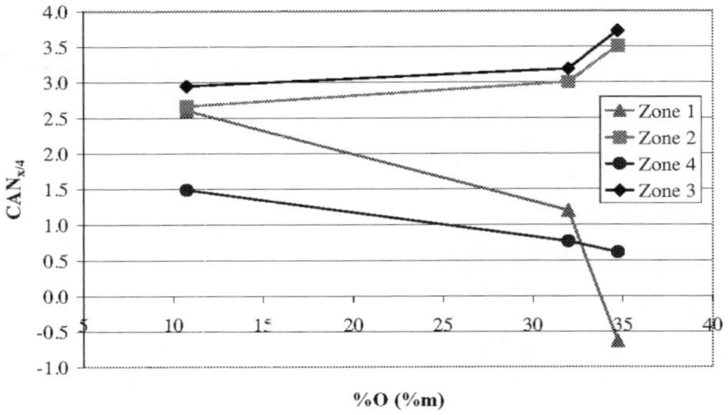

Fig. 12 Correlation between fuel oxygen content and $CAN_{x/4}$

This figure shows that the impact of the oxygen content is dependant from the engine load.

- *At low load* (zones 1 and 4), an increase in the oxygen content leads to a narrowing of the CAI running zone.
- *At high load*, on the contrary, an increase in the fuel oxygen content leads to an enlargement of the CAI running zone.

The analysis of zone 1 shows a specific behavior: ethanol (O) induces a negative CAN. The blend with ethylal (O2) induces a strong increase in the CAN. These evolutions are not due to LHV, but more to the cetane number increase induced by ethylal.

DISCUSSION

All the correlations that have been calculated can be brought together in a table to get a global view of the most influencing fuel parameters. In **Table 3** below, a positive correlation is noted "+" and a negative correlation "-".

According to these results, it can be determined that an enlargement of the CAI zone at low speed / low load (zone 1) can be obtained through:

- an increase in fuel cetane,
- a low oxygen content,
- a low latent heat of vaporization.

On the opposite, in zone 2 (low load, high speed), to be efficient a fuel should have:

- a high oxygen content,
- a high latent heat of vaporization,
- a low cetane index.

	RON	MON	IC	Density kg/m^3	Heating value kJ/kg	IBP °C	FBP °C
Zone 1	-		+		+	+	
Zone 2			-	*	-	-	
Zone 3				*			
Zone 4	-		+		+	+	

*: dependant on fuel cetane number

	%C	%H	%O	LHV	FBP-IBP	E 50	E 100
	%m	%m	%m	kJ/kg	°C	%v	%v
Zone 1	+	+	-	-			
Zone 2	-	-	+	+			
Zone 3	-		+			-	
Zone 4	+	+	-	-			-

Table 3. Correlations between CANx/4 and fuel parameters

As a consequence, the optimal fuel has to be a compromise between all these parameters.

Moreover, the results obtained here are only preliminary ones and deserve closer analyses. A set of fuels has been tested and some correlations have been calculated, without claiming at any formulation optimization or trying to decorrelate the fuel properties.

So far, it remains difficult to identify precisely which fuel parameter is a truly important one. For instance, fuel volatility is positively correlated with initial boiling point and the CAI range for zones 1 and 4, which means that the optimal fuel at low load condition appears to be a heavy fuel. This observation is quite surprising, as the main issue at low load is to obtain a satisfying auto-ignition. It would have been more logical if heavy fuels, with their high initial boiling points and low volatility would not rank as the best ones. In fact, these surprising results seem to be mostly due to the auto-ignition properties, as most of our fuels with high initial boiling points were "diesel-like" fuels (diesel fuel, Fischer-Tropsch...) with high cetane numbers, while, on the opposite, most of the fuels with low initial boiling points were "gasoline like" fuels (ethanol, gasoline...) with low cetane numbers.

To obtain a more comprehensive evaluation of fuel impact on CAI running zone, the ability to discriminate between the fuels properties will be an unavoidable need. It should be considered early, when determining design of experimental method for instance, in order to be able to determine which parameter is the most influent in each zone: auto-ignition delay, volatility, density, chemistry of the fuel, combustion speed...

CONCLUSION

The IFP-developed CAI™ (Controlled Auto-Ignition) combustion system has been applied to a conventional gasoline automotive 4-stroke engine. This engine has been used to undertake a detailed study of the influence of the fuel formulation on the engine operating range that can be covered in CAI combustion mode. From the information and results generated by this study, the following main conclusions can be stated:

- A repeatable and discriminating testing procedure was developed, able to show the fuel effect on the CAI running zone.
- The development of this method leads to the definition of four indices, according to the engine speed and load. The fuel impact has to be studied separately for each zone.
- The study of several fuel formulations has shown that the fuel selective formulation could be one of the relevant ways to enlarge strongly the CAI running zone in some of the four areas previously defined.
- Correlation calculations were achieved to try to link the fuel physical properties and the CAI running zone. It has been shown that the engine load is one of the main parameters modifying the fuel impact: at low load, fuel requirements would be a short auto-ignition delay and a low LHV, while at high load a longer auto-ignition delay and a high LHV would be preferred. A compromise between theses characteristics will be needed to determine the optimal fuel.
- Further works are necessary to determine the true impact of each of the fuel physical parameters, and to be able to rank their influence on the CAI running area: auto-ignition delay, volatility, chemical composition...

Acknowledgments. The authors would like to acknowledge Christophe Gunkel who performed all the engine tests.

REFERENCES

1. DOUAUD A. Vers des technologies de combustion "ZERO" émissions. EET Paris 20-21 Novembre 2001
2. DURET P., DOUAUD A. New Combustion Processes: Homogeneous Diesel and Auto-Ignition Gasoline towards Zero Emission with Major Engine and Fuel Technology Challenges. Global Powertrain Congress, Detroit June 5-7, 2001
3. A New Generation of Engine Combustion Processes for the Future, IFP International Congress, November 26-27, 2001, Editions Technip
4. DURET P., VENTURI S. Automotive Calibration of the IAPAC Fluid Dynamically Controlled Two-Stroke Combustion Process - SAE 960363
5. ISHIBASHI Y., ASAI M. Improving the Exhaust Emissions of Two-Stroke Engine by Applying the Activated Radical Combustion - SAE 960742
6. DURET P., DABADIE J.C., LAVY J., ALLEN J., BLUNDELL D., OSCARSSON J., EMANUELSSON G., PEROTTI M., KENNY R., CUNNINGHAM G. The Air Assisted Direct Injection ELEVATE Automotive Engine Combustion System - SAE 2000-01-1899
7. LAVY J., ANGELBERGER C., GUIBERT P., MOKHTARI S. Towards a Better Understanding of Controlled Auto-Ignition (CAI™) Combustion Process from 2-Stroke Engine Results Analyses - SAE/SETC 2001-01-1859/4276
8. LAVY J., DABADIE J.C., ANGELBERGER C., WILLAND J., JURETZKA A., SCHAFLEIN J., MA T., LENDRESSE Y., SATRE A., SCHULZ C., KRAMER H., ZHAO H., DAMIANO L. Innovative Ultra-low NOx Controlled Auto-Ignition Combustion Process for Gasoline Engines: the 4-SPACE Project - SAE 1000-01-1837
9. DURET P., LAVY J. Gasoline Controlled Auto-Ignition (CAI™): potential and prospects for future automotive application – IMechE Conference – 21st Century Emissions Technologies – 4 & 5 December 2000 – London
10. LAVY J., DABADIE J.C., DURET P., ANGELBERGER C., LE COZ J.F., CHEREL J.

Controlled Auto-Ignition (CAI): a New Highly Efficient and Near Zero NOx Emissions Combustion Processes for Gasoline Engine Application - IFP International Congress - November 26-27, 2001 - Editions Technip
11. DURET P., LAVY J. Process for Controlling Self-Ignition in a 4-Stroke Engine - US Patent 6 082 342
12. LI J., ZHAO H., LADOMMATOS N., MA T. Research and Development of Controlled Auto-Ignition (CAI) Combustion in a 4-Stroke Multi-cylinder Gasoline Engine - SAE paper 01FL-393, Detroit, March 2001
13. ALLEN J. et al. Advanced Combustion Using a Lotus Active Valve Train. Internal Exhaust Gas Recirculation Promoted Auto Ignition, IFP International Congress, Nov 26-27 2001
14. DOREAU J. Contribution of New Generation Variable Valve trains and Engine Management Systems for the Control of New Combustion Processes, IFP International Congress, Nov 26-27 2001.D. E. FOSTER - An Investigation into the Effect of Fuel Composition on HCCI Combustion Characteristics - SAE 2002- 01-2830
15. STEIGER W. Volkswagen Strategy to High Efficient Powertrains - 22nd International Vienna Motor symposium - April 26-27, 2001
16. MONTAGNE X., DURET P. What Will Be the Future Combustion and Fuel-related Technology Challenges? - IFP International Congress - November 26-27, 2002 - Editions Technip
17. OLSSON J.O., TUNESTAL P., HARALDSSON G., JOHANSSON B. A Turbo Charged Dual Fuel HCCI Engine - SAE 2001-01-1896
18. DURET P. Gasoline CAI and Diesel HCCI: the Way towards Zero Emission with Major Engine and Fuel Technology Challenges - SETC Kyoto October 2002 - JSAE 20024280 & SAE 2002-32-1787
19. JEULAND N., MONTAGNE X. Engine and fuel related issues of gasoline CAI (Controlled Auto-Ignition) combustion, JSAE 2003-03-49, SAE 2003-01-1856 - Society of Automotive Engineers of Japan congress, Yokohama, 19-22 May 2003, Proceedings
20. JEULAND N., MONTAGNE X. New approach to characterize the engine/fuel couple in CAI™ combustion mode. Impact of fuels. Direkteinspritzung im Ottomotor IV. 5 Internationale Tagung, Essen, 1-2 July 2003, Proceedings, pp. 249-272

Contact. Nicolas Jeuland (nicolas.jeuland@ifp.fr) Fuel, Lubricant and Emission Department IFP

Which Fuels For Low CO_2 Engines?
© P. Duret, X. Montagne (Eds.) and Editions Technip, Paris, 2004, pp. 99 – 106
27 rue Ginoux, 75015 Paris

SYNTHETIC FUELS – KEY FOR FUTURE POWERTRAINS

Dr. Wolfgang Steiger
Volkswagen AG

ABSTRACT – *The development of motor vehicles continues to be determined by increasingly stringent emissions standards including fuel consumption and CO_2 emissions. In addition, the dominance of crude oil as the primary energy source for mobility gives rise to considerable future risks. So, there is a strong need for new fuels and diversified energy sources including renewables.*

Synthetic, liquid fuels play a major role in this, as they enable the use of a multitude of diverse primary energies for generating high quality fuels. In an first step GTL-fuels will come in the market, providing an high-quality, environmentally-friendly diesel fuel. In comparison with conventional diesel, GTL-Diesel revealed significant lower particle emissions and NO_x emissions in various vehicles. In a second step, biomass may also be used to manufacture fuels. Synthesis gas out of biomass, combined with Fischer-Tropsch synthesis, leads to CO_2-neutral BTL-fuels ("SunFuels") with properties similar to GTL-fuels. On a well-to-wheels basis, this dramatically reduces CO_2 emissions from vehicle operation whilst consumption remains the same.

In a third step, thanks to its high flexibility, the GTL/BTL process can be used to manufacture tailor-made fuels which optimally satisfy the high demands of modern engine generations and of novel combustion schemes such as Homogenous Combustion Engines as well.

In the long-term, it is assumed that the current hydrogen storage and infrastructure problems will be solved. Synthetic, liquid fuels meanwhile offer an ideal transition from hydrocarbon to hydrogen management. However, this is not anticipated within the next 20 years.

INTRODUCTION

From an environmental standpoint, rising exhaust emission standards in the world, above all as in California, where practically zero emissions are required even from conventional powertrains, continue to have a determining influence on the development of motor vehicles and powertrains. However, the importance of measures to reduce consumption and CO_2 emissions in the optimisation of vehicle and powertrain concepts is steadily growing also. With the dramatically diminishing effect of vehicle emissions in the immission situation, i.e. air quality, the balance between typical local exhaust-gas emissions such as CO, HC and NOx and the global CO_2 emission will presumably shift towards an increasing significance of CO_2 . The main reason for this is that global warming is now largely accepted as a reality although objective evidence has yet to be shown. Considerable pressure is being brought to bear on the automobile industry because CO_2 emissions will, contrary to the trend of other pollutants, continue to rise until the year 2010. In Europe, however, under a clear technical consideration this fact would be of secondary importance as the CO_2 emissions produced by road traffic account for only approx. 11.5% of total anthropogenic emissions.

Another important facet in this picture is the rising global demand for energy services allied to the anticipated decrease in the availability of inexpensive, low-cost fossil primary energy sources, particularly mineral oil. Above all the extreme concentration on petroleum as a primary energy source especially in the mobility sector holds considerable risks for the future. The mobility economy is greatly affected by this. Securing a long-term supply of energy sources for public and private transport, especially against the background of political instability in oil producing regions, requires not only that existing fuel resources be utilised as economically as possible, but also that the energy sources used in fuel production be diversified in the medium and long term to include alternative and finally renewable resources in particular.

STRATEGY

It is from substantial interest for the automotive industry to ensure the availability of all needed energy as base for future business. In the same time continuous demands for emission reduction makes the availability of high quality fuels necessary. So the Volkswagen AG and DaimlerChrysler AG set the joint task of devising a strategy for a step-by-step transition from present-day powertrains and conventional mineral oil based fuels to future powertrains and related fuels and primary energy sources needed for their operation. Just to concentrate on long term high risk technology is not enough to ensure future mobility. This strategy encompasses three areas

- A systematic further increase in the efficiency of powertrain units
- The incorporation of alternative energy sources in the fuel production process
- The development of CO_2 neutral paths for vehicle operation

Already remarkable successes have been achieved in the field of fuel consumption reduction. According to the voluntary commitment of the European automotive industry for an overall fleet CO_2 reduction of about 25% from 1995 to 2008 the ACEA fleet is full in line as shown in Fig. 1.

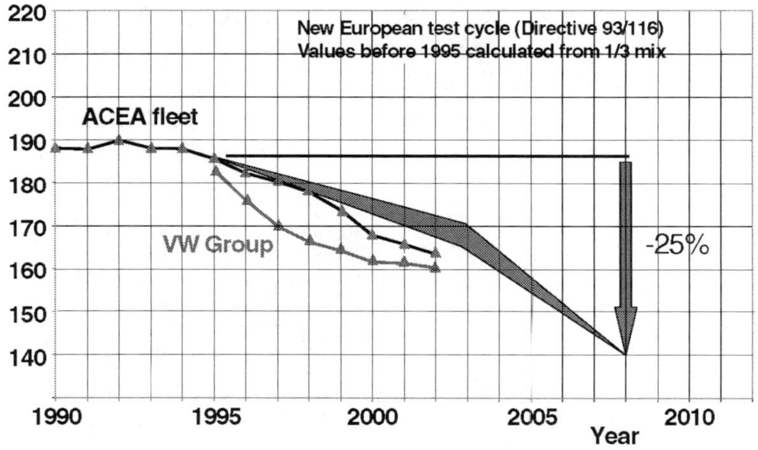

Fig. 1 ACEA Commitment

In a comparison of the various means of transport, the modern-day passenger car is already definitely competitive. Nevertheless, development continues towards even more economical powertrain systems.

From today's point of view, the hydrogen fuel cell with conducted electric motor has the highest tank-to-wheel efficiency of all powertrain units. Ongoing development is still necessary to reach the systems maturity for market introduction. However, a precondition for its use is the availability of hydrogen. Hydrogen is not an primary energy, but must be produced out of different sources. If this fuel production is included in an well-to-wheel analysis the comparison between fuel cell and combustion engine powered vehicles show no big difference in the overall energy efficiency. (Fig. 2) This means, that a major step towards CO2 reduction and saving fossil energy must come from the fuel production.

In this context Hydrogen can only contribute to the reduction of CO_2 emissions if it is produced on base of renewable energy. However, there are three critical technological barriers that must still be overcome: the lack of a mobile fuel storage acceptable to the customer, the lack of infrastructure and, finally, the lack of an economical viable technology for the renewable production of

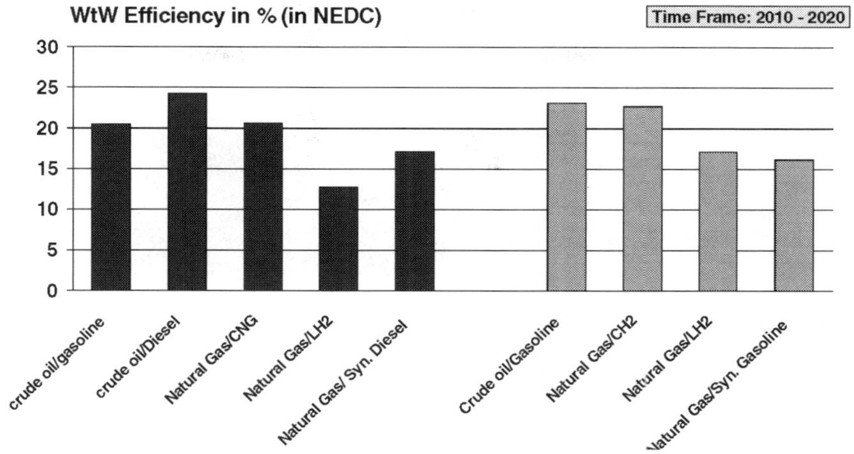

Fig. 2 well to wheel analysis of different powertrains

hydrogen. As no technological solution to any of the three barriers is available yet, hydrogen can only be regarded as a long term solution.

At present, no singular energy source, not even hydrogen, is able to meet the four key demands on a future fuel for mobility industry:

- Safe supply
- economic viability
- Fulfilment of environmental and climate protection requirements
- High volume and weight specific energy density.

Given this background, a diversification of fuels would seem to be the next logical step. However, the simultaneous market availability of diesel, petrol, methanol, ethanol, natural gas and other fuels is not an economically viable solution as a separate powertrain would have to be developed for each of these fuels and our customers will heavily be confused on the filling station.

It is therefore important to look for a way to diversify primary energy sources whilst concentrating the energy sources for mobile applications on as few variants as possible.

SYNTHETIC FUELS

In the next few years, increased use will be made of natural gas in particular, which is of advantage with regard to specific CO_2 emissions in comparison to an gasoline engine. Natural gas can, and will, be used directly as a means of automotive propulsion. However, natural gas is expected to only supplement today's fuels rather than replace them, because of the known drawbacks of all gaseous fuels regarding their range and tank space requirements, not to mention the rising costs of exhaust treatment in order to meet strict exhaust emission limits such as Euro IV. However, other secondary energy sources can

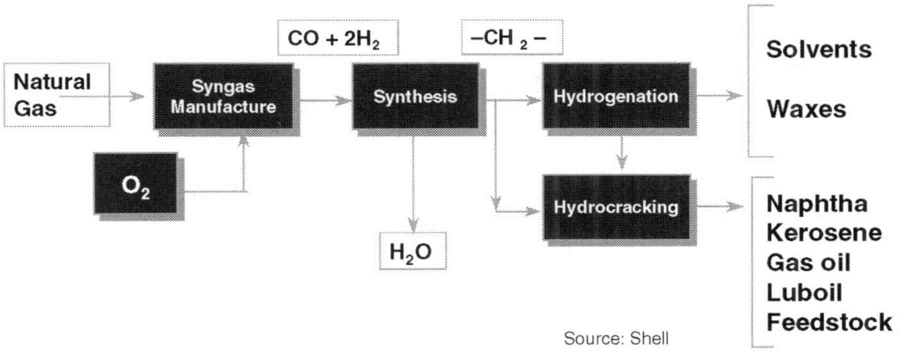

Fig. 3 Shell SMDS process

also be produced from natural gas by using known and industrially approved processes such as the Shell Middle Distillate Synthesis (SMDS, Fig. 3).

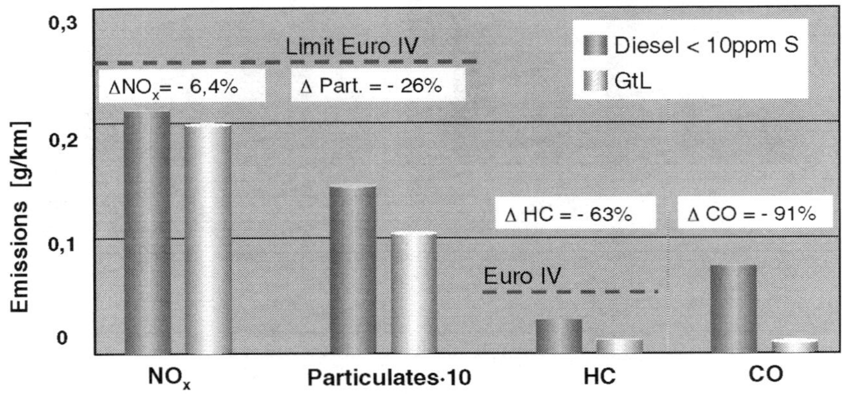

Fig. 4 Synthetic fuel in the diesel test fleet

In the first process step, natural gas is converted to a synthetic gas - a blend of manly hydrogen and carbon monoxide - by means of vapour reformation. Conventional fuels, above all a high-grade diesel fuel containing no sulphur or aromatic compounds, can be produced from this synthetic gas by means of Fischer Tropsch Synthesis.

This so-called Gas To Liquid Technology (GTL) is highly economical given the current price level of petroleum in many regions of the earth where natural gas or petroleum gas can be produced cheaply. It will, no doubt, take another 5 to 8 years, which are needed for investments and the construction of synthesis plants, before we see a secure and relevant supply of these synthetic fuels. This, therefore, represents a short to medium-term solution.

These synthetic fuels have a huge potential for improving engine combustion systems. The synthetic diesel fuel specification is particularly impressive because of its high cetane number and the absence of any aromatic compounds or sulphur.

In may 2003 under the presence of the German chancellor Gerhard Schröder and Volkswagen CEO Dr. Pischetsrieder a fleet test has been started in Berlin with 25 Golf vehicles equipped with 1.9ltr 74 kW Diesel engines already fulfilling Euro IV emission standards without active after threatment systems. The aim of the test was to demonstrate the viability and benefits of Shell GTL Diesel in daily use. The final results of this test has been shown on 27th of October 2003. Fig. 4 shows the emission results of those test vehicles. Even with those latest technology demonstrating engines a clear advantage with SynFuel cold be shown. The mean value of Particulate emission could be further reduced by about 26% with any increase in any other emission component. Stability of the emission levels could be demonstrated as extreme stabile over the running time of the test. This demonstrates clear, that the technology of synthetic fuels is ready for the market.

SUNFUEL

The intermediate stage of this technology - synthetic gas - now also allows the use of renewable energy sources such as waste wood, straw, energy plants or organic waste. The decisive factor is that quality of

Fig. 5 CO_2 cycle with SunFuel

the end product is not depending on the primary energy used. This solution eliminates the finite availability and CO_2 emission of synthetic fuels. The energy content stored in annual plant growth on earth is equivalent to many times the energy consumption of the human race, i.e. there exists an enormous potential for substitution. From a political viewpoint, the supply situation can be improved by using biomass since it is distributed relatively evenly across the earth, unlike the fossil energy sources.

This will not lead to zero CO_2 exhaust emission but will see the creation of a CO_2 neutral cycle powered by solar energy, illustrated in Fig.5. By doing so, we integrate the fuel cycle into the natural CO_2 cycle, which accounts for approx. 98% of overall CO_2 emissions. For this reason, we refer to a SynFuel produced in this way as a "SunFuel".

In Germany and Europe, biomass would mainly be derived from three sources: waste wood, herbal waste and the cultivation of energy plants. All in all there is a secure medium-term technical potential to meet more than 20% of current total fuel demand in Europe and acceding countries. This is not taking into account that there is an huge potential of increasing the biomass output per square km by new and modified plants. Up to now agricultural people and industry are not fully aware of this new business field, but is starting research and development under the new aim of maximising the mass output of the fields. In parallel there is strong research and development of the so called biomass-to-liquid BTL production process.

Fig. 6 represents the currently most developed production system schematically. The diagram shows the CHOREN CarboV® process. In the first process step, the biomass is broken down into a gaseous constituent and a solid constituent (organic coke) by means of low-temperature gasification. The second process step involves producing the synthetic gas. The arising slag contains, among other things, all mineral constituents absorbed by the plants during their growth. The synthetic gas is then converted to fuel in a Fischer-Tropsch Synthesis process with downstream fuel optimisation by Hydrogen aftertreatment i.e. isomerisation or hydro cracking.

Biomass-based SunFuel can be seen as a medium-term solution, as it is not yet economically viable. The production costs of SunFuel, excluding taxes (based on a plant size of 200 MW_{th}), are approx. 25 Cent/litre higher than those of petroleum-based fuels or natural-gas-based fuels (approx. 25 Cent/litre). However, the production costs are well below today's pump prices in Europe, so the onus is on the politicians to adopt suitable fiscal legislation promoting the development and introduction of SunFuels until such time as these fuels become economically feasible.

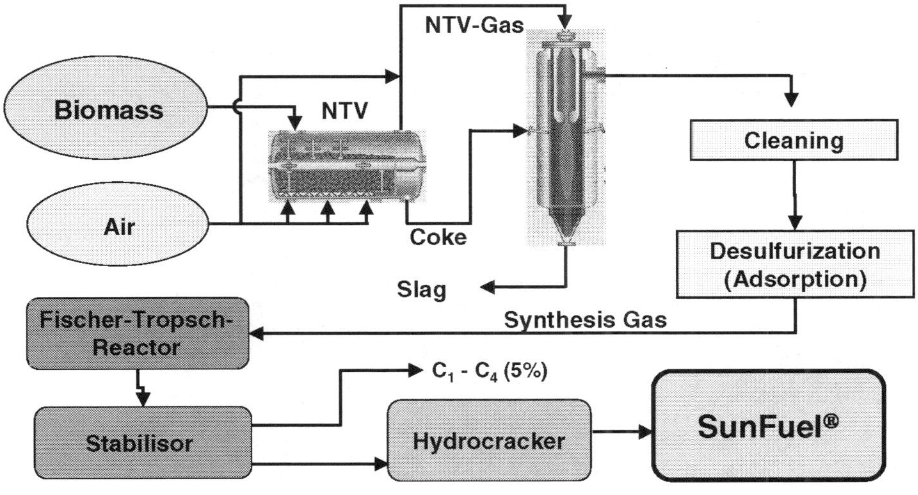

Fig. 6 Schematic diagram of the CarboV process

In the long term, subject to the availability of an inexpensive and renewable hydrogen source, this hydrogen could be added to the BTL process, doubling nearly the fuel output of the process. The implementation of a hydrogen economy, therefore, will not necessarily result in the use of hydrogen in the mobility economy. Biomass-based synthetic fuels could, on the whole, prove to be the better option from a sustainability viewpoint. Fuels of this kind are CO_2-neutral as they do not produce additional carbon dioxide emissions.

As we approach the technical limits to petroleum production, and with rising global energy demand, alternative energy sources will in future become increasingly important. Also, the demands on the quality and purity of conventional fuels are rising. The rises in cost involved will pave the way for generally more expensive alternatives. This is why VOLKSWAGEN and DaimlerChrysler foresees in the coming years an evolution in fuels away from conventional petroleum-based fuels towards synthetic fuels produced from natural gas and biomass-based SunFuels. Only in the distant future, once all technological barriers have been overcome, will hydrogen come, eventually, into play as an energy carrier for mobile use as shown in Fig.7.

Huge Investment is necessary to realise sufficient substitution of mineral oil based transportation fuels by renewables. So time is needed for the transition. Fig. 8 shows an time wise optimistic scenario for this transition. It can be expected, if the political willingness is stabile for at least 20 years, resulting in supporting activities, that in 2020 the goal of the European parliament to substitute 20 % of the European transportation fuels by alternative fuels will be reached.

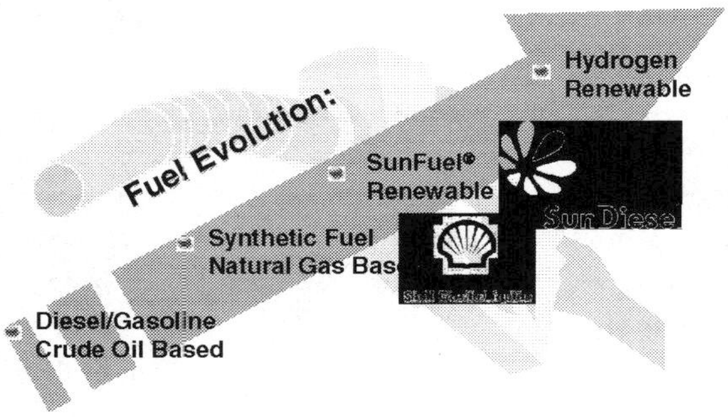

Fig. 7 Fuel Evolution

NEW COMBUSTION SYSTEMS

Over the next 30 years, we can expect to see the availability of high quality liquid hydrocarbons as fuels. Synthetic fuels also offer the possibility to optimally adapt fuel properties to combustion parameters. A further reduction in pollutant emissions or in the considerable costs of exhaust treatment will only be achievable if, in particular, the raw emissions of NOx of stratified combustion systems can be lowered. In other words, NOx production must be suppressed during the combustion system without impairment to engine

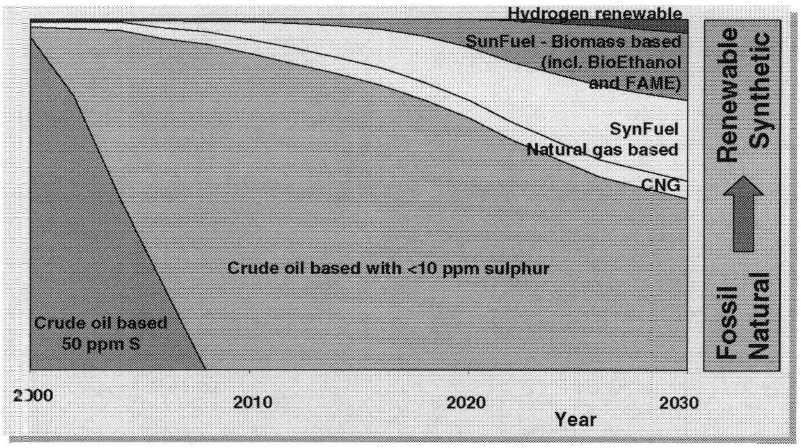

Fig. 8 Optimistic Fuel Scenario for Europe

efficiency. To this end, quality regulation by direct injection (DI, TDI or FSI system) must be retained. The objective is, therefore, to marry the advantages of the petrol engine and the diesel engine in a new process.

The combustion systems of both engine concepts are now much more closely related now that direct injection system is also used in petrol engines. The next stages of combustion system development will serve to reinforce this trend. The "partly homogenized diesel combustion system" and the "auto ignition petrol engine" currently being researched and developed are already based on hardware that is essentially similar. The next logical step is to contemplate the development of a new combustion system that unites the key attributes of both systems. Engine developer refer to such an system as the Homogeneous Charge Compression Ignition System HCCI.

The best basis of this process is a new synthetic fuel. As far as can be seen at this time (the basic requirements of such a process are currently being determined), the decisive factors are the evaporation and ignition response, i.e. the composition, of the fuel. The requirements for increased homogenisation of the atomised mixture without premature commencement of spontaneous ignition are a fuel with low initial and final boiling points, as well as a reduced tendency to spontaneous ignition. This is enabling the base systematic of the HCCI process, the extreme early injection with long evaporation time while still keeping the main combustion after top dead centre. These properties increase in the following order: diesel ⇒ kerosene ⇒ naphtha.

Fig. 9 shows, by way of example, the reduction of NOx particulate trade-offs for Diesel Kerosene and Naphtha at an engine speed of 1600 rpm and at a torque of 100 Nm in HCCI combustion. As can be clearly seen, kerosene offers a much greater reduction in emissions in combination with HCCI combustion, while Naphtha enables a NOx-free combustion in this point. In other conditions Naphtha is unfortunately more problematic due to instability and boiling problems in the injection system. So an Kerosene-like fuel seems to have the best fundamental specifications.

Implementing the HCCI process entails crossing a number of hurdles. In stationary operation, the process was shown to be highly stable. However, dynamic operation requires the development of entirely new control concepts, sensors and actuators. A market launch can not be expected before the year 2010.

SUMMARY

Fig. 10 shows the most probable scenario for the development of future powertrains and associated fuels, and represents schematically the resulting development of specific CO_2 emissions. With today's mineral oil based fuels and conventional powertrain units, specific CO_2 emissions will continue to decrease given by the self-commitment of the automobile industry and technical progress. Engines with direct fuel injection will play a key role in this process.

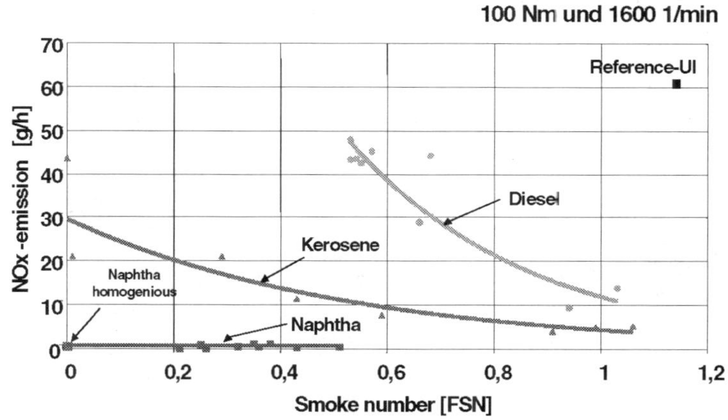

Fig. 9 Fuel potentials in the HCCI process (engine displacement 1.9 liter)

In addition to mineral oil based fuels, conventional synthetic fuels, principally natural gas based fuels, will come onto the market during this decade. For our customers (common customers of the automobile industry and the fuel industry), the introduction of synthetic fuels will change nothing as all functional characteristics and the infrastructure will remain unchanged. Synthetic fuels are free of sulphur and aromatic compounds and their properties can be specified more stringent than today's fuels. These advantageous properties will enable automobile manufacturers to continue developing their products, resulting in reduced consumption and, particularly with diesel engines a further improvement in emission quality.

Producing synthetic fuel on the basis of CO_2 free or CO_2 neutral energy rather than from fossil primary energy will reduce specific CO_2 emissions during vehicle operation even if there is no change in fuel consumption. This applies regardless of fuel type, i.e. it also applies to synthetic fuels derived from renewable raw materials ("SunFuel"). The great advantage of taking this route is that our present-day fuel infrastructure can be preserved during this phase.

As already shown, new engine combustion systems combining the consumption advantages of modern-day diesel engines with the emissions potential of petrol engines will come into use in the medium term. Suitable fuels must be adapted to these hybrid combustion systems. Synthetic fuels (SynFuel, later SunFuel) offer the best potential.

The development of the combustion engines itself will be accompanied by the development of more intelligent powertrain systems. Starter-Generator systems will be further upgraded to mild hybrid systems. This technology offers the potential for a better linkage between the powertrain and the user requested power output leading to an more comfortable and less fuel consuming solution.

In the long term, it is to be expected that the persisting problems of hydrogen storage and infrastructure will be solved. The green light for the hydrogen economy can then be given subject to a sufficiently positive overall evaluation. However, that is unlikely to happen in the next 20 years.

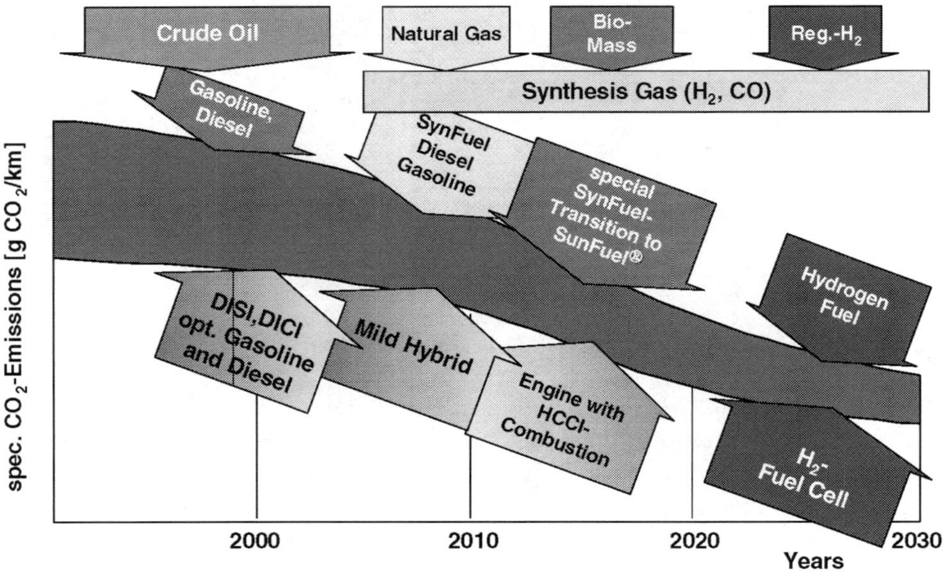

Fig. 10 Fuel and Powertrain Strategy

ENERGY CONVERSION IN ENGINES FROM NATURAL GAS

Which Fuels For Low CO_2 Engines?
© P. Duret, X. Montagne (Eds.) and Editions Technip, Paris, 2004, pp. 109 – 121
27 rue Ginoux, 75015 Paris

NATURAL GAS ON THE ROAD FOR LOWER CO_2

Rik S.G. Baert, J. de Rijke, R.T.M. Smokers

TNO Automotive, Delft, The Netherlands

ABSTRACT – *Different pathways for meeting the EU ambitions of greenhouse gas reduction and energy resources diversification are reviewed and the need for a shift towards CO_2-neutral fuels respectively low-carbon fossil fuels is illustrated. Based on data from different well-to-wheel studies, the case for natural gas vehicles as a timely contribution towards lowering automotive CO_2-emissions is evaluated and motivated.*

Both technological and non-technological barriers for market introduction and government acceptance of these vehicles are reviewed. A corresponding technology roadmap for natural gas vehicle solutions is presented and critical gas vehicle research and (aftertreatment, fuelling, storage,..) technology developments are identified. Results from recent and ongoing research development projects are used to illustrate the technical feasibility of competitive low-CO_2-emitting natural gas engines.

Finally, it is shown how natural gas vehicle development and application can be part of different gaseous-fuel-based pathways towards sustainable mobility.

CLEAN NATURAL GAS ON THE ROAD

Gas fuelled vehicles have been with us for some time. In fact, one of the earliest vehicles was powered with a Lenoir engine running on so-called "town gas".

Historically, engine manufacturers and vehicle owners have turned to gaseous fuels to solve problems with the supply of conventional fuel. This was the case in Europe in the period before respectively during and shortly after the second World War. This happened again during the 1970's when oil prices soared and supply was short.

For the last decades however, gaseous fuelled vehicles have been promoted especially because of their lower emissions level of both regulated and unregulated emissions. In that period natural gas vehicle power train, storage and fuelling technology has been further developed and its potential for quiet and very clean urban transport has been time and again demonstrated. In fact, one of the first vehicles to meet the Californian NZEV emission level targets was a Honda natural gas vehicle [1]. Since then, also regular gasoline fuelled cars have demonstrated similar emission levels. To achieve this however, much more technical effort was needed. Of course, the light-duty diesel engine is far from meeting such low emission levels.

A similar situation is gradually emerging for heavy-duty applications. Until now, the emissions advantage of the natural gas driven vehicles has remained unchallenged. However, because the competitor diesel engine is set to become a lot cleaner than before (but also more expensive), also for these applications the "emissions" gap is closing.

Of course, the natural gas vehicle (NGV) retains its non-regulated emissions advantage. This was confirmed again recently in an evaluation study, some results of which are shown in Figure 1 [2].

Nonetheless, it goes without saying that the advent of cleaner vehicles on conventional fuels is not making life easier for the NGV industry.The good news for NGV's in Europe is that, as the emissions problem gets solved, automotive transportation is confronted with the need to improve sustainability and the need for energy resources diversification.

ON THE ROAD FOR LOWER CO_2

Over the years, automotive vehicle transportation in Europe has increased considerably. In recent years road transport vehicles have taken up more than one quarter of all energy resources in Europe [4]. Furthermore, the large majority of these vehicles run on fossil fuel. Automotive transportation is therefore one of the major contributors to the overall anthropogenic (i.e. man-made) production of greenhouse gases, i.e. of carbon dioxide.

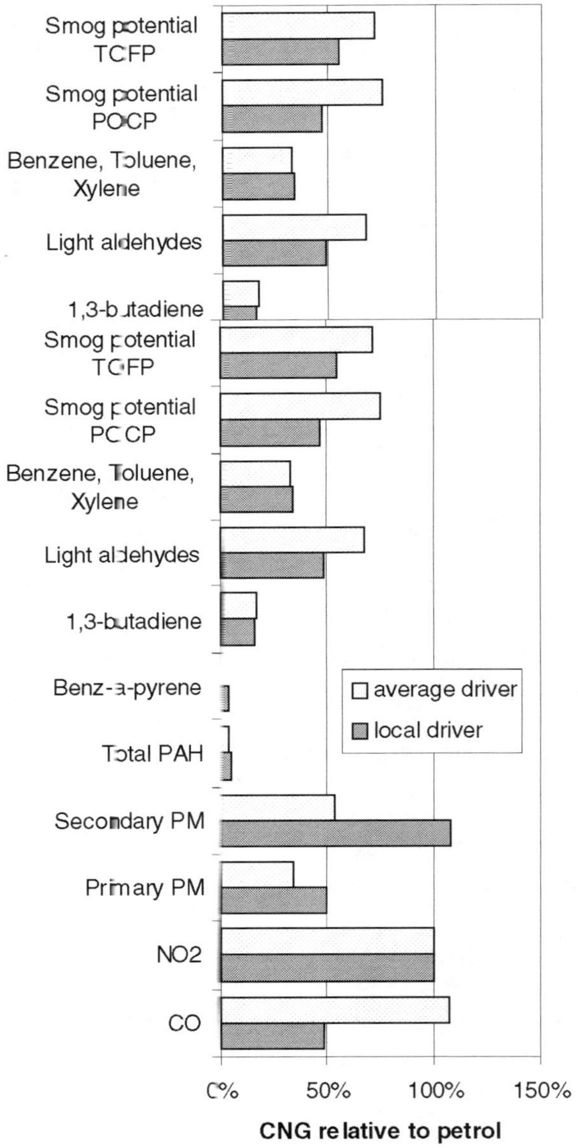

Fig. 1 Comparison of various (non-regulated) emissions of modern (Euro3) passenger cars on gasoline and CNG (OEM-equipment) [2]. Results are based on vehicle tests on a real-world driving cycle (Common Artemis Driving Cycle [3]) and on representative weighting of emissions on the urban cold, urban warm, rural road and highway parts of the cycle for typical urban driving and for average Dutch driving. POCP and TOFP are (alternative) indicators of ground-level ozone forming potential.

Of course, as shown in Figure 2, different measures can be taken to reduce carbon dioxide production in transportation. Reducing the need for mobility, the introduction/implementation of smarter (intelligent / guided) driving behaviour and a shift towards different (lower carbon dioxide emitting) fuels could all be key contributors. All of these measures will however need time for development and for getting introduced respectively accepted. On a short term, the biggest effect is to be expected from the introduction of more fuel-efficient conventional vehicles. This, and the Kyoto objective to realize sizeable reductions in greenhouse gas production by 2008-2012 has therefore lead to the setting of targets for the reduction of CO_2 emissions from conventionally fuelled passenger cars in Europe.

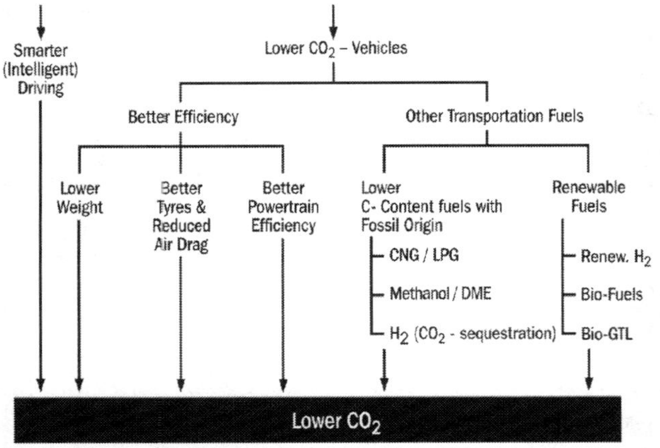

Fig. 2 Different pathways for automotive CO_2 reduction

Back in 1995 the fleet average CO_2-emission for new (diesel and gasoline) cars introduced on the road was 185 g/km. In that year ACEA, the association of car manufacturers in Europe, voluntarily set the goal of achieving 140 g/km by 2008. Similar covenants between the European Commission and JAMA and KAMA respectively have been arranged with 2009 as target year. As shown in Figure 3 the industry is on track for achieving this target : by 2001 CO_2 emission had already decreased to 167 g/km [5]. Until now, this has been achieved primarily by increasing the market share of diesel engines and by the further optimisation of petrol engines. On a somewhat longer term, the EU itself has set a target of 120 g/km to be achieved by 2010 at the latest through complementary (fiscal/regulatory) measures [6]. Expectations are that this latter target will not be reached. Currently discussions are therefore going on between EU and industry on the feasibility of achieving new fleet average CO_2 emissions of 120 g/km by 2012.

At the same time the feasibility of achieving emission levels of 80 g/km has been the subject of demonstration projects both in the US (PNGV target of 80 miles per gallon of gasoline for a conventional middle class vehicle) and in Europe (3-liter/100 km for a considerably smaller VW Lupo). Although these are very low levels indeed, they have been mentioned as realistic targets both in the UK Foresight study (80 g/km by 2020,[7]) and in the FURORE roadmap study (95 g/km by 2020 respectively 80 g/km by 2030) [8].

Currently, best-of-class mid-range (1300 kg weight / 75 – 95 kW) diesel vehicles have a fuel consumption in the order of 5.2 l/100 km (or 140 g CO_2/km) when measured in the European NEDC cycle. Obviously, with conventional powertrain technology, even if all vehicles were to achieve diesel-like efficiency, and if diesel engine efficiency were to be further improved with 20 %, new fleet average emissions could at best reach 110 g/km. In practice therefore, the CO_2-reduction trend shown in Figure 3 implies considerable additional savings on vehicle weight and the introduction of considerable amounts of very fuel efficient hybrid and/or fuel cell vehicles.

- With this scenario, some critical observations are to be made :currently the trend for vehicle weight is not to decrease but to increase, both because of increasing (active and passive) safety demands and because of consumer demands for more comfort and more performance (airco, 4-wheel drive, SUV),
- when running on conventional gasoline, fuel cell vehicles show little (or only a limited) efficiency benefit when compared to hybrid vehicles; in recent years fuel cell vehicle development has therefore shifted almost exclusively towards application of hydrogen as an energy carrier,
- weight reduction and hybridisation (and certainly the application of fuel cell vehicles) will result in a considerable increase of vehicle production costs.

It is clear that very strong (fiscal and other) incentives will be needed to meet with the CO_2-reduction trend shown in Figure 3.

Of course, this somewhat negative/grim picture changes when we allow fossil fuels to be replaced with climate-neutral fuels such as biomass derived fuels and renewable hydrogen. This route not only helps with solving the GHG problem, but also helps in diversifying the energy options, another important European objective.

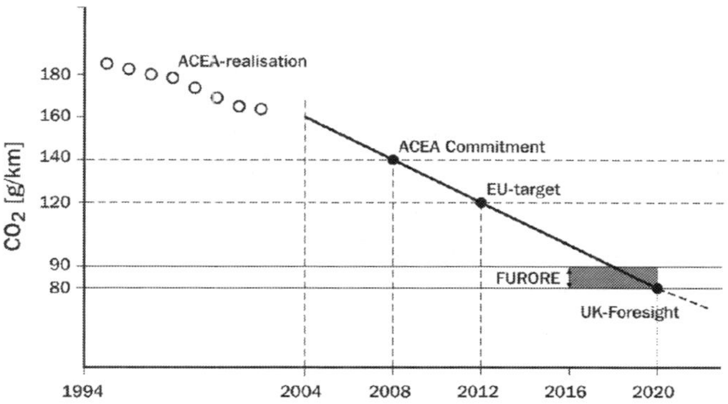

Fig. 3 New fleet average CO_2 emissions and future target levels; New European Driving Cycle (NEDC) [6,7,8]

At the same time the feasibility of achieving emission levels of 80 g/km has been the subject of demonstration projects both in the US (PNGV target of 80 miles per gallon of gasoline for a conventional middle class vehicle) and in Europe (3-liter/100 km for a considerably smaller VW Lupo). Although these are very low levels indeed, they have been mentioned as realistic targets both in the UK Foresight study (80 g/km by 2020,[7]) and in the FURORE roadmap study (95 g/km by 2020 respectively 80 g/km by 2030) [8].

Currently, best-of-class mid-range (1300 kg weight / 75 – 95 kW) diesel vehicles have a fuel consumption in the order of 5.2 l/100 km (or 140 g CO_2/km) when measured in the European NEDC cycle. Obviously, with conventional powertrain technology, even if all vehicles were to achieve diesel-like efficiency, and if diesel engine efficiency were to be further improved with 20 %, new fleet average emissions could at best reach 110 g/km. In practice therefore, the CO_2-reduction trend shown in Figure 3 implies considerable additional savings on vehicle weight and the introduction of considerable amounts of very fuel efficient hybrid and/or fuel cell vehicles.

With this scenario, some critical observations are to be made :

- currently the trend for vehicle weight is not to decrease but to increase, both because of increasing (active and passive) safety demands and because of consumer demands for more comfort and more performance (airco, 4-wheel drive, SUV),
- when running on conventional gasoline, fuel cell vehicles show little (or only a limited) efficiency benefit when compared to hybrid vehicles; in recent years fuel cell vehicle development has therefore shifted almost exclusively towards application of hydrogen as an energy carrier,
- weight reduction and hybridisation (and certainly the application of fuel cell vehicles) will result in a considerable increase of vehicle production costs.

It is clear that very strong (fiscal and other) incentives will be needed to meet with the CO_2-reduction trend shown in Figure 3.

Of course, this somewhat negative/grim picture changes when we allow fossil fuels to be replaced with climate-neutral fuels such as biomass derived fuels and renewable hydrogen. This route not only helps with solving the GHG problem, but also helps in diversifying the energy options, another important European objective.

However, although biomass-to-liquid (BTL) synthetic fuels and renewable hydrogen are considered the long term automotive energy carriers, they will take many years and considerable investment to build up sufficient production capacity[1]. The arguments for this statement are :

- fuel production technology is still subject of research and development; this will be followed with testing for scale-up potential and optimisation in pilot plants. Only then are the first large scale BTL plants expected to appear where they will first meet extra demand and later on gradually replace "old" refineries. Because these fuels will be considerably more expensive than current conventional fuels and the because of the (resulting) limited market drive, significant BTL production is not expected before 2015.
- Until recently there was considerable enthusiasm towards large scale automotive application of hydrogen. Earlier scenario studies "guestimated" hydrogen to reach 10 % of the final energy carrier market by the year 2010 (BP, 2002, [9]) respectively 2025 (Shell, 2001, [10]). A recent report by the US National Research Council however estimates that the impact of (hydrogen fuelled) FCV is likely to be minor for the next 25 years [11]. This would be the time needed to set up the needed infrastructure, to develop the carbon sequestration technology (for production of hydrogen) and to make the FCV meet with market demands for autonomy, reliability and affordability.
- Finally, it should be pointed out that when (renewable or CO_2-neutral) hydrogen becomes available, it should be used for those applications where it has the biggest benefit. And these are not necessarily automotive applications. In fact, forced introduction of limitedly available renewable hydrogen in the transport sector may even lead to a net increase of the CO_2-emissions of the overall energy system [12].

For the next 15 years therefore other fuels have to be considered. In Europe, the EU has the ambition to replace 20 % of the traditional transportation fuels by alternative fuels by 2020.

In an optimistic development scenario communicated by the EU, the possible contribution of biofuels (RME / ethanol), natural gas and hydrogen towards meeting this target was identified [13]. This scenario is repeated in Table 1. In view of the previous observations the proposed contribution of hydrogen is somewhat optimistic. This was also the consensus view of the FURORE members [8].

Year	Biofuel	Natural Gas	Hydrogen
2005	2		
2010	6	2	
2015	(7)	5	2
2020	(8)	10	5

Table 1. EU optimistic scenario target levels for substitution of total transport fuel consumption by alternative fuels [13]

Between the remaining fuel candidates, natural gas has some additional interesting properties / advantages :

[1] Unless some calamity occurs

- contrary to the use of biofuels produced from food crop (such as biodiesel made from soyabeans / rapeseed, or ethanol made from wheat / sugar beet / sugar cane) there are no ethical objections towards use of natural gas,
- production of biofuels such as FAME (from rapeseed) and ethanol from sugar beet is energy-intensive (equivalent to between 20 to 40 % of its energy content); this reduces their GHG reduction potential,
- contrary to biofuels, natural gas is readily available in very large quantities, no production capacity has to be built,
- natural gas production costs are low and linked to those of crude oil; on the short term natural gas will remain the cheapest alternative fuel,
- there already exists a natural gas distribution grid in a large part of the EU; this grid will be extended in response to the increased use of natural gas for electrical power generation, extra investment costs for automotive will therefore be focusing on the installation of refuelling stations,
- market ready automotive vehicles have been developed that can run on natural gas.

Clearly, for the short and medium term, natural gas has the most potential to give a timely contribution towards meeting the CO_2 reduction targets.

NATURAL GAS FOR LOWER CO_2

Of course the interest for NG as an energy carrier originates with its composition. Most of the fuel molecules in any natural gas are methane molecules. This results in a very low C/H ratio and correspondingly in a low production of CO_2 per unit of energy when burnt. Typically a natural gas will produce 56 g CO_2 for every MJ of energy released in the combustion process. This is approximately 25 % less than with diesel or gasoline.

The effective CO_2 emission with natural gas will depend not only on the fuel composition but also on the powertrain efficiency and on the way this efficiency is affected by the change in fuel composition. Modern CNG vehicles have a similar energy consumption and therefore a considerable CO_2 benefit compared to comparable gasoline fuelled vehicles. And although a diesel powered passenger car is much more energy efficient, a comparable NGV will produce similar amounts of CO_2. Of course, amongst current production vehicles, the lowest CO_2 output comes from the hybrid vehicles.

Similar comparisons have been made for the vehicles that are expected on the road around 2010. Figure 4 shows such a comparison for different future gasoline, diesel and natural gas fuelled vehicle options. These TTW numbers originate from the recent CONCAWE/EUCAR/JRC study [14]. All options shown are for a compact size, 5-seater sedan (1250 kg reference weight for the advanced gasoline version). All diesel options are assumed to be equipped with DPF.

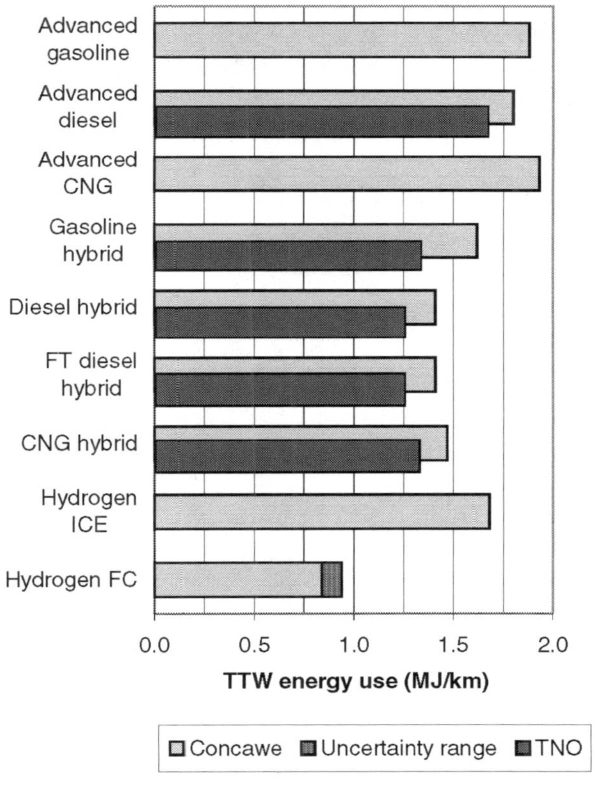

Fig. 4 Tank to Wheel energy use of different vehicle configurations as expected around 2010 [14]

The light coloured bars in Figure 4 represent the outcome of [14]. Some of these numbers, however, can be questioned :

- 2010 hybrid vehicle fuel consumption is considered too high; the current Toyota Prius already has a substantially better energy efficiency of 1.37 MJ/km (4.3 l / 100 km),

- from observations on the potential for better fuel efficiency with a diesel engine, an energy consumption of 1.28 MJ/km (3.6 l / 100 km) should be possible by 2010 for diesel hybrid,
- in line with observation in the Furore report [8], a diesel efficiency of 1.71 MJ/km (4.7 l/100 km) is considered a realistic target (to be achieved a.o. through advanced downsizing / friction reduction)
- although somehat conservative, the Concawe study CNG TTW assumption was retained.

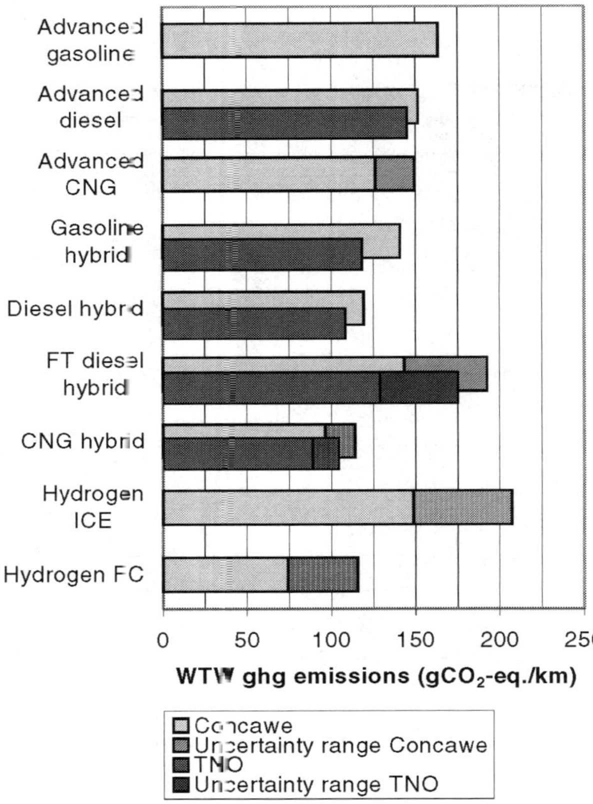

Fig. 5 Well-to-Wheel CO_2 emission of different vehicle configurations as expected around 2010 [14]

From these observations, alternative TTW estimates (dark-coloured bars) were added to Figure 4.

When estimating GHG impact, it is clear that also the GHG/CO_2 emitted during production and distribution of the fuel should be taken into account. In line with the time frame of this study, it is assumed that both FT-diesel and hydrogen are produced from natural gas. Then the WTW CO_2 emission of these vehicle options are as shown in Figure 5.

For Figure 5, in general, the WTT-data from [14] were used. For CNG however it was assumed that for distances up to 4000 km, transportation of natural gas will be with pipeline; for larger distances transportation will be as LNG with no extra CO_2-penalty.

Notwithstanding these amendments to [14], the same conclusions are arrived at :

- CNG hybrids have WTW CO_2 emissions that can compete with those of a hydrogen FCV (with hydrogen produced from natural gas),
- for automotive applications, direct use of natural gas in an ICE produces WTW less CO_2 than when it is first converted to diesel (FT-synthesis) or to hydrogen. There are indications that more energy efficient FT-production techniques are being developed [15], but even then, direct use of CNG gives lower CO_2.

To conclude, some general remarks can be made on the subject of well-to-wheel analysis :

- most WTW studies take the fuel consumption of a vehicle as measured in the emissions test cycle as being representative of real-life fuel consumption; recently alternative test cycles have been suggested that better simulate actual driving patterns [3],
- none of the studies in the literature take into account the trend of increasing energy demand (e.g. linked to the introduction of airconditioning).
- future regulation should target overall (well-to-wheel) CO_2-reduction.

LD-CNG POWERTRAIN TECHNOLOGY ROADMAP

Given the fundamental combustion properties of natural gas, current NG vehicles in production are provided with engines that have been derived from conventional multipoint timed port fuel injection (MPFI), naturally aspirating (NA), spark ignition (SI) gasoline engines.

In recent years, development on natural gas engines focused primarily on meeting better than gasoline engine emissions requirements with competitive reliability, durability and with acceptable performance. Main areas of research were :

- development of dedicated combustion concepts for natural gas,
- development of dedicated (λ = 1)-catalyst technology (catalyst formulation and corresponding lambda sensor evaluation),
- solving the practical problems of fuel quality variation,
- development of compact, light-weight on-board fuel storage,
- development of dedicated fuelling systems compatible with future OBD requirements.

TNO worked together with other partners in a number of European research projects targeting aftertreatment and fuel quality problems [Methacar,1994; Levings, 1996; NG components and fuels, 1996]. On the fuelling system side, TNO was involved with the development of a multipoint sequential PFI fuelling system. For this system, both hardware and software were developed. (It is worth while mentioning here that the fuelling system injectors were designed such that they are scale-able and can be used for CNG, but also for LPG, and – in the future – for H2 dedicated engines.) The corresponding control system was furthermore designed to be compatible with future OBD requirements.

With this or similar new technology, competitive natural gas light-duty vehicles can and have been developed.

In practice, most of these vehicles are still bi-fuel (gasoline and CNG) vehicles. Contrary to monofuel dedicated vehicles they are not able to make full use of the higher octane quality of natural gas. This and their higher vehicle weight (due to the use of heavy gas storage tanks) results in a higher energy consumption. Furthermore, most of these vehicles have naturally aspirating engines. With PFI of natural gas instead of gasoline, the effective air breathing capacity of these engines is reduced. This in turn results in loss of maximum power and torque, as illustrated in Figure 6.

Of course, with supercharged or turbocharged engines such performance loss no longer occurs. (Alternatively, direct injection of natural gas would also remedy this.)

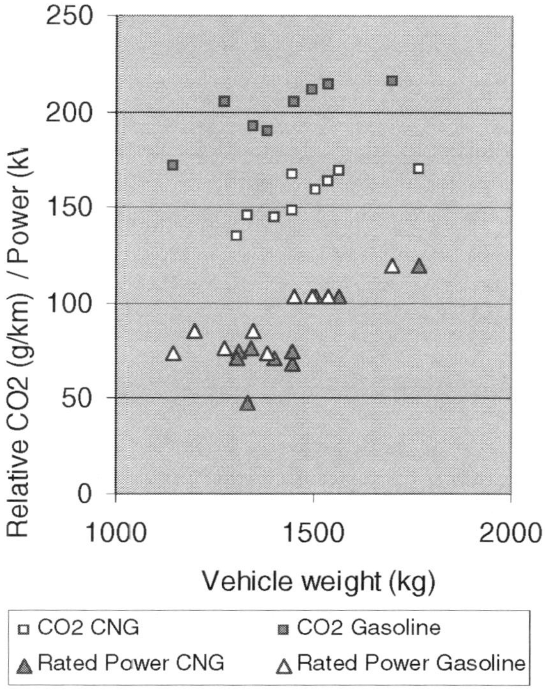

Fig. 6 Comparison of current NGV with comparable (related) gasoline vehicles produced by the same OEM (power in kW)

At the same time there is need for further research activities [16] :

- the higher costs for a dedicated (higher noble metal content) catalyst and (OEM-quality) fuelling systems are a disadvantage; there is a need to identify alternative and cheaper solutions [16].
- Future (SULEV-like) emission levels put high demands on fuelling control and on fuel system reliability and durability. An in-house comparative study of different fuelling systems demonstrated a need for further improvement with some natural gas components.
- Future OBD and OBM requirements also demand additional research and development,
- For highest engine efficiency, a combination of fuel quality and combustion (knock) sensing is needed.

It is expected that all of the above challenges can and will be solved in the next few years.

Apart from the development of light-weight, compact (and conformable) fuel storage solutions, none of these technologies will result in important improvements of CNG vehicle fuel efficiency. Such energy saving technologies will, necessarily, come from the mainstream (gasoline) world.

To identify them we can look at the future LD gasoline technology roadmap. Figure 7 is a (simplified) version of such a roadmap, where focus is on engine technology (of course also advanced transmissions - e.g. CVT - are expected to be used more and more).

From this figure, the following observations can be made :

- the main drive for new technology with the LD gasoline engine is efficiency improvement,
- new technologies such as mechanical systems for valve time changing, direct injection of gasoline and engine downsizing (through super- and turbocharging) have already been implemented in a limited number of vehicles; their use is expected to grow significantly in the coming years,
- other, even more expensive, technologies such as fully flexible variable valve control and variable compression ratio are subject of current investigations; given their high level of cost and complexity their implementation is likely to remain limited for some time to the high end of the market,
- the success of lean burn engine concepts will depend on the availability of corresponding, efficient (more efficient than at present) and affordable aftertreatment technology,
- new combustion concepts such as CAI are also currently being investigated; unless expensive enabling technology such as VVT and VCR is also implemented, application of CAI is likely to remain limited to the lower bmep part of the engine operating range, where it will result in some additonal efficiency gain.

As most CNG applications are expected in the midsize market segment, the primary candidate technologies for application on natural gas engines are downsizing and DI-CNG, possibly – on a somewhat longer term – in combination with CAI.

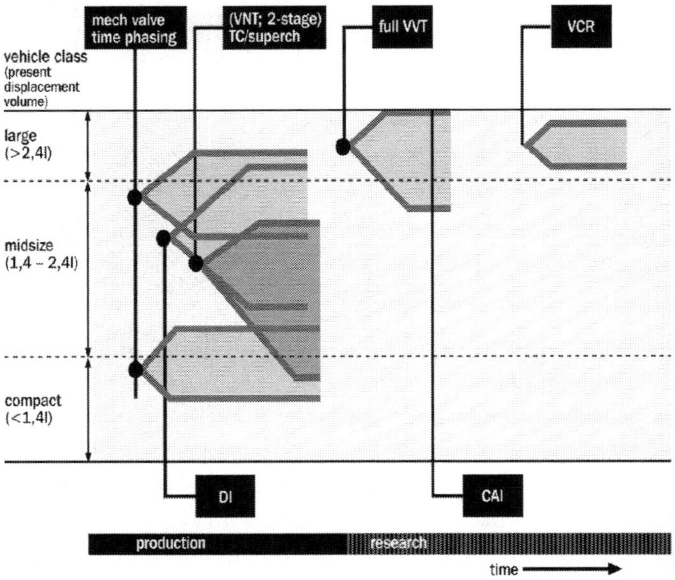

Fig. 7 LD gasoline engine technology roadmap

In line with this observation, Figure 8 presents some results of on-going investigations into the potential of downsizing in combination with exhaust gas recirculation and/or lean mixture combustion to reduce CNG fuel consumption.

- the success of lean burn engine concepts will depend on the availability of corresponding, efficient (more efficient than at present) and affordable aftertreatment technology,
- new combustion concepts such as CAI are also currently being investigated; unless expensive enabling technology such as VVT and VCR is also implemented, application of CAI is likely to remain limited to the lower bmep part of the engine operating range, where it will result in some additonal efficiency gain.

As most CNG applications are expected in the midsize market segment, the primary candidate technologies for application on natural gas engines are downsizing and DI-CNG, possibly – on a somewhat longer term – in combination with CAI.

In line with this observation, Figure 8 presents some results of on-going investigations into the potential of downsizing in combination with exhaust gas recirculation and/or lean mixture combustion to reduce CNG fuel consumption.

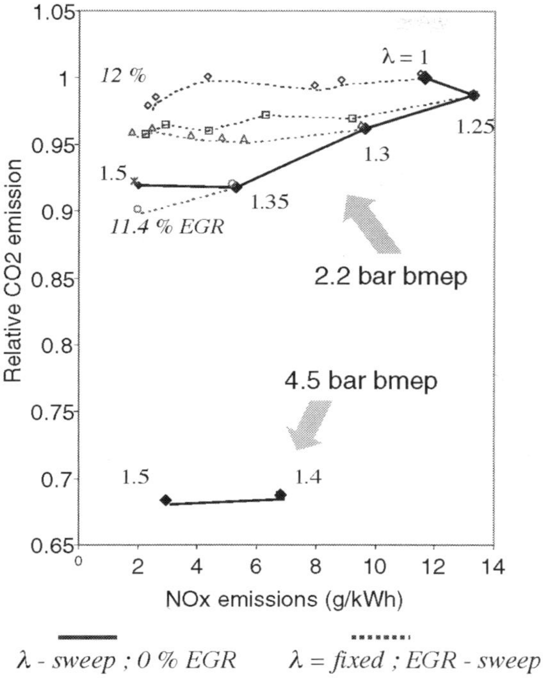

Fig. 8 Effects of charge dilution (LB and or EGR) and downsizing on part load fuel efficiency.

Tests were done on a MPFI 0.5 l/cylinder multicylinderTC natural gas engine. Work concentrated on part load engine conditions that are indicative of the engine when operating in the NEDC test procedure. For EGR both a long route and a short route approach were tested. Results shown are mainly for 1 engine working point : 1600 rpm, 2.2 bar bmep. The results shown illustrate that, (in this TC engine configuration) at low bmep levels, EGR is limited and therefore brings only a limited efficiency improvement. Increasing air-fuel-ratio is clearly more effective. A much bigger effect is obtained with downsizing (increasing bmep) : at 4.5 bar bmep energy efficiency is improved by almost 35 %. Similar results were obtained at other working points. Further work is going on aimed at identifying efficient engine/aftertreatment combinations.

HD-CNG POWERTRAIN TECHNOLOGY ROADMAP

Over the past 25 years different heavy-duty (HD) CNG vehicles have been built and used (mainly) for public transport and/or for special applications (such as refuse collection) in areas with a high population density. The introduction of these vehicles were financially supported by local and national governments because of their considerably lower emissions and noise level.

Year	1998	1997	1997
Concept	$\lambda = 1$	Lean Burn	Lean Burn
Pmax (kW)	170	190	213
Tmax (Nm)	940		960
Displ. Vol. (l)	8.66	10.4	10.52
Fuel system	MPFI	SPFI	SPFI
Fuel type	Dutch NG	Dutch NG	Jap. NG
MN	85	85	60
Net cal. value (MJ/kg)	32.61	32.61	40.14
AFR control	Cl. Loop	Cl. Loop	Cl. Loop
E-gas	Yes	No	Yes
Aftertreatment	3-W cat	Oxycat	Oxycat
Emission test procedure	ECE R49 13-mode	ECE R49 13-mode	Jap. G13 13-mode
NOx (g/kWh)	0.22	2.07	1.0
THC (g/kWh)	0.017	0.83	1.5
NMHC (g/kWh)	< 0.003		
CO (g/kWh)	0.68	0.29	0.0

Table 2. Typical characteristics of current λ=1 and lean-burn HD CNG engines.

This is illustrated in Table 2. This table gives an overview of the main characteristics of some typical turbocharged and aftercooled HD CNG engines developed at TNO between 1996 and 1998.

HD diesel emission limits (in Europe) at that time were : NOx < 7 g//kWh, THC < 1.1 g/kWh, CO < 4.0 g/kWh (measured in the ECE R49 13-mode test procedure). Clearly, these HD CNG engines had no problem in achieving considerably better emission levels than their diesel competitor.

Notwithstanding their clear emissions advantage, the succes of the HD natural gas engine has, until now, been limited. There are a number of reasons for this:

- fuel storage takes up a lot of space and does not conform easily,
- SI engines in general need maintenance more frequently than CI engines.
- Durability is not comparable to that of a diesel. This should be no surprise as almost all HD gas engine fuelling and ignition system components are derivatives from the less durable (smaller

lifetime) LD gasoline/gas engine world. With only a fraction of the effort that goes into regular diesel (fuelling) technology development at their disposal, it is no surprise that some/most natural gas components do not match diesel component lifetimes. On the other hand, mileage with the typical urban transport NGV application is much smaller than with typical diesel-fuelled vehicles.

- Natural gas HD vehicles were almost exclusively developed for application in fleets. Investments into the rather costly refuelling systems are at the expense of the fleet owner.
- Energy consumption is higher than with HD diesel vehicles; because most fleet owners get diesel at a reduced price, this also works against CNG.

This has given the HD NGV a bad press and created doubts about the (future) viability of HD CNG application. However, a lot of the negative rumours refer to earlier versions of HD NGV's with old (e.g. carburettor type) fuelling technology and without closed loop fuelling control.

For the typical urban transport application (buses /refuse collection) durability and reliability issues have since been dealt with in a satisfactory way, and acceptable solutions have been demonstrated.

Nevertheless some challenges remain.

One of these challenges is the future clean diesel engine.

Table 3 illustrates the extent to which HD diesel engine emissions will have to decrease in Europe and the US in the coming years.

The NOx emission levels mentioned for the second LB + oxycat configuration in Table 2 represent a lower limit that could only be achieved through a combination of E-gas, carefull fuelling control and clever calibration. Although MPFI will give some extra margin, the conventional (premixed charge, SI) PFI LB HD CNG powertrain will have difficulty in matching future diesel emissions (below 1 g/kWh NOx) without NOx aftertreatment of some kind. Stoichiometric natural gas engines on the other hand can already now meet the US2010 emission, but suffer from a lower fuel efficiency.

	EU III	EU IV	EU V	US 2007
Implemented in year/period	2000	2005	2008	2007 -2010
Test	ESC	ETC	ETC	HDDTC
NOx (g/kWh)	5	3.5	2.0	0.27
HC (g/kWh)	0.66	0.55	0.55	0.188
CO (g/kWh)	2.1	4.0	4	-
PM (g/kWh)	0.1	0.03	0.03	0.0134

Table 3. HD Emission limits (vehicles with > 3.5 tons in EU, resp. > 8500 lbs in US). ESC = European Steady state test Cycle; ETC = European Transient test Cycle; HDDTC = US HD Transient test Cycle.

At the same time, HD diesel maximum bmep levels are expected to further increase (towards 28 bar [25]). For CNG engines to keep pace with this trend will be a challenge. Especially for the stoichiometric engine concept, where pre-turbine exhaust temperature levels are currently already at the practical limit. And although LB engines are not likely to run into these problems, a higher torque density goes hand in hand with higher charge density and higher maximum pressure, both of which make succesfull spark ignition of lean mixtures increasingly difficult. Some solutions to solve this ignition problem have been suggested (for instance micro-pilot diesel ignition) but they still need further prove/confirmation.

Finally, as illustrated in Figure 9, current HD CNG engines use more energy. This is of course a result of part load throttling but also of their typical lower bmep. (An exception to this is the recent MB Citaro CNG engine that achieves up to 18 bar bmep).

Different alternative engine concepts have been suggested for improving gas engine thermal efficiency. One such concept is that of a conventional (PFI, SI, homogeneous charge combustion) engine with add-on variable valve timing technology [18]. Alternatively, direct injection of natural gas (for SI stratified charge combustion) has been suggested to improve (mainly) part load efficiency [19]. A somewhat special solution is that of direct injection of natural gas aimed at diesel-like combustion (i.e. combustion controlled by rate of fuel injection and the ensuing fuel-air turbulent mixing process) with combustion initiated by either glow plug heating respectively micro-pilot diesel injection [20].

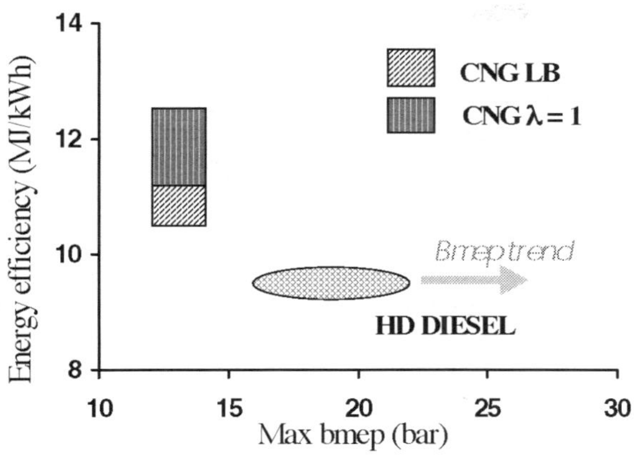

Fig. 9 Overview of typical current HD CNG engine efficiency and maximum bmep characteristics. Energy efficiency in MJ of energy input per kWh of energy delivered.

Evidence shows however that - at any moment of time - the most likely NGV configuration will not necessarily be the one that is theoretically most energy efficient. It will be the vehicle concept that combines a competitive efficiency (fleet owner combined investment and running cost) with the lowest (manufacturer) development cost. As with the LD CNG engine, TNO therefore believes that the HD CNG engine is likely to feed on new mainstream HD diesel technology developments such as increased bmep (through advanced VNT or 2-stage TC), flexible valve operation (VVT) and the – possible -introduction of flameless combustion (HCCI/CAI).

For the next years, most is expected of downsizing. Already in 1994, TNO demonstrated the potential of downsizing for better efficiency [21]. This was confirmed in a recent study. For this study a model for a typical modern bus configuration was built using the ADVANCE model library [22]. Key elements in the corresponding drive-line model are the ZF 4HP500 4-speed automated transmission and the vehicle mass. Using this model it is possible to calculate engine speed and torque as a function of time for any imposed vehicle-time trajectory. By combining this with available (steady-state) efficiency maps, it is possible to make a projection of the corresponding fuel consumption.

For this study 3 different engine configurations were simulated : a modern diesel engine, a stoichiometric (no EGR) CNG engine with 3-way catalyst and a lean burn CNG engine with oxycat. From results obtained in previous and ongoing engine development and in-use-compliance monitoring projects, engine maps were constructed that met with the following engine characteristics :

- bmep at maximum torque : 18 bar,
- maximum power 225 kW,
- comparable engine speed range,
- maximum torque between 1000 and 1200 rpm.

For all 3 engine configurations fuel consumption was calculated in the European Transient Cycle (ETC).

Figure 10 shows the results of this calculation for all three parts of the ETC : the urban part, the ural part and the motorway part. For comparison the CO_2 level is shown of a hypothetical HD vehicle that would combine diesel like energy efficiency with NG fuel CO_2 production levels (in CO_2 per MJ of fuel).

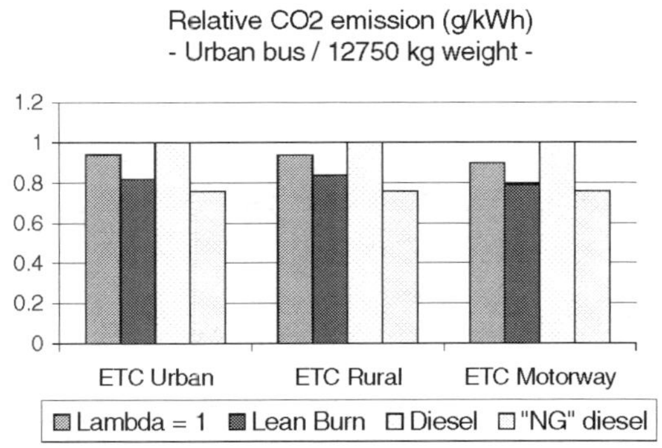

Fig. 10 Comparison of different urban bus vehicle configurations. "NG" diesel represents a vehicle with diesel-like fuel efficiency and with natural-gas-like CO_2-production.

Clearly, both CNG vehicles have a lower CO_2 emission than the diesel version, and the LB version comes a long way towards meeting the NG-diesel reference.

(As a final remark it should be mentioned however that, although a bmep of 18 bar is typical for current diesel bus engines, this level is high for a CNG engine, and possibly low for future diesel engines.)

ADDITIONAL CRITICAL ISSUES WITH NGV INTRODUCTION

For a succesfull market penetration of NGV's, the development of very clean and energy-efficient powertrain technology alone will not be sufficient. Other (both technical and non-technical) issues have to be dealt with in a satisfactory way. They are :

- development of light-weight, compact and preferably conformable on-board gaseous fuel storage technology,
- development of energy efficient refuelling station technology with acceptable refuelling times,
- setting-up of a widespread and harmonized refuelling infrastructure,
- (possibly) development of affordable and reliable home-fuelling equipment,
- installing of a widespread, well-trained, servicing organisation,
- promotion of NGV public acceptance by positive discrimination of gaseous fuelled vehicles, for instance in terms of parking privileges,
- promotion of low-CO_2 emitting vehicles in general, through regulatory or fiscal measures.

PATHWAY TOWARDS LONG TERM SUSTAINABLE TRANSPORT SOLUTIONS

From the WTW comparison discussed before it is clear that natural gas vehicles can achieve very low CO_2 emission levels. Eventually however a transition will have to be made towards climate-neutral energy carriers such as hydrogen or biofuels. The introduction NGV's on the road can help with this transition in a number of ways.

1. First of all, natural gas vehicles are able to run on biogas (from waste) and on synthetic methane (if it can be produced efficiently and at competitive cost).
2. NGV's are also able to run on mixtures of natural gas and hydrogen. Both LD and HD natural gas vehicles will accept such mixtures with up to 20 % (vol) of hydrogen. In fact, adding a limited amount of hydrogen can have a positive effect on NGV performance. Furthermore, with some modifications, even higher H2/NG ratio's are possible.
3. When natural gas, biogas and hydrogen all are available in a refuelling station, it is possible to blend them towards improved fuel specifications (improvements are in terms of energy content and combustion/ignition characteristics).
4. Of course, production costs with climate-neutral fuels will be considerably higher than with current fossil fuels. For such fuels maximum efficiency vehicle concepts will be selected. For hydrogen these will first be hybrid ICE powered vehicles. As more hydrogen becomes available, automotive FCV will become affordable and H2-FCV will take over.
5. If succesfull, NGV's will pave the way for hydrogen in terms of public acceptance of gaseous fuelled vehicles.

CONCLUSION

1. For the short and medium term, the introduction of natural gas vehicles on the road is the route that has the most potential to give a timely contribution towards meeting the CO_2 reduction targets in automotive transport.
2. When focused on urban transport applications, such a strategy will in addition further reduce local regulated and unregulated emissions.
3. For both LD and HD applications, the natural gas powertrain development strategy is most likely to be one of implementation (at low cost) of generic new technology developed for the mainstream (gasoline or diesel) vehicle concepts.
4. Over the next few years development efforts will focus on downsizing in combination with optimised diluted charge combustion concepts.
5. For succesfull market penetration of NGV's not only technical but also non-technical supporting measures are needed.
6. Because they are able to run on (blends of different) future climate-neutral gaseous fuels such as biogas, syngas and hydrogen, efficient NGV's can be a pathway towards long term sustainable transport solutions.

REFERENCES

1. Honda Civic GX, http://www.hondacars.com
2. P. Hendriksen et el., *Evaluation of the environmental impact of modern passenger cars on petrol, diesel,*

automotive LPG and CNG, TNO report 03.OR.VM.055.1/PHE, December 2003.
3. ARTEMIS Assessment and Reliability of Transport Emission Models and Inventory Systems. http://www.trl.co.uk/artemis/introduction
4. Eurostat (2000)
5. ACEA and the Commission Services joint report (2002) *Monitoring of ACEA's commitment on CO_2 emission reduction from passenger cars. Final report, 25 June* http://www.acea.be/ACEA/20020712PublicationEmissions.pdf
6. COM(95)689.1995 *A community strategy to reduce CO_2 emissions from passenger cars and improve fuel economy*. European Commission – DG XI, Brussels, 1995.
7. N.N. (1999) *The Foresight Vehicle Strategic Plan*, UK Department of Trade and Industry (DTI), http://www.foresightvehicle.org.uk
8. FURORE (2003) *Future Road Vehicle Research. R&D Technology Roadmap*. FP5 Thematic Network supported by the European Commission. October 2003, ISBN : 3-200-00017-1
9. BP (2002).www.bp.com (17-12-2002)
10. Shell (2001). Energy Needs, Choices and Possibilities. Scenarios to 2050. London, Shell International.
11. N.N. (2004) *The Hydrogen Economy: Opportunities, Costs, Barriers, and R&D Needs*. US National Research Council, February 2004.
12. S. Rahmesohl et al., *Bedeutung von Erdgas als neuer Kraftstoff im Kontext einer nachhaltigen Energieversorgung*, Wuppertal Institut für Klima, Umwelt, Energie, September 2003.
13. COM(2001)547 On alternative fuels for road transportation and on a set of measures to promote the use of biofuels. Brussels, 7.11.2001.
14. Concawe/ Eucar/ JRC (2003) *Well to Wheels analysis of future automotive fuels and powertrains in the European context*. http://ies.jrc.cec.eu.int/Download/eh
15. C. Daey Ouwens, member of the BIG-FiT team *(Climate neutral transport fuels from biomass)*. Private communication, June 2004.
16. Lox, E., Lindner, D., Carelli, G., Miletto, G., Ferrera, M. (2003) *The aftertreatment system and the NMHC issue*. Int. Conf. "The urban ecological transportation : the gaseous fuel option", October 23-24, Mantova.
17. TNO CNG Powertrain technology workshop, Delft, January 2003.
18. Uitenbroek, J., Cremer, P., Kraemer, H.P., Minnaert, J., *High efficiency natural gas engine with throttle-free load control and Miller-cycle valve timing*. IANGV Congress.
19. Miura, A., Honjou, F., Nakamura, A., Harayama, N., Sasaki, M. (2000) *Further Development of Fuel Consumption for Heavy-Duty CNG Engine*. FISITA World Automotive Congress, June 12-15, Seoul, Korea.
20. Harrington, J., Munshi, S., Nedelcu, C., Quelette, P., Thompson, J., Whitfield, S. (2002) *Direct injection of natural gas in a heavy-duty diesel engine*. SAE paper 2002-01-1630.
21. van der Weide, J., de Haas, J., van Ling, J., de Rijke, J., van der Steen, M. (1994) *CNG city bus engine with optimised part-load efficiency, high mean effective pressure and low emissions*. 4th Biennial Int. Conf. & Exh. On Natural Gas Vehicles, Toronto, October 3-6.
22. E. van den Tillaart, J. Eelkema, W. Vink (2002) *ADVANCE, a modular simulation environment for vehicle and powertrain design and analysis*, 19th International Battery, Hybrid and Fuel Cell Electric Vehicle Symposium & Exhibition (EVS 19), Busan, Korea.
23. Hoeijmakers, M.J., Rondel, M. (2004) *The Electric Variable Transmission in a city bus*. PESC 04, Aachen, Germany.
24. Electromechanical Converter, Patent WO 03/075437 A1.
25. Sturm, W.L., Kruithof, J. (2000) *Development and Testing of a HD Diesel Engine with Two-stage Turbocharging*. 9. Aachener Kolloquium Fahrzeug- und Motorentechnik.

Which Fuels For Low CO_2 Engines?
© P. Duret, X. Montagne (Eds.) and Editions Technip, Paris, 2004, pp. 123 – 132
27 rue Ginoux, 75015 Paris

UTILIZATION OF NATURAL GAS FOR INTERNAL COMBUSTION ENGINES

Yukihiro Tsukasaki, Satoshi Taniguchi, Akio Yasuda, Toshiaki Tanaka
Toyota Motor Corporation, Japan

ABSTRACT – *Natural gas is expected as one the promising alternative energies for the future. This study focused on CNG and GTL as an automobile fuel for internal combustion engines. The investigation results reveal that CNG direct injection engine has good potential to improve the full load performance and to reduce the partial load CO_2 emissions by stratified charge lean combustion. The GTL (FT Diesel) fuel showed the possibility of low emission diesel engine due to its advantage of zero aromatics and zero sulfur. But it was suggested that low volume change of elastomer dipped into the FTD fuel might cause fuel leakage in the fuel system.*

INTRODUCTION

Automotive industry has been strongly required to develop the technologies of clean and low fuel consumption for air quality improvement, global warming gas reduction and energy security. As an ultimate ecological powertrain, a fuel-cell technology has been developed around the world. However, there still remains a lot of issues to be solved for the widespread application of fuel-cell vehicles. Therefore, it is considered that the internal combustion engines with the conventional fuels from crude oil would be mainly utilized for a few decades. From viewpoints of alternative energy and clean fuel, natural gas has been applied to automobiles in many countries. A typical use of natural gas is a CNG. The CNG vehicle with an Otto-type engine, which is equipped with mixer or multi-point fuel injection systems, has already been introduced into many countries [1,2]. The disadvantages of current CNG vehicles are poor engine performance and short driving range, compared to the gasoline vehicle. These disadvantages of the current CNG engines are fundamentally caused by low charging efficiency and low energy density due to gaseous fuel. The bi-fuel system of CNG and gasoline, which has been popular in Italy, and so on., is practical method for the customer to enjoy the long driving range with CNG vehicle.

Fischer-Tropsch (FT) technology is more attractive for the flexibility of energy source and the realization of clean fuel [3-7]. FT Diesel fuel is one of promising automotive fuel because of its characteristics of sulfur free, aromatics free and high cetane number, and has been already commercialized by a few oil makers. But, the present application of FT Diesel is mainly for the quality improvement of market diesel fuel as one of blending oils.

In this study, the potential of engine performance and fuel economy in the CNG direct injection (DI-CNG) system and the application of pure FT Diesel fuel to DI diesel engine have been investigated as the utilization of natural gas to internal combustion engines.

CNG DIRECT INJECTION ENGINE

A prototype direct injection CNG (DI-CNG) engine was used to investigate the advantages of engine performance and fuel economy against a conventional CNG engine with a multi-point fuel injection system. The schematic diagram of DI-CNG engine is shown in Fig. 1. The test engine was based on an in-line 4, 2 liter, gasoline direct injection engine. The compression ratio was increased from 9.8 to 13. The CNG injection nozzle was a hole type one with high flow rate and anti-wear surface treatment on its needle and seat (Fig. 2). The injection pressure of CNG was set at 5 MPa. The combustion chamber with a narrow cavity type piston was designed for CNG (Fig. 3). The specification of test engine was shown in Table 1.

The natural gas property of tested CNG was the same as 13A city gas in Japan. As a reference fuel, a regular gasoline was also tested in this study. The specifications of test fuels are shown in Table 2.

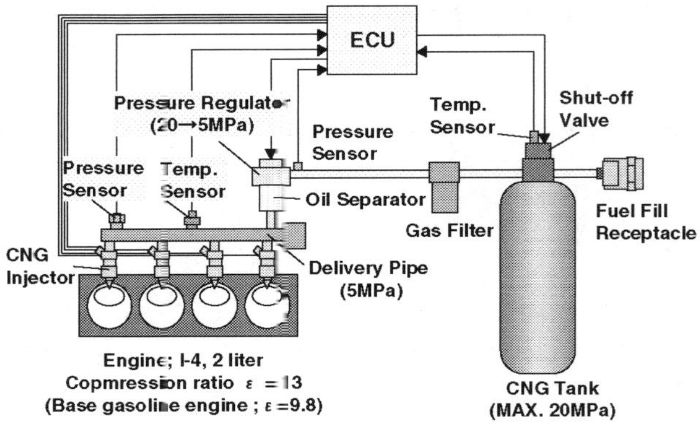

	CNG injector	Gasoline injector
Fuel pressure	5 MPa	6 - 15 MPa
Static flow rate	480 NL/min. (@ 5 MPa, 503 cc/min. gasoline equivalent)	840 cc/min. (@ 12 MPa)
Nozzle type	1 hole type	Slit type
Spray type	Solid cone type	Fan spray

Fig. 1 Schematic diagram of DI-CNG engine system

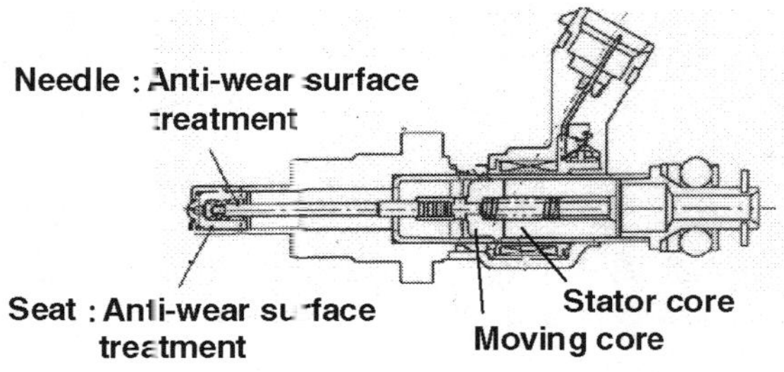

Fig. 2 DI-CNG injector

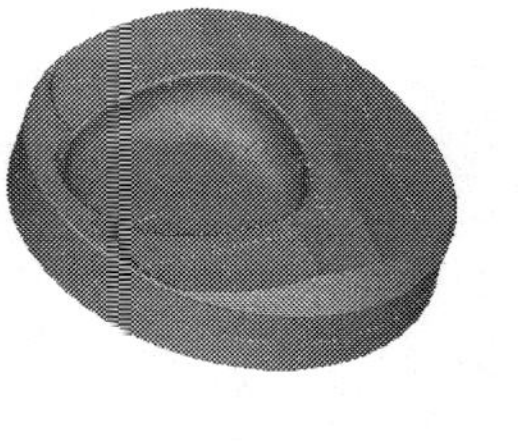

Fig. 3 Piston cavity in DI-CNG engine

Table 1. Specification of DI-CNG engine

Engine type	Direct injection, 4 stroke CNG engine (prototype)
Number of cylinders and valves	In- line4, 4 valves per cylinder
Displacement (Liter)	1.998
Bore x Stroke (mm)	86 x 86
Compression ratio	13 : 1 (Base gasoline engine 9.8 : 1)

Table 2. Test fuel properties

		CNG *1	Gasoline (Regular)
Density		0.84 kg/Nm3	0.74 kg/L (@15 deg.C)
RON		120 - 130	91
Lower heating value	MJ/kg	50.2	42.9
	MJ/L	0.042	31.7
Properties		CH4 88% C2H6 8% C3+ 4%	Aromatics 29% Orefins 14% Oxygenates 0%
Stoichiometric air-fuel ratio		16.8 : 1	14.7 : 1

*1 13A city gas in Japan

FULL LOAD PERFORMACE OF DI-CNG ENGINE

The full load engine performances in DI-CNG and multi-point injection CNG (MPI-CNG) engine are shown together with DI-gasoline in Fig. 4. The engine torque in DI-CNG was higher than that of MPI-CNG and could reach to the same level as the base gasoline engine at the engine speed below 2400 rpm.

But the torque advantage of DI-CNG against MPI-CNG was little over 3200rpm. It is considered that this is caused by the long injection duration including the suction stroke due to the insufficient fuel flow rate of tested CNG injector. Fig. 5 shows the comparison of charging efficiency between DI-CNG and MPI-CNG engine. The charging efficiency (η_v) of DI-CNG at 2400 rpm was about 5% higher than that of MPI-CNG. On the other hand, little difference of charging efficiency between DI-CNG and MPI-CNG was observed at 4800 rpm. This suggests that the CNG injector with larger fuel flow rate would be able to reduce the injection duration and to increase the charging efficiency at higher engine speed conditions.

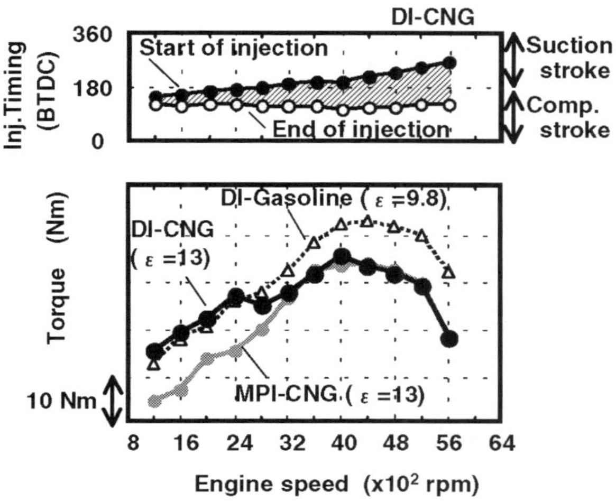

Fig. 4 Full load engine performances

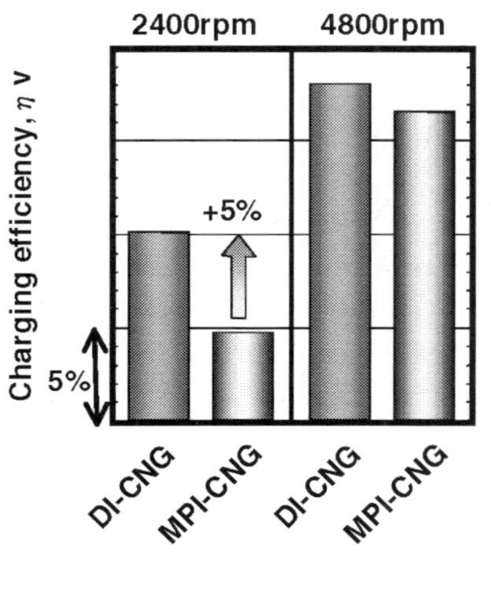

Fig. 5 Comparison of charging efficiency between DI- CNG and MPI-CNG engine

FUEL ECONOMY AND CO_2 EMISSIONS OF DI-CNG ENGINE

The Spark Ignition Direct Injection (SIDI) engine with stratified charge combustion can improve fuel economy and reduce CO_2 emissions compared to conventional MPI engines with homogeneous stoichiometric combustion. In DI-gasoline engines, the stratified mixture is formed through fuel vaporization process, and mixture formation is one of key issues to realize the stratified charge combustion under the wide engine operation conditions. In the case of DI-CNG engine, the fuel is naturally gaseous and methane, which is main component of CNG, has characteristics of extended inflammable limit and low CO_2 generation in the combustion compared to a conventional gasoline. Therefore, DI-CNG engine is expected to be a promising internal combustion engine for reducing CO_2.

Mixture formation in DI-CNG

Comparison of mixture formations between CNG and gasoline in a SIDI engine was conducted by means of an optical method [9]. The experimental apparatus is shown in Fig.6. The HC concentration in the vicinity of spark plug was measured by an infrared (IR) absorption method with a He-Ne Laser (3.39μm) and an IR detector. The measurement was conducted under the engine motoring operation of 1200rpm. The HC concentration (equivalent ratio) traces in compression and expansion stroke are shown in Fig.7. In the case of gasoline, the equivalent ratio in the vicinity of spark plug steeply increased and rich mixture over the inflammable equivalent ratio was formed at 20 degree CA after the injection. Then, the equivalent ratio decreased rapidly.

On the other hand, DI-CNG did not form so rich mixture as DI-gasoline and the inflammable mixture in the vicinity of spark plug was maintained longer than that of gasoline. This suggests that CNG is suitable for the stratified charge combustion from viewpoints of mixture formation. Fig.8 shows the comparison of the mixture formations between CNG and gasoline by CFD simulation. This result indicates that the velocity of CNG spray is very fast and CNG mixture can diffuse quickly in the combustion chamber.

Fuel economy and CO_2 potential in DI-CNG

The combustion characteristic and fuel economies of DI-CNG were investigated in the engine bench tests. The heat release rate changes of DI-CNG and MPI-CNG engine are shown in Fig.9. It is confirmed that the DI-CNG engine had a rapid combustion characteristic although the MPI-CNG engine showed a moderate combustion, which is a general property of CNG (methane). Especially, the initial combustion duration (0% to 10% heat release) became short in the DI-CNG (Fig.10). This is caused by rapid formation of inflammable mixture in the vicinity of spark plug, as shown in Fig.7. The fuel consumption and CO_2 emissions under the partial loads are shown in Fig.11. At the light load condition, the stratified charge lean combustion could attain remarkable reductions of fuel consumption and CO_2 in the CNG engine. These reductions decreased at higher load condition, but the NOx emissions were greatly lower than that of the stoichiometric combustion.

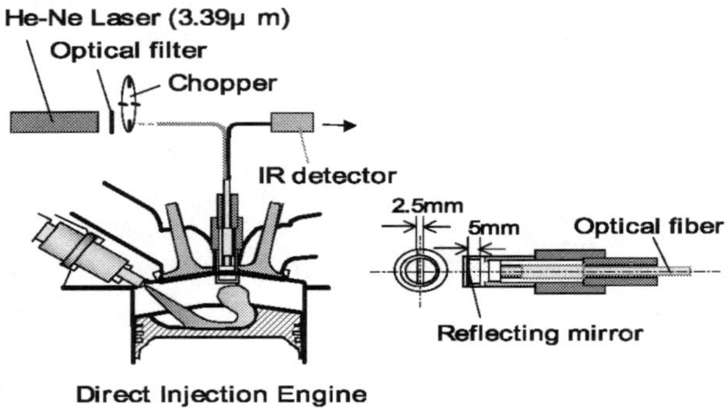

Fig. 6 Experimental apparatus of mixture concentration measurement

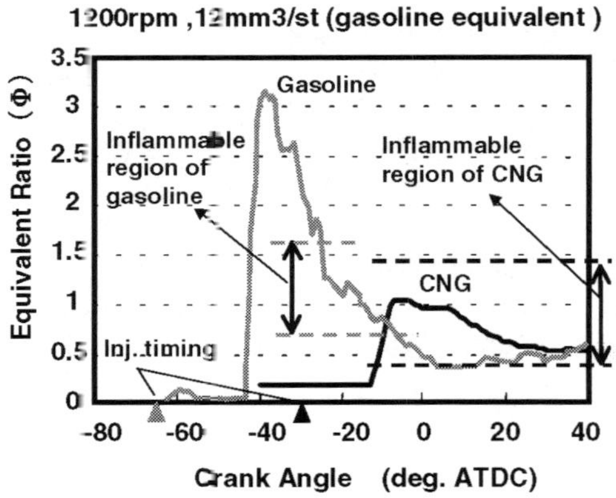

Fig. 7 Equivalent ratio traces in compression and expansion stroke

Therefore, it is considered that the DI-CNG engine is one of promising technologies to realize the environmental friendly vehicle with internal combustion engine.

As mentioned above, the DI-CNG technology has a good potential to improve engine performance, fuel economy and exhaust emissions in the CNG engines. However, it is necessary to solve a lot of technical issues for the establishment of this technology. It is considered that the reliability and durability of fuel injector are the most important issue in the DI-CNG engine. This study has just focused on the potential of engine performance and fuel economy in the DI-CNG, and the durability test has not been conducted yet. In future, the DI-CNG injector should be further developed for the practical application with enough reliability and durability performance.

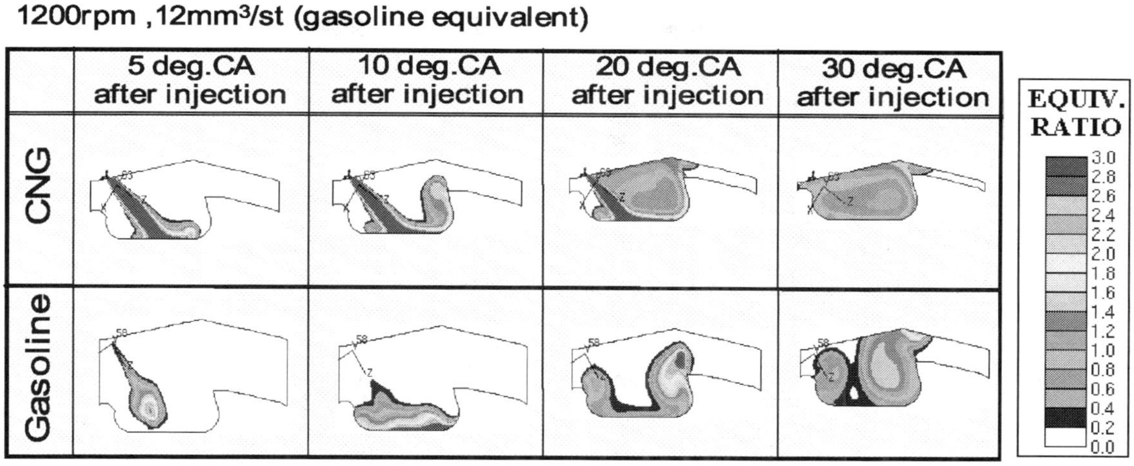

Fig. 8 Comparison of mixture formation process between DI-CNG and DI-gasoline engine

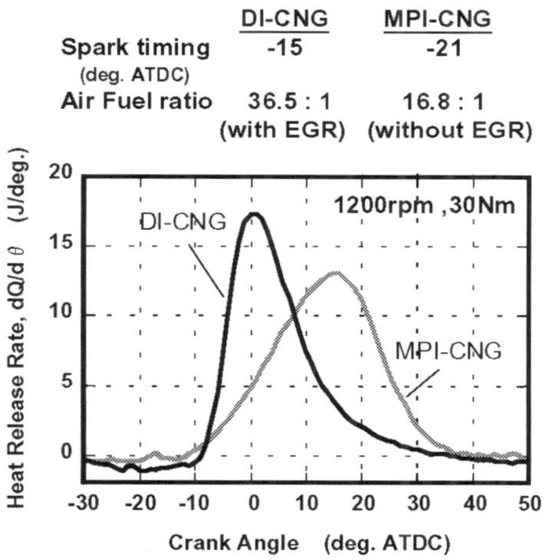

Fig. 9 Comparison of heat release rate between DI-CNG and MPI-CNG engine

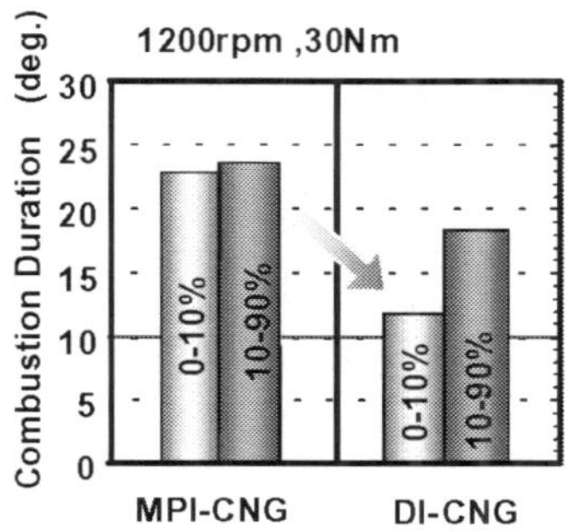

Fig. 10 Comparison of combustion duration between DI-CNG and MPI-CNG engine

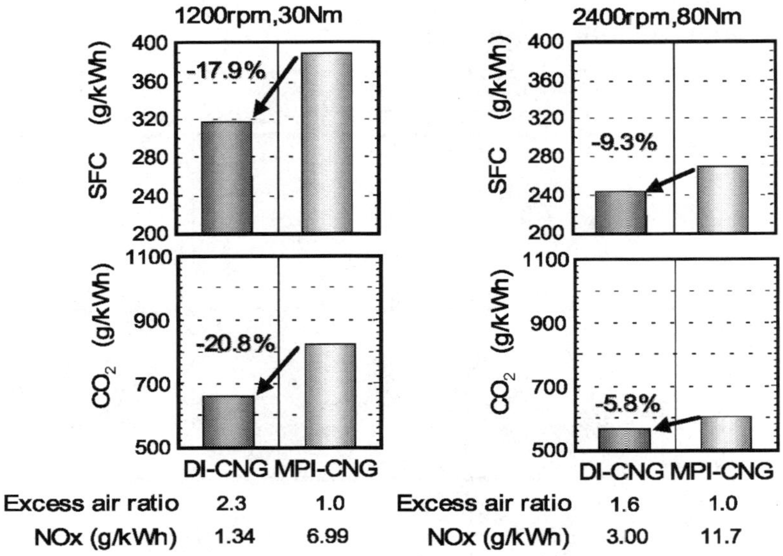

Fig. 11 Fuel economy and CO_2 emissions under partial load conditions

GTL (FT DIESEL) FUEL FOR AUTOMOBILE

Effects of FT Diesel fuel on the diesel combustion characteristic, the engine performance and the exhaust emissions were investigated. The test fuel specification is shown in Table 3. The FT Diesel fuel had low density, high viscosity and high distillation temperature in the range of T10 to T70 compared with those of a reference diesel fuel in this study. The specifications of test engine and test vehicle are shown in Table 4. The engine was DI diesel with a turbo-charger, an inter-cooler and a common rail fuel system.

Table 3. Test fuel properties

Property		FT Diesel fuel	Diesel fuel
Density @15℃	g/cm³	0.7826	0.8330
Viscosity @30℃	mm²/s	4.145	3.496
Cetane number		88	55
Sulfur	ppm w	<1	350
Aromatics	vol.%	0	26.7
IBP	deg.C	170.5	174
T10	deg.C	257	216
T30	deg.C	274.5	253
T50	deg.C	292	277
T70	deg.C	308	300
T90	deg.C	327.5	333
EP	deg.C	338.5	360
Pour point	deg.C	0	-7.5
Lower heating value	MJ/kg	43.52	42.95
	MJ/L	34.06	35.78

Table 4. Specifications of test engine and test vehicle

Engine type	Direct injection diesel with turbocharger, inter- cooler
Number of cylinders and valves	In- line4, 4 valves per cylinder
Displacement (Liter)	2.446
Bore x Stroke (mm)	92 x 92
Compression ratio	17 : 1
Fuel supply system	Common rail
Vehicle type	Passenger car
Transmission	5 speed manual
Vehicle weight (kg)	1,360
Engine type	Direct injection diesel with turbocharger, inter-cooler
Displacement (L)	1.995
Number of cylinders and valves	In-line4, 4 valves per cylinder
Bore x Stroke (mm)	82.2 x 94
Compression ratio	17 : 1
Fuel supply system	Common rail

ENGINE PERFORMANCE AND EXHAUST EMISSIONS WITH FT DIESEL FUEL

Fig.12 shows the comparison of full load engine performances between the FT diesel fuel and the reference diesel fuel. In this evaluation, smoke level under the full load conditions was limited at 10%. The test result shows that FT diesel fuel could improve the engine output about 13% compared to that of the reference diesel fuel. This was realized by the characteristic of less smoke emission, which enabled the increase of fuel injection quantity under the limited smoke level. The changes of heat release rate in the partial load are shown in Fig. 13. Generally, injection timing retarded is used in order to decrease NOx emissions in diesel engines. In the case of FT diesel fuel, its high cetane number property resulted in little change of heat release rate even at the retarded injection timing.

On the other hand, the reference fuel with the cetane number of 53 showed the steep rise of heat release with ignition delay at the retarded injection timing. This suggests that the effect of injection timing retarded on combustion characteristic is small in the FT diesel fuel and the NOx reduction may be achieved without the combustion noise increase. Fig. 14 shows the comparison of exhaust emissions under the Japanese 10.15 mode. The test vehicle had no after-treatment catalyst and the engine was calibrated with the reference diesel fuel. The emission test result indicates that PM emissions were reduced about 40% by using the FT diesel fuel. Moreover, the FT diesel fuel could reduce HC and CO emission remarkably. With regard to PM emissions, the particle numbers in FT diesel fuel were smaller than those of reference diesel fuel over all range of particle size but the particle size distributions had little difference in both fuels, as shown in Fig.15. The CO_2 emission and the fuel economy under the Japanese 10.15 mode are shown in Fig.16. The CO_2 emission and the fuel consumption in the FT diesel fuel were lower than the reference fuel.

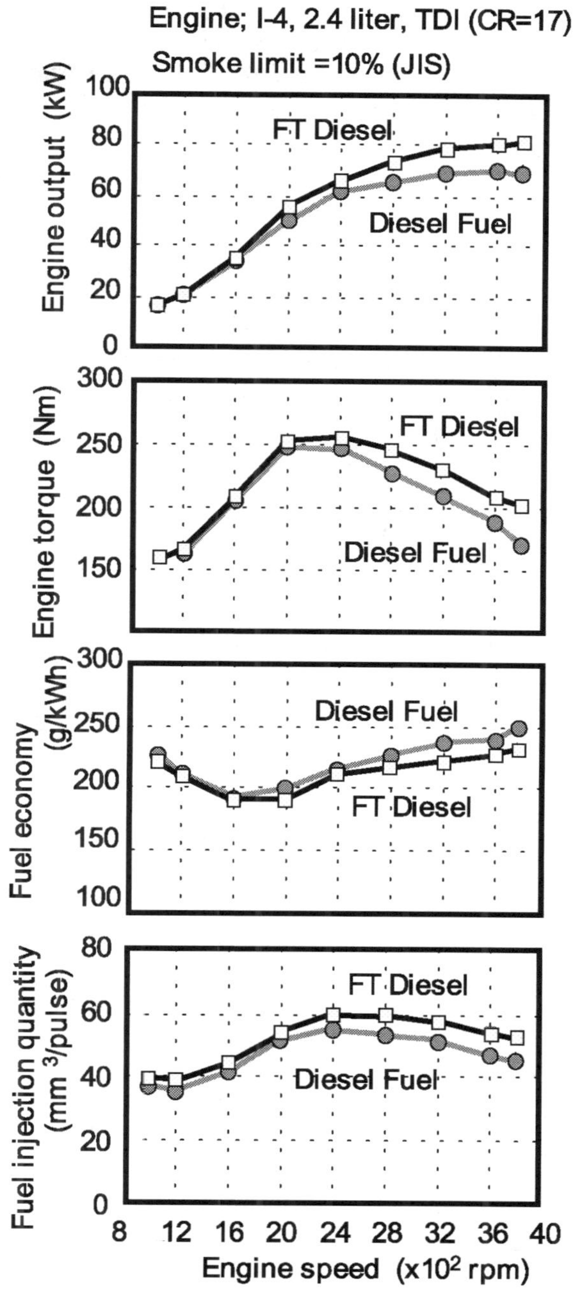

Fig. 12 Comparison of full load engine performances between FT diesel and reference diesel fuel

The FT diesel fuel had lower heating value on volume basis but resulted in better fuel economy than the reference diesel fuel. It is considered that the high cetane number of FT diesel fuel can suppress the combustion deterioration by the retarded injection timing in engine calibration for reducing the engine-out NOx. With regard to CO_2 emission, the amount of CO_2 relates to the carbon to hydrogen ratio (C/H) of fuel. The FT diesel fuel has no aromatic components and its C/H is lower than the reference diesel fuel with 26.7% aromatics. Therefore, FT diesel fuel is effective to reduce CO_2 emission during the combustion (Tank to Wheel CO_2). However, it is well known that the synthetic fuels generate much CO_2 in the current production process (Well to Tank CO_2) and FT diesel fuel is inferior to a conventional diesel fuel in CO_2 reduction. It is necessary to improve the energy efficiency in GTL (FT diesel) fuel production and to develop higher efficiency engine for GTL (FT diesel) fuel in future.

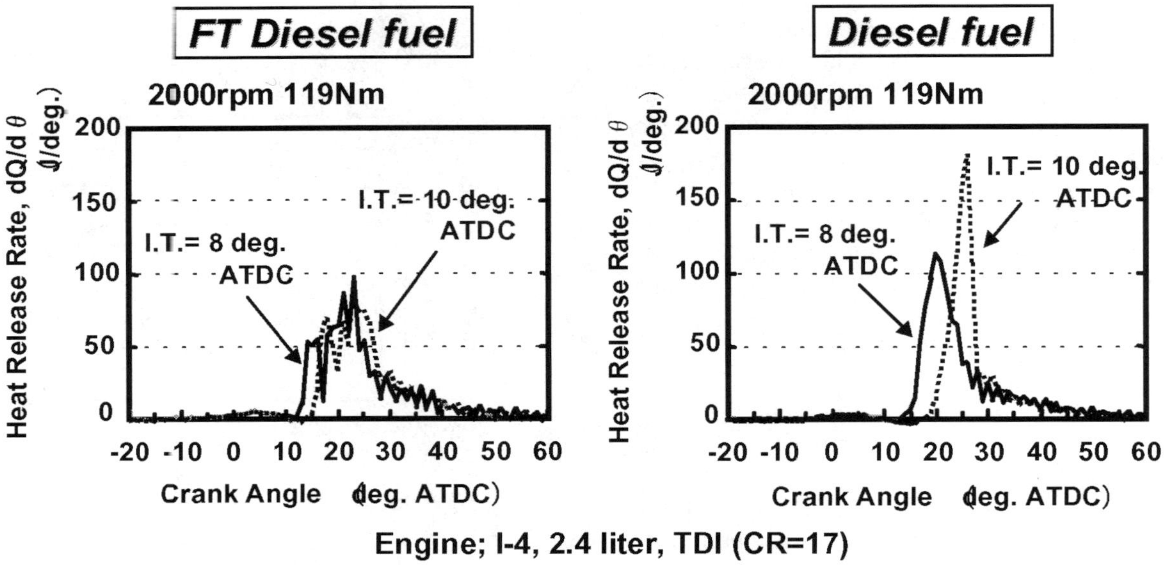

Fig. 13 Comparison of heat release rate changes between FT diesel and reference diesel fuel

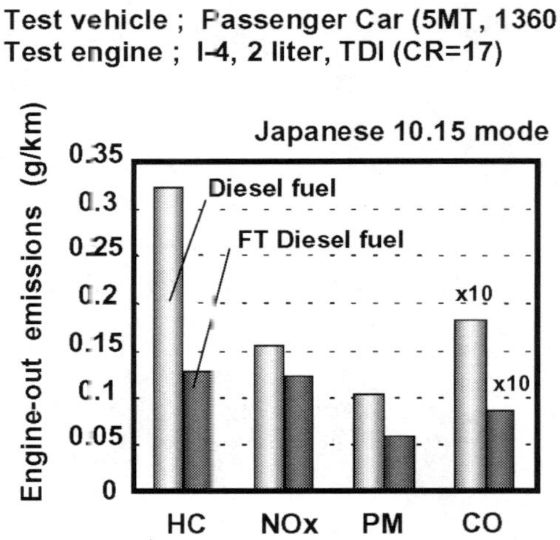

Fig. 14 Exhaust emissions under the Japanese 10-15 mode

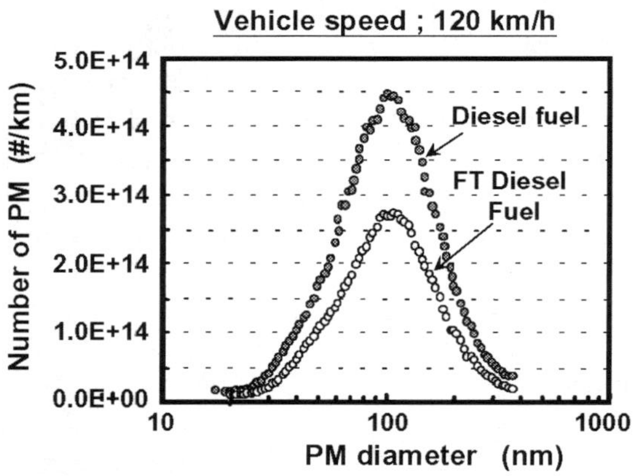

PM	FT Diesel Fuel	Diesel fuel
Engine-out emissions (g/km)	0.038	0.050
Mean Diameter (nm)	108	110

***PM diameter was measured by SMPS.**

Fig. 15 Particle size distributions under vehicle cruising condition

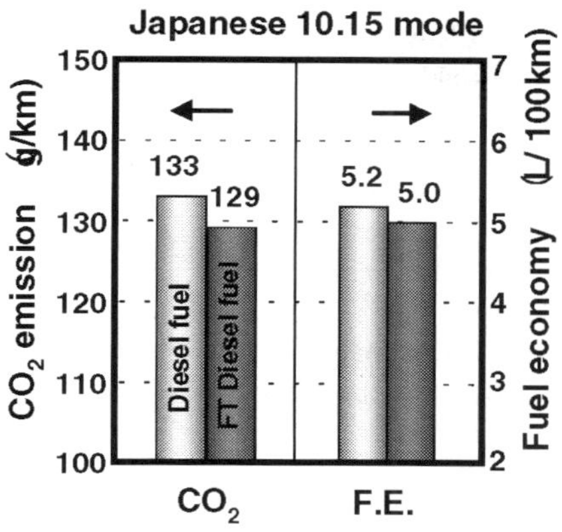

Fig. 16 CO_2 emission and fuel economy under the Japanese 10-15 mode

MATERIAL COMPATIBILITY IN FT DIESEL FUEL

Material compatibility is one of requirements for automotive fuels. Effects of FT diesel fuel on the physical property of elastomer in fuel system were investigated. Three test fuels, that is, the FT diesel fuel, the reference diesel fuel and the blended fuel of 50% FT diesel and 50% reference diesel fuel were evaluated in this study. The NBR was selected for the test piece because of a typical elastomer in a conventional diesel fuel system. Fig.17 shows the changes of physical properties in the NBR dipped in each test fuel. It is found that little change of those occurred in the case of 100% FT diesel fuel. Particularly, the reference diesel fuel resulted in 10% increase in the volume of NBR after the test time of 72 hours, but a little volume decrease of NBR was observed in the FT diesel fuel. It is considered that this phenomenon was caused by zero aromatics in the FT diesel fuel, because the elastomer swelling is related to the concentration of aromatic contents in the fuel. The volume of tested NBR slightly increased due to the blends of FT diesel and reference diesel fuel. The design of sealing parts such as an O-ring in a fuel system is usually based on the characteristic of elastomer swelling in a conventional diesel fuel with aromatic contents. Therefore, it is necessary to pay attention to the fuel leakage in the sealing parts for the application of FT diesel fuel.

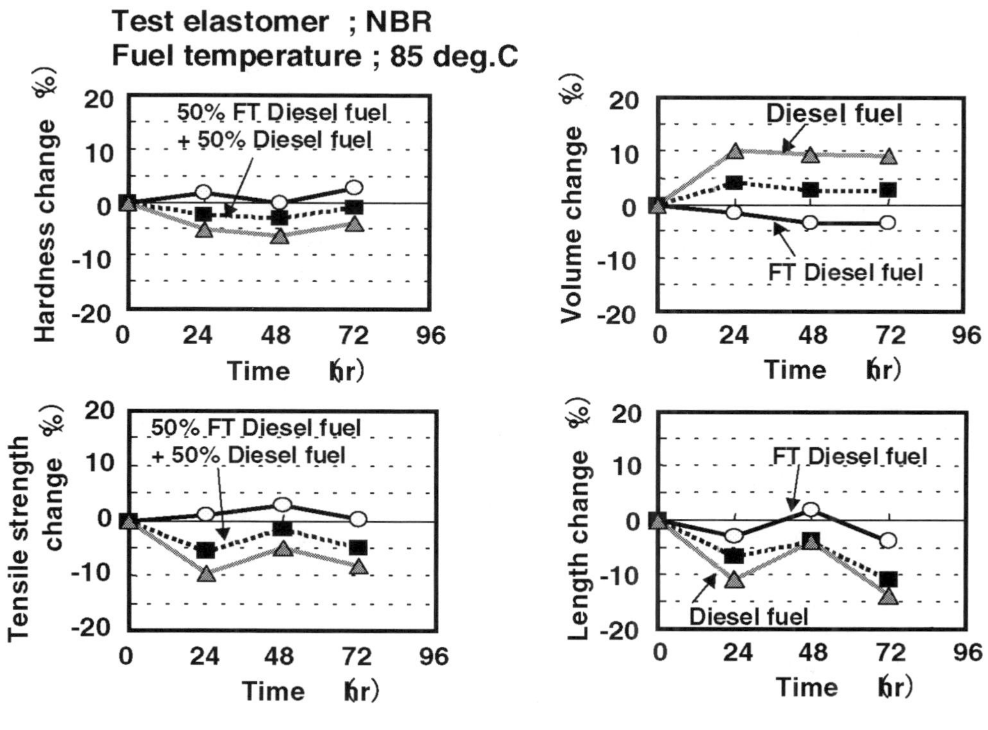

Fig. 17 Physical property changes in NBR

CONCLUSION

As application methods of natural gas to internal combustion engines, a direct injection CNG (DI-CNG) system in an Otto-type engine was developed. Meanwhile, a Fischer-Tropsch diesel (FTD) fuel was applied to a conventional diesel engine. The engine performance, the exhaust emissions, and the fuel economy in these applications were evaluated in this study. The evaluation results were as follows:

1. DI-CNG engine can improve the engine full load performance by recovering the charging efficiency (□v) and can further reduce the fuel consumption and the CO2 emission by applying the stratified charge lean combustion, compared to those of a conventional CNG engine with a multi-point fuel injection system and stoichiometric combustion.

2. For the establishment of DI-CNG engine technology, it is most important issue to develop the high performance DI-CNG injector with its durability and reliability.

3. The DI diesel engine operated on FTD fuel emits less smoke because of the fuel properties of zero aromatics, and can improve the full load performance compared to that with a conventional diesel fuel.

4. The high cetane number characteristic of FTD fuel is advantageous for suppressing the deterioration of diesel combustion and fuel economy under the low NOx conditions with injection timing retarded.

5. However, CO_2 emission is emitted in the production process of FTD (GTL) fuel. It is necessary to improve the energy efficiency of synthetic fuel production and to develop high efficiency diesel engines for FTD fuel in future.

It is considered that internal combustion engines will still remain the main powertrain of automobile for a few decades. From a viewpoint of GHG (CO_2) and exhaust emission reduction, natural gas is an effective energy for automobile. The utilization of natural gas would be more important and would depend on the development of fuel and engine technology in future.

Acknowledgments.

The authors would like to express our deepest appreciation to Dr. K. Kawamura and Mr. H. Inagaki of Toyota Central Research & Development Laboratories, Inc. and Mr. M. Kawauchi and Ms. C. Sasaki of Toyota Motor Corporation for their analysis works on DI-CNG mixture formation in this study.

REFERENCES

1. Suga T., Knight B., Arai S., (1997) Near Zero Emissions Natural Gas Vehicle, SAE Paper No. 972643
2. Kato K., Igarashi K., Masuda M., Otsubo K., Yasuda A., Takeda K., Sato T., (1999) Development of Engine for Natural Gas Vehicle, SAE Paper No. 1999-01-0574
3. Tsukasaki Y., (2000) Technical Trend of GTL Fuels for Automobiles (in Japanese), Journal of Society of Automotive Engineers of Japan, Vol.55, No.5, PP.67-72
4. Johnson J.W., Berlowitz P.J., Ryan D.F., Wittenbrink R.J., Genetti W.B., Ansell L.L., Kwon Y., Rickeard D.J., (2001) Emissions from Fischer-Tropsch Diesel Fuels, SAE Paper No.2001-01-3518
5. Oguma M., Goto S., Konno M., Sugiyama K., Mori M., (2002) Experimental Study of Direct Injection Diesel Engine Fueled with Two Types of Gas To Liquid (GTL), SAE Paper No. 2002-01-2691
6. Schaberg P.W., Zarling D.D., Waytulonis R.W., Kettleson D.B., (2002) Exhaust Particle Number and Size Distributions with Conventional and Fischer-Tropsch Diesel Fuels, SAE Paper No.2002-01-2727
7. Myburgh I., Schnell M., Oyama K., Sugano H., Yokota H., Tahara S., (2003) The Emission Performance of a GTL Diesel Fuel – a Japanese Market Study, SAE Paper No.2003-01-1946
8. Ueda S., Mori Y., Iwanari E., Oguma Y., Minoura Y., (2000) Development of a New Injector in Gasoline Direct Injection System, SAE Paper No.2000-01-1046
9. Kawamura K., Suzuoki T., Saito A., Tomoda T., Kanda M., (1997) Development of instrument for Measurement of Air-Fuel Ratio in Vicinity of Spark Plug – Application to DI Gasolin Emgine, The 14th Internal Combustion Engine Symposium

Which Fuels For Low CO_2 Engines?
© P. Duret, X. Montagne (Eds.) and Editions Technip, Paris, 2004, pp. 133 – 143
27 rue Ginoux, 75015 Paris

LEAN BURN HIGHLY BOOSTED SI ENGINES FOR GASOLINE, GAS AND REFORMULATED FUELS

Stefan Pischinger[1], Fabian Fricke[1], Gerhard Lepperhoff[2], Olivier Lang[2], Knut Habermann[2]

1. Institute for Combustion Engines (VKA), RWTH Aachen
2. FEV Motorentechnik GmbH

ABSTRACT – *Increasingly growing demands on fuel consumption reduction combined with ACEA´s commitment to decrease the new vehicles fleet CO_2-emissions below 140 g/km until 2008 are distinctly pushing the development of new passenger car engines and improved fuel formulations.*

In the last decade, the fleet consumption could be reduced by strongly increasing the share of modern direct injection, turbocharged Diesel engines. However, a further reduction requires the efficiency improvement also of spark ignition engines. Gasoline direct injection and variable valve trains, both in series production, have already shown a distinct reduction potential. New combustion systems are currently under development which also include potentials regarding alternative fuels like natural gas (NG).

This paper describes a new engine concept which aims to further use the potentials linked to CNG. In a first step, a dedicated (monofuel) natural gas vehicle based upon a stoichiometric, highly boosted engine with a specific power of 80 kW/l is presented. Further fuel saving potential is given by lean operation. Conclusively, a new lean burn combustion system, achieving Diesel-like fuel efficiencies, capable for different fuels, as gasoline and CNG has been developed. With gasoline, particularly the mixture formation optimization was of primary focus. The extreme lean burn plus EGR operation may permit the EU5 limits to be achieved without nitrogen oxide (NOx) aftertreatment.

INTRODUCTION

Considering continuously tightened emission regulations, as well as increasing shortage of energy resources, the development of new highly efficient and environmentally friendly combustion systems, associated with the use of alternative fuels becomes increasingly important. Since CO_2 emissions of passenger car traffic and its greenhouse potential is in the public interest, natural gas is discussed as an attractive alternative fuel. Due to the comparably lower c/h ratio the CO_2 emission is approximately 25% lower than with gasoline or Diesel at constant efficiency. Considering the greenhouse effect, which derives from the CO_2 emissions, one also has to take into account the other emissions, as far as they contribute to that climatic phenomenon. Hydrocarbon (HC), which consists of greenhouse active methane (CH_4) under natural gas operation, is here particularly significant. Despite the 75 times higher global warming potential (GWP) of CH_4 compared to CO_2, due to the considerably shorter duration of existing and the significantly lower absolute emission output, natural gas operation allows about 20% benefit of greenhouse effect compared with gasoline [1]. It is also well known that the fuel composition has a considerable influence on the emission behavior (NOx, HC, CO), as reported in [2, 3, 4]. Also, latest activities within the energy supply, but also the automotive industry are focusing on reformulated fuels from biomass, resulting in a neutral CO_2 balance [5, 6].

Nevertheless, independent on which fuel is finally used, the increase of overall efficiency of SI combustion systems by taking advantage of the still remaining improvement potential is absolutely necessary to meet the agreed CO_2 fleet targets [7]. One promising approach for future SI engines to reach Diesel-like fuel economy and at the same time obtain lowest engine out emissions, is homogeneous lean operation in the whole engine map, which will be discussed in the paper.

CNG OPERATION IN SI ENGINES

COMBUSTION SYSTEM

Engine concepts, which have been applied to date, are mainly based on common gasoline engine technology, and are carried out as bifuel vehicles, due to the few number of natural gas stations [8, 9, 10]. Only mono-fuel application can make use of the better RON potential of CNG ($RON_{NG} \approx 130$), in order to enable increased compression ratios (C.R.), and hence obtain thermodynamic benefits. The lower mixture heating value of natural gas, caused by reduced mixture density, leads to a theoretical reduction of full load performance between 7 and 10%, which can only partly be compensated by better efficiency, when C.R. is increased.

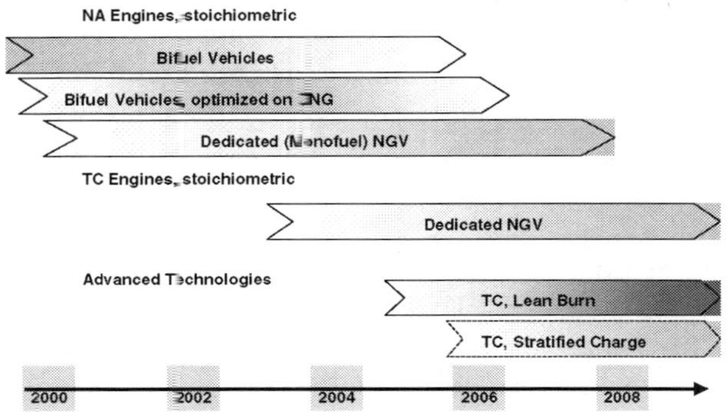

Fig. 1 Roadmap of CNG Powertrains (Passenger Cars) [11]

Based on the high knock tolerance, CNG offers high potential to be utilized in turbocharged engine concepts. Therefore, the before mentioned lack of performance will be overcompensated. In addition, comparatively low part load efficiency of today's common boosted SI engines can be improved by about 3 to 6%, when increased C.R. is applied. Moreover, the combination of turbo-charging and lean burn technology is a perfect match for CNG engines. Main development areas here are the ignition system, the boosting system, and also the mechanical engine characteristics. In addition, lean operation has to focus on lowest nitrogen oxide and hydrocarbon emission, in order to minimize efforts for aftertreatment systems. In **Fig. 1** the prognosis of future CNG powertrains for passenger cars is presented. The expected market penetration of passenger cars especially optimized for CNG will not only be dependent on improved engine technology, but also on tax incentives, and also on future filling station density. The announcement of a sufficient CNG station network through different fuel supply companies in Germany, Europe and worldwide in medium term will definitely support the launch of CNG fuelled vehicles.

As next development step, turbocharged CNG vehicles are already planned by some manufactures. In future, the combination of boosting and lean operation modes will be the consequent development step to make use of the fuel immanent benefits.

EXHAUST GAS AFTER TREATMENT

Whereas today's, stoichiometrically operated SI-engines are capable to fulfill the most stringent emission standards, e.g. in California, the major challenge for the Diesel engine is to reach an equal emission level, especially facing the immanent NOx-particulate trade-off, without sacrificing the excellent fuel economy. Generally, NG operation allows to significantly reduce the exhaust gas emissions (THC, CO, NOx), when compared with gasoline [12]. The fundamental difference between gasoline and natural gas operation is caused by the composition of the emitted hydrocarbons. Catalytic oxidation of the very compact methane molecule compared to long chained hydrocarbons is more difficult and requires either higher temperature levels or increased catalyst loading. Additionally, catalysts show a limited adsorption capability in the case of methane, which results in less tolerance of lambda variations. In contrary, CNG operation produces fewer total HC emissions immediately after cold start and during warm-up, which derives from the absence of fuel wall wetting and also good mixture homogeneity.

Real progress can be achieved when the aftertreatment system has been consistently designed for CNG operation, among other things by managing the exhaust gas temperature, for example with close-coupled catalysts, air-gap-insulated exhaust manifolds and down pipes, as well as more methane focused catalyst formulations and optimized rel. AFR control.

DEDICATED TC CNG ENGINE

A stoichiometrically operated, highly charged engine concept, achieving a very high specific power output, seems to be the most attractive short-term CNG application for passenger cars. In the following, FEV´s dedicated CNG demonstrator vehicle, powered by a stoichiometrically operated, highly boosted 1.8 L engine with a specific power of 80 kW/L is presented [11].

The gasoline base vehicle, chosen for conversion from gasoline to NG operation is the 110 kW version of the Audi A6 Variant with manual gear shift, for the European market, as presented in **Fig. 2**.

Fig. 2 CNG Demonstrator A6 1.8 l Turbo

This concept, featuring a close-coupled catalyst positioning, meets the EU-IV-emission targets (see **Fig. 3**), and achieves a specific power of 80 kW/L, while giving a clear benefit in full load fuel consumption, due to the stoichiometric operation even at high speeds. Regarding the composition of HC emission, it has to be taken into account that the biggest portion is consisting of methane, not considered as toxic, as represented in the Californian emission legislation.

The CO_2-benefit in comparison to a natural aspirated gasoline engine, having the same power, which is attained by combining the positive effects of engine downsizing and NG, is approximately 35% to 37%. It has to be mentioned that this reduction is not only achieved in the NEDC, but will also be feasible under real driving conditions. E.g., at rated speed, a CO_2 reduction of 50%, compared to a TC gasoline engine of the same power can be realized (see **Fig. 4**).

Further development steps, such as improved charging devices, if applicable with additional low-end-boost support, optimized combustion chamber design, with increased C.R., and optimization of the exhaust aftertreatment strategy offer further potential for additional CO_2 benefits.

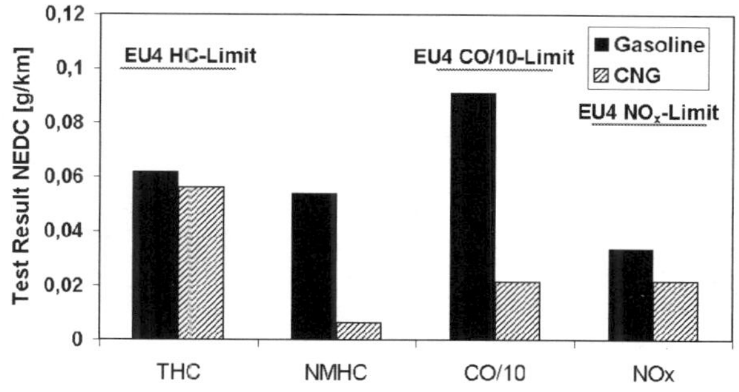

Fig. 3 EU4 Test Result

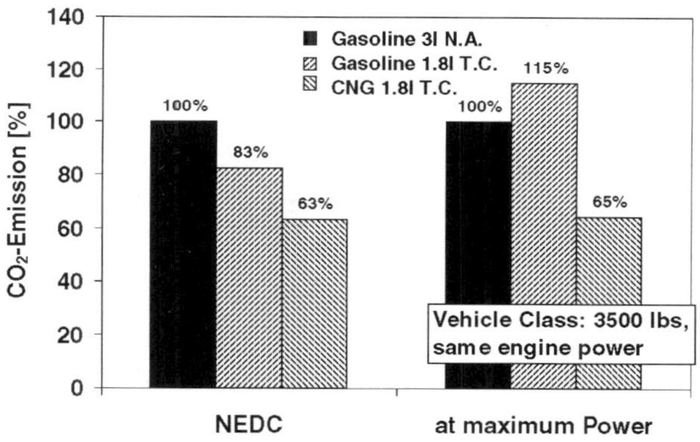

Fig. 4 CO_2 Reduction in NEDC and at max. Power

THE LEAN BURN HIGH BOOSTED SI ENGINE

In order to reach Diesel-like fuel economy and at the same time obtain lowest engine out emissions, the most promising solution for future SI engines is homogeneous lean operation. To avoid comparatively high costs for lean NO_X aftertreatment, target is to reach lowest possible NO_X raw emissions, by high degrees of charge dilution, considering lean operation and exhaust gas recirculation. Also, a high homogeneity of the mixture is basic demand, because homogeneous premixed combustion nearly avoids soot formation. In **Fig. 5** the efficiency, typical trends of hydrocarbon, carbon monoxide and nitrogen oxide emissions versus the rel. AFR are shown. The stability of combustion, shown in the graph in the middle of **Fig. 5** is representing the limiting factor for charge dilution. To obtain best efficiency as well as lowest engine out emissions a rel. AFR target of 1.7 to 1.9 is desired, and consequently a combustion layout for highest operation stability is required. To ensure reliable inflammation and subsequent complete combustion, taking the lean ignition limit into account, measures as increased compression ratio, increased in-cylinder turbulence and increased ignition energy are essential.

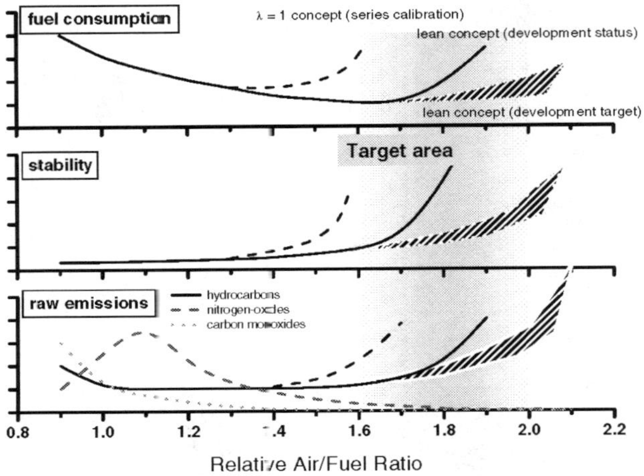

Fig. 5 Efficiency and Emissions vs. rel. AFR

Applying basic thermodynamic relations, the process efficiency of any combustion process can ideally be improved by increasing the compression ratio (C.R.) as well as the rel. AFR. In this chapter an approach of a homogeneous lean burn, highly boosted SI combustion concept allowing high compression ratios, focusing on high charge turbulence will be introduced. In Fig. 6, the ideal cycle efficiency as a function of C.R. and rel. AFR for conventional engine concepts as well as for the proposed lean burn concept is shown. Compared to state-of-the-art, stoichiometrically operated gasoline engines a cycle efficiency improvement potential of at least 8% points is realistic, by lean operation (rel. AFR in the range of 1.6) and increased C.R. (in the range of 13). Compared to Diesel engines, featuring higher compression ratios, and lean part load operation a theoretical efficiency gap of 4% points is still remaining. With regard to the trend of real cycle losses versus C.R., lower mechanical efficiencies, and special application requirements of Diesel engines to meet current emission targets, the realistic disadvantages of the lean burn boost system compared to the Diesel engine will be even lower.

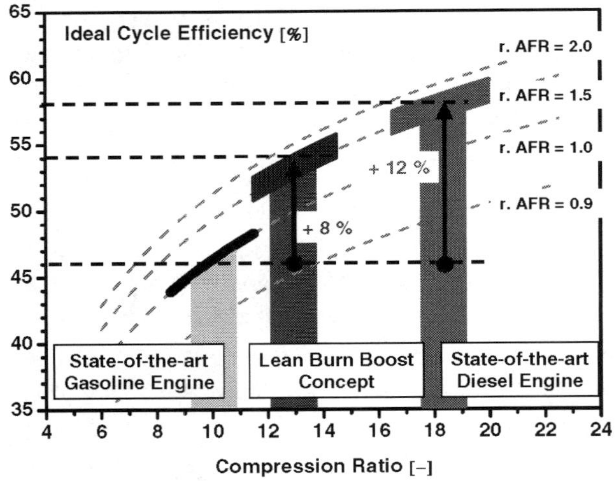

Fig. 6 Ideal Cycle Efficiency, comparing the SI and the Diesel Combustion Process (based on constant-volume cycle; high part load operation)

Fig. 7 the simulation of real cycle losses is shown, and it reveals that the improvement potential is even higher, as the ratio of effective to thermal efficiency is approximately 10% higher. This efficiency gain is mainly a result of drastically reduced heat and gas exchange losses, whereas the losses caused by non-ideal combustion are increased, but highly depending on the actual realization of the concept. For the given part load point at 2000 rpm, IMEP = 6.0...7 bar the total efficiency improvement of lean operation with high compression ratio (C.R. = 14.5), compared to stoichiometric operation with C.R. = 11.4 sums up to approximately 27%. The detailed analysis of the real cycle losses also proves that the lean burn gasoline engine is capable to reach Diesel-like fuel efficiencies in the medium load range.

Because fuel consumption is not only of interest under test cycle conditions, such as NEDC or FTP, but also under real driving conditions [13], concepts focusing on efficiency improvements in the complete engine operation map are gaining more attention. Lean operation even at high loads allows significant efficiency improvements compared to common SI engine full load operation modes. Additionally, due to reduced exhaust gas temperature, as a result of lean operation and high C.R., enrichment for component protection (turbocharger, close-coupled catalyst) can be avoided, resulting in an additional fuel efficiency benefit. Reduced exhaust gas temperatures at the same time allow to consider the application of a VTG charging system also for SI engine concepts.

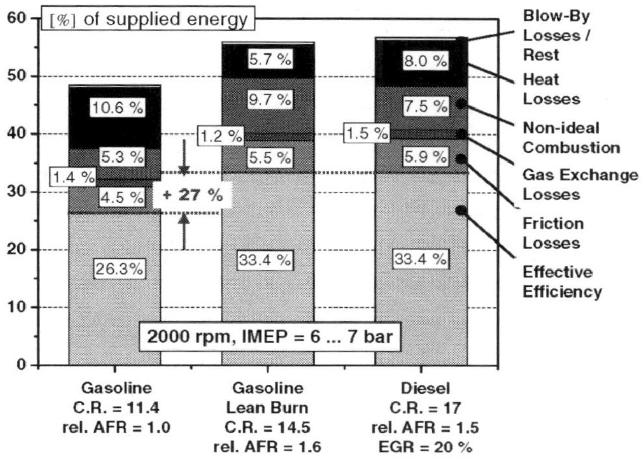

Fig. 7 Comparison of Split of Losses for different Engine Concepts

In future, high boosting, with high low-end torque and good transient behavior will be a key technology for future SI engines, as it is today for the Diesel engine. With regard to lean operation in the entire engine map, to ensure compensation of reduced mixture heating value, the lean high boost concept especially requires a well-considered charging technology. The resulting comparatively high specific performance characteristics at the same time open up further potential to reduce the in-use fuel consumption by consequent downsizing. Hence, the proposed lean burn highly boosted concept in total offers a high potential to make a contribution to the CO_2 discussion.

Since the CO_2 emission is not only depending on the efficiency of fuel conversion, but also influenced by the chemical composition of the fuel, combustion systems capable to allow the application of alternative fuels, e.g. natural gas or in future also hydrogen, have to be considered within future SI engine research and development.

ENGINE LAYOUT AND TESTING RESULTS

GENERAL CONCEPT CONSIDERATIONS

Basically, lean burn highly boosted combustion is limited by lean operation capability on the one side, and knocking on the other side [14]. The proposed combustion system features a high charge turbulence level in order to overcome reduced combustion velocities, a comparatively high C.R., a well-considered combustion chamber design and increased spark energies, and hence enables an extremely high lean burn capability. An additional effect of high turbulence level is the acceleration of the flame propagation, reducing the knock sensitivity, and therewith further increased C.R. and boost pressure can be applied. As consequence of the above, this design involves comparatively high cylinder peak pressures, and for this reason the applicable boosting level has to be chosen with regard to the mechanical capability of the engine.

Beside high turbulence levels, a high degree of charge homogeneity is required, in order to avoid flame distinction, due to lean regions, and also for lowest NOx emissions, by avoidance of enriched regions, close to stoichiometric operation. Low local variations of the rel. AFR are also basic necessity to surely avoid high cycle-to-cycle variation, in worst case resulting in misfiring or in engine knocking. Conclusively, apart from the generation of a sufficient in-cylinder mixture formation, the mixture preparation, depending on the injector and the injection process has to be optimized. For reliable ignition and early flame development under lean conditions also increased spark energies and extended spark duration offer certain improvement potential.

Engine Layout

The presented testing has been performed on a single cylinder engine, based on a modified Diesel engine design. A special ignition system was developed for testing purposes, allowing ignition voltages up to 70 kV. The high pressure boosting was realized by a 2-stage boosting device, enabling boost pressure levels higher than 2 bar. Base engine design parameters are given in **Table 1**.

Table 1. Engine Design Specifications

Bore [mm]	88
Stroke / bore ratio	1
Displacement [dm^3]	0.54
Compression ratio [-]	11.4 , 13, 14.5
# of valves per cylinder	4
Integral swirl ratio c_u/c_a [-]	approx. 2.3
▪ Central spark plug position ▪ Capacity discharge ignition system ▪ External boosting device ▪ Injection system: PFI and DI applicable	

TESTING RESULTS

In a first stage, testing with propane ($RON_{Propane} \approx 100$) is performed, in order to exclude the effect of liquid fuel atomisation and vaporization. Based on the measurement results and analysis of the heat release rates, optimisation steps regarding the generated charge motion level and the piston top land design are introduced in the concept. In **Fig. 8** the flow coefficient vs. integral swirl ratio of the original and the optimized intake port layout is depicted. In comparison to state-of-the-art four valve Diesel engines, the modified port design features a high swirl motion, but a still tolerable flow coefficient, finally effecting the layout of the required charging device.

To generate a high degree of in-cylinder turbulence close to TDC, a special piston crown layout, illustrated in **Fig. 9** is designed. The design is based on a bowl-in-piston layout, with enlarged valve pockets. Taking the flat cylinder head into account, a relatively high share of squish area can be realized.

Subsequent testing shows that based on these optimisations, the given performance targets, e.g. at 2000 rpm, IMEP = 23 bar, with rel. AFR = 1.7 are achievable with propane. The developed lean burn high boost combustion system demonstrates the capability for very high loads, however for the overall concept the boosting system performance has to be taken into account.

A subsequent test phase focuses on the carry-over of these promising results to gasoline operation. Due to the different inherent characteristics of the considered fuels, main research focus is to optimize the mixture formation, aiming at a locally and temporally highly homogeneous lean fuel/air mixture. For that reason, different possibilities of low pressure port fuel injection, beside others also taking air-supported injectors and mixture heating measures into account, as well as direct injection is investigated. Best results are achieved in part load as well as in full load operation using a high pressure injection system, with fuel pressures of up to 120 bar integrated as PFI system. According to [15], this system enables fuel droplet sizes in the range of 20 µm. Variations of injector positioning and injection timing show best results for injector positions close to the intake ports and injection timings simultaneous to the intake stroke. As expected the mixture formation can be positively influenced by a high degree of fuel atomization, and high flow velocities.

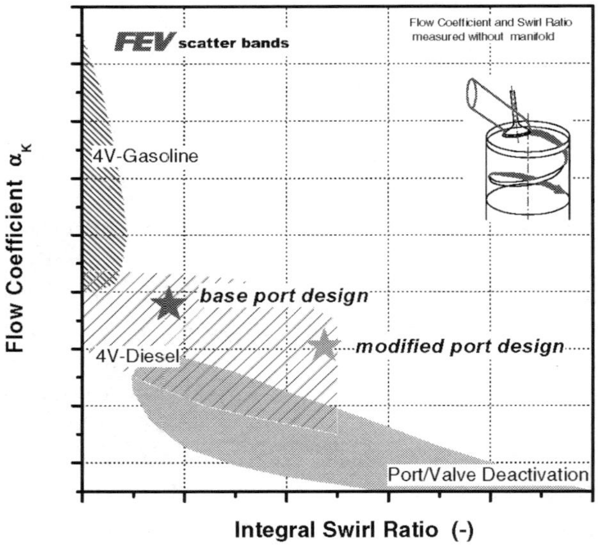

Fig. 8 Intake Port Flow Characteristics

The part load testing results at 2000 rpm, IMEP = 6.0 bar, comparing propane and optimized gasoline operation are given in **Fig. 10**. It can be stated that optimized gasoline operation shows only minor drawbacks in lean burn capability, compared to propane operation, and allows lean operation of up to rel. AFR´s of approximately 1.65 at C.R. = 13 and 1.75 when further increasing the C.R. to 14.5. The stability limit is defined as the standard deviation of IMEP of 0.15 bar, COV = 2.5%, respectively. The efficiency benefit gained by lean operation, in comparison to stoichiometric operation for all depicted tests is about 10%.

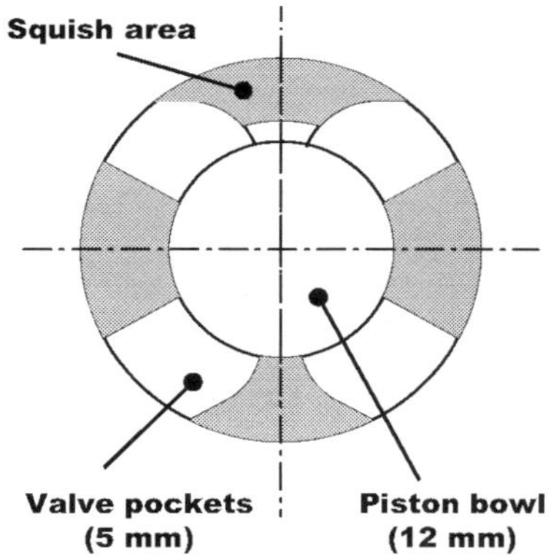

Fig. 9 Piston Crown Layout

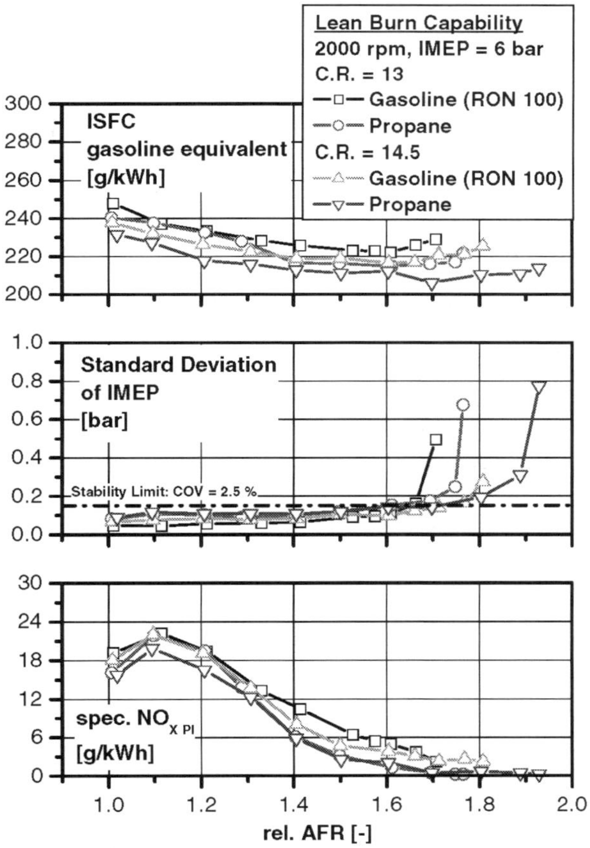

Fig. 10 Lean Burn Capability at 2000 rpm, IMEP = 6.0 bar

An assessment of the part load results focusing on fuel efficiency and nitrogen oxide emissions, compared to state-of-the-art gasoline as well as Diesel engines is given in Fig. 11. Due to the high C.R. and lean operation the proposed concept shows an outstanding fuel consumption, even compared to Diesel engines, but keeping in mind that the given scatter bands are consisting of engine concepts with series calibration, for Diesel engines especially meaning an application focusing on the NOx emission target. When considering a CO_2 related discussion, the lower c/h ratio of gasoline, propane and especially natural gas compared to Diesel, results in an additional advantage. Special attention has to be given to the NOx raw emission level, determining the requirement of additional NOx aftertreatment systems, resulting in fuel consumption as well as overall cost disadvantages. As a result of high in-cylinder turbulence the NOx emissions of the lean boost concept range in the upper part of the scatter band of stoichiometrically operated SI engines. As shown, lean operation allows a reduction of the NOx raw emission level of more than 90%. Comparing propane and gasoline operation, it is obvious, that using a gaseous fuel, a NOx advantage is still apparent, which is caused by the more homogeneous mixture quality, and also by reduced in-cylinder temperatures, as a result of the different fuel composition [3, 4]. Based on the high stability of the combustion system, further reduction of the NOx emission can be obtained by application of EGR. Particularly an internal EGR, realized by valve timing strategies, seems to offer high improvement potentials, as also the mixture formation can be positively influenced.

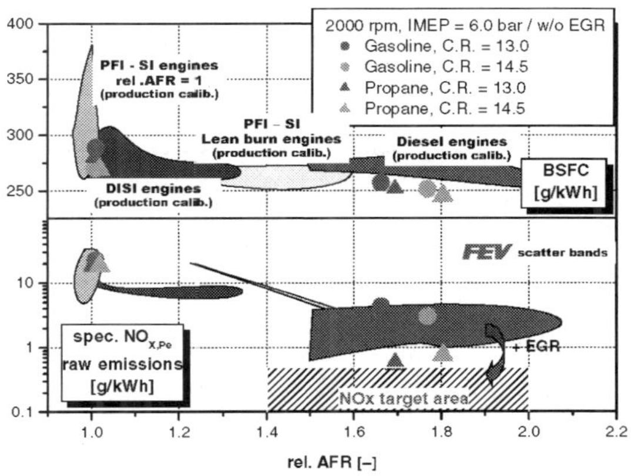

Fig. 11 Part Load Fuel Consumption and NOx Emission

Moreover, the proposed combustion system offers high potential for integration of controlled auto ignition operation (CAI) in wide part load ranges. Therewith, as described in [16] lowest NOx emissions are achievable, and no additional NOx aftertreatment will be required.

As discussed initially, essential goal of this concept is to realize lean operation not only at part load, but also at full load operation. The performed full load testing with propane demonstrates indicated mean effective pressure levels of above 21 bar when operated lean with rel. AFR of 1.6 to 1.7 (see **Fig. 12**).

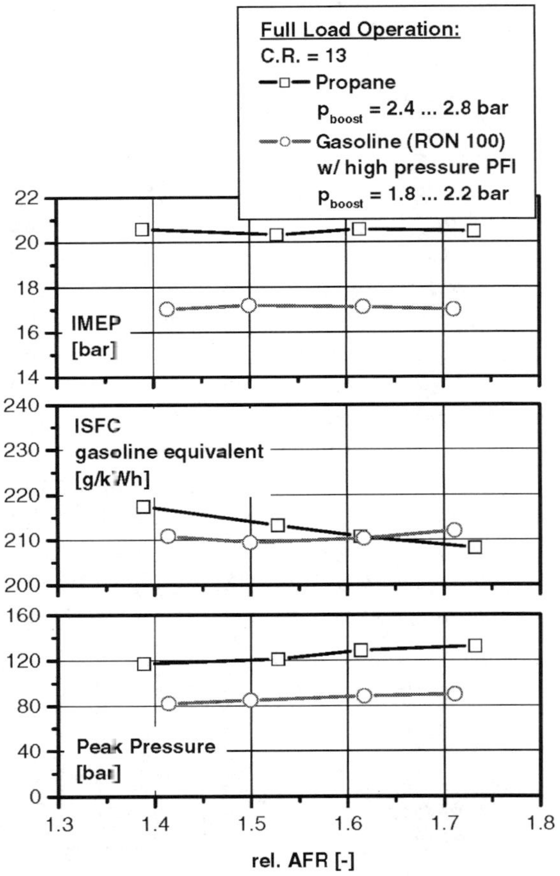

Fig. 12 Full Load Operation at 2000 rpm

The comparison of full load results from gaseous and liquid fuel shows a reduction of the operable range between misfiring and knocking limit. With gasoline maximal loads of 17 to 18 bar IMEP at maximal rel. AFR's of approximately 1.6 can be achieved with acceptable stability (see **Fig. 13**). In the load range between IMEP = 12 and 18 bar, very good indicated specific fuel consumptions of approximately 195 to 210 g/kWh are obtained, corresponding to efficiency values of 41 to 43%. In addition a positive influence on combustion stability by increased electrode gaps, allowing to enlarge the amount of mixture exposed to the spark, can be stated. Concerning the ignition system, it has to be considered that increased spark gaps again lead to an increase of ignition voltage demand.

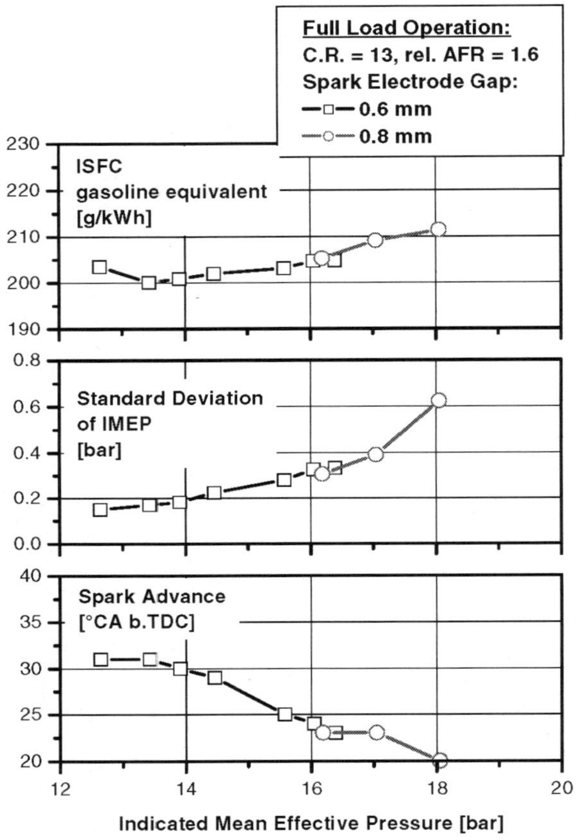

Fig. 13 Full Load (Gasoline) at 2000 rpm, rel. AFR = 1.6

To assess and compare the achieved full load results, in **Fig. 14** and **Fig. 15** scatter bands of modern gasoline and Diesel passenger car engines are given, depicting BMEP and BSFC versus the engine speed range. It can be stated that the lean burn boosted concept in operation with gaseous fuel, is capable to reach Diesel-like full load performance. Therewith, the performance of the final concept is mainly determined by the adopted charging device. The loads achieved in gasoline operation modes, ranging in the middle of the scatter band of charged SI engines. In **Fig. 15** it is shown that the lean boost concept allows to obtain full load fuel consumptions of up to approximately 13% below the lower end of scatter band of boosted SI engines in the range of modern DI Diesel engines.

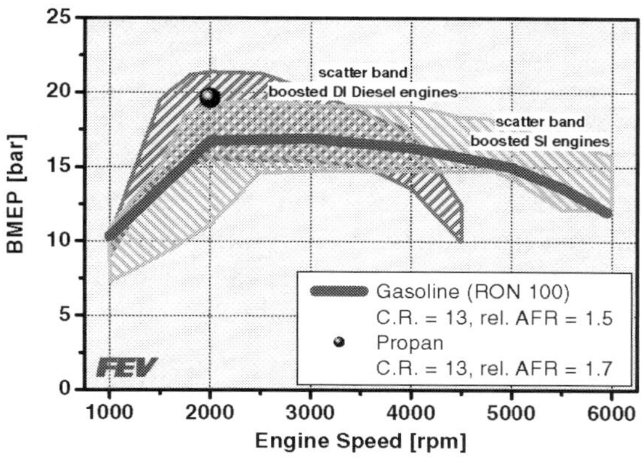

Fig. 14 Full Load Performance Scatter Band

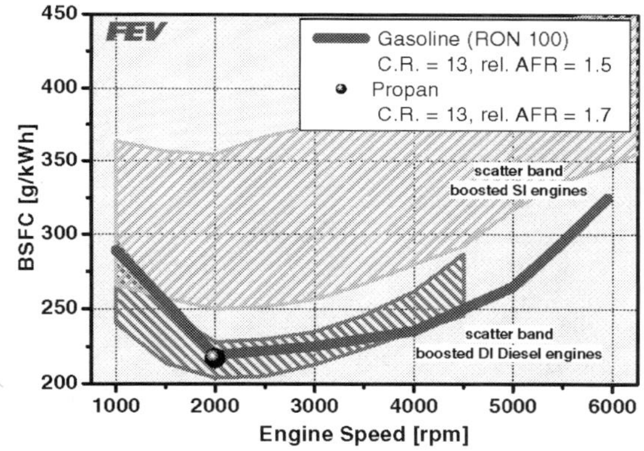

Fig. 15 Full Load Fuel Consumption Scatter Band

ASSESSMENT OF IGNITION VOLTAGE DEMAND

In the context of the high boost lean combustion concept the increased ignition voltage demand has to be discussed in more detail. For testing purposes a capacity discharger ignition system (CDI), capable of ignition voltages of up to 70 kV was prepared. The recording of the resulting ignition voltages at high loads (IMEP = 23 bar, rel. AFR = 1.4) shows maximum breakdown voltages of up to 45 kV (see **Fig. 16**).

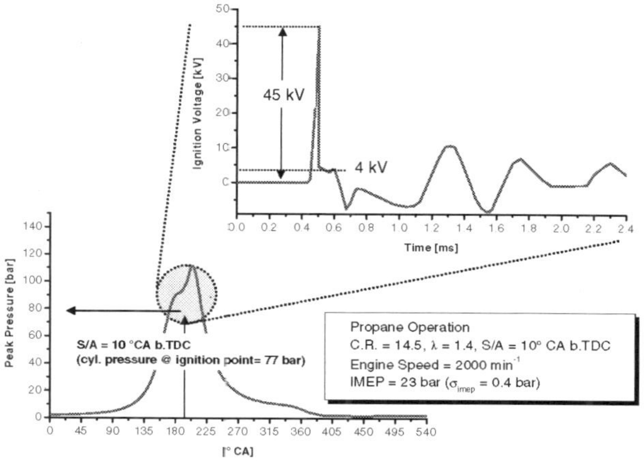

Fig. 16 Ignition Voltage Requirement at Full Load

With regard to the used CDI system, comparatively short spark burning durations of only 0.1 ms with burning voltages of 4 kV are recorded. Fundamental development focus conclusively is to increase the spark burning duration, leading to further improved lean burn capability, as described in [17, 18], while maintaining a high ignition voltage, exceeding current state-of-the-art ignition systems by up to 20 kV. Concerning the high ignition energy level, a higher spark plug electrode wear can be expected, again resulting in increased spark voltages, in order to ensure a reliable ignition. Based on the performed investigations, the following basic demands for the ignition system can be concluded:

- ignition voltage higher than of state-of-the-art TC engines (> 40 kV)
- spark burning duration of approximately 2 ms [18]
- reduced energy stored in the system
- high reliability under all operation conditions

ALTERNATIVE FUEL CAPABILITY

Based on the presented results, it can be assumed that the lean burn high boost concept offers high potential for the use of alternative fuels. Especially, natural gas can easily be combined with a lean combustion layout because of its extended ignition limits. Also, essential advantages of this fuel lies in its high knock tolerance and good ability to homogenate. An alternative approach could also be the use of hydrogen, favorable due to its outstanding lean ignition ability.

The basic requirements for an optimized fuel for a lean operated and boosted SI engine concept can shortly be summarized as follows:

- High octane number
- High energy density
- Good ability to homogenate
- High h/c ratio
- Low aromatic content

CONCLUSION

The proposed lean burn high boosted SI engine concept demonstrates a high potential for highest fuel efficiency in the complete engine map. It is shown that in part load as well as in full load operation fuel consumptions close to today's Diesel engines are obtainable. In addition, the good full load performance characteristic, mainly effected by the charging device, enables further improvement, due to the effect of engine downsizing.

High turbulence levels and high pressure port fuel injection allow gasoline operation almost similar to the operation with gaseous fuels. But especially the use of fuels having good ability to homogenate seems to offer highest potentials for future high efficient, low CO_2 emitting combustion systems.

Subsequent research will focus on the transfer of the concept to SI engine typical tumble systems and on an optimization of mixture preparation. Based on the results shown, fuel droplets diameters in the range of 20 µm or lower should be achieved in order to improve the mixture homogeneity in gasoline operation. So-called low pressure fuel injectors, developed for SULEV concepts, demonstrating a good fuel atomization ability, are assumed as an interesting alternative to the applied DI injection systems [19, 20]. Based on the performed testing, a detailed layout for an appropriate charging system has to be conceived with regard to the individual application demands. Because the concept allows a reduction of exhaust temperatures, the variable turbine geometry system could be one key technology, but also two stage charging should be considered, when aiming at loads higher than BMEP = 20 bar. Additional potential to reduce the NOx emission by the application of an internal EGR are also part of future test series, finally leading to a statement concerning the demands of lean exhaust gas aftertreament.

A next step, including work on the mentioned remaining tasks will be the development of a lean burn high boosted full-size engine concept, capable to run on different fuels, e.g. gasoline and natural gas. In future also hydrogen, as regenerative secondary energy resource should be taken into consideration, offering high potential for ultra-lean operating combustion systems.

Acknowledgments. The performed testing, presented in this study was partly funded by: Deutsche Forschungsgemeinschaft (DFG).

REFERENCES

1. Umierski, M.; Baumgarten, H.; Geiger, J.; Wolters, P.: 'Neue Motorkonzepte für erdgasbetriebene Fahrzeugmotoren', 23. Internationals Wiener Motorensymposium, 2002
2. Pischinger, F. ; Adams, W.: 'Abgasemissionen bei Verwendung alternativer Kraftstoffe für Kraftfahrzeug-Ottomotoren', VDI-Berichte Nr. 531, 1984
3. Kleinschmitt, W.: ,Untersuchung des Arbeitsprozesses und der NO-, NO_2- und CO-Bildung in Ottomotoren', Disseration, RWTH Aachen, 1974
4. Harrington, J. A.; Shishu, R. C.: 'A single-cylinder engine study of the effects of fuel type, fuel stoichiometry, and hydrogen-to-carbon ratio on CO, NO, and HC emissions', SAE paper 730476
5. Bartsch, C.: ' Synthetischer Dieselkraftstoff als Weg in die Zukunft', MTZ 65, 4/2004
6. Friess, W.; Herrmann, H.-O.; Maly, R. R.; Schnell, M.; Schaberg, P.: 'Emissionsverbresserung durch GTL-Dieselkraftstoff', Tagungsband 20. VDI/VW Gemeinschaftstagung , 12/2003
7. 'Monitoring of ACEA´s Commitment on CO_2 Emission Reductions from Passenger Cars (2002)', Final Report, 05.09.2003, www.acea.be

8. Brochure Opel Zafira 1.6L CNG, Opel Astra 1.6L CNG
9. Brochure Volvo S60, S80, V70
10. Brochure VW Golf Variant Bi Fuel
11. Pischinger, S.; Rütten, O.; Umierski, M.: 'Erdgas als Kraftstoff für die Zukunft, Effiziente Brennverfahren und Applikation', Tagungsband 20. VDI/VW Gemeinschaftstagung , 12/2003
12. Hupperich, P.; Dürnholz, M.: 'Exhaust Emissions of Diesel, Gasoline and Natural Gas Fueled Vehicles', SAE paper 960857
13. Lake, T.; Stokes, J.; Murphy, R.; Osborne, R.: 'Turboaufgeladene Konzepte für Downsized Benzindirekteinspritzung Ottomotoren', 12. AC Kolloquium, 2003
14. Pischinger, S.; Umierski, M.; Hüchtebrock, B.: 'New CNG Concepts for Passenger Cars: High Torque Engines with Superior Fuel Consumption', SAE paper 2003-01-2264
15. Preussner, C.; Döring, C.; Fehler, S.; Kampmann, S.: 'GDI: Interaction between Mixture Preparation, Combustion System and Injector Performance', SAE paper 980498
16. Wolters, P. ; Salber , W. ; Geiger, J. ; Duesmann, M. ; Dilthey, J.: 'Controlled Auto Ignition Combustion Process with an Electromechanical Valve Train', SAE paper 2003-01-0032
17. Pischinger, S.: 'Effects of Spark Plug Design Parameters on Ignition and Flame Development in an SI-engine' Doctoral Thesis, MIT 1989
18. Böwing, R.: 'Der Einfluss von Zündung und Zylinderinnenströmung auf die Verbrennung in Ottomotoren mit hoher Ladungsverdünnung', Dissertation, RWTH Aachen, 2000
19. Samenfink, W. ; Albrodt, H. ; Frank, M. ; Gesk, M. ; Melsheimer, A. ; Thurso, J. ; Matt, M.: 'Strategies to Reduce HC-Emission during Cold Starting of Port Fueled Injected Gasoline Engine', SAE paper 2003-01-0627
20. Posselt, A. ; Samenfink, W. ; Frank, M. ; Huppertz, T.: 'Strategien zur Reduzierung der HC-Emissionen des Verbrennungsmotors mit Saugrohreinspritzung im Kaltstart unter Berücksichtigung unterschiedlicher Katheizkonzepte', 9th Symposium ,The Working Process of the Internal Combustion Engine', Graz , 2003

Contact.

Fabian Fricke
Institute for Combustion Engines
RWTH Aachen
Schinkelstraße 8
52064 Aachen
Fricke@vka.rwth-aachen.de

DEFINITIONS, ACRONYMS, ABBREVIATIONS

ACEA: Association des constructeurs européen d'automobiles
BMEP: brake mean effective pressure
BSFC: brake specific fuel consumption
b.TDC: before top dead center
°CA: degree crank angle
c: carbon mass fraction
CAI: controlled auto ignition
CDI: capacity discharge ignition
CH_4: methane
CNG: compressed natural gas
CO: carbon monoxide
CO_2: carbon dioxide
COV: coefficient of variation
C.R.: compression ratio
DI: direct injection
DISI: direct injection spark ignition
EGR: exhaust gas recirculation
FTP: federal test procedure
GWP: global warming potential
h: hydrogen mass fraction
HC: hydro carbon
Hu: calorific value
IMEP: indicated mean effective pressure
ISFC: indicated specific fuel consumption
NEDC: new European driving cycle
NG: natural gas
NGV: natural gas vehicle
NOx: nitrogen oxides
PFI: port fuel injection
rel. AFR: relative air-to-fuel ratio
RON: research octane number
S/A: spark advance
SI: spark ignition
SULEV: super ultra-low emission vehicle
σ_{imep}: standard deviation of IMEP
TC: turbo charged
THC: total hydrocarbon
VTG: variable turbine geometry

Which Fuels For Low CO_2 Engines?
© P. Duret, X. Montagne (Eds.) and Editions Technip, Paris, 2004, pp. 145 – 155
27 rue Ginoux, 75015 Paris

DEVELOPMENT OF NATURAL GAS DEMONSTRATOR BASED ON AN URBAN VEHICLE WITH A DOWN-SIZED TURBOCHARGED ENGINE

Richard Tilagone, Stéphane Venturi
Institut Français du Pétrole

ABSTRACT – *The main objective of this project is to confirm the high efficiency of CNG fuel when used in urban conditions. For this purpose, a demonstrator vehicle using only CNG fuel has been developed when modifying an existing gasoline urban vehicle. The target is to achieve low CO2 emissions with a maximum level of 100 g/km on MVEG driving cycle, while keeping pollutant emissions below EURO IV level.*

In addition, partners involved focus to maintain vehicle behaviours as maximum vehicle speed and acceleration close to standard gasoline version. For these reasons, a Smart car has been chosen offering, on one hand, a small displacement engine useful to obtain low CO2 emissions (downsizing approach), and on another hand, turbocharger characteristic necessary to maintain torque and power outputs in accordance with expectation.

The first step was dedicated to CNG specific engine parts design (pistons, connecting rod, crankshaft, camshaft, etc...) in order to optimise CNG configuration. During the following steps, the engine has been optimised on a test bench. Valve timing and turbocharger selection have been optimised in order to obtain the best trade-off in terms of power consumption and pollutants. The final step has been spent to calibrate the engine, firstly on the test bench and secondly on the vehicle. This work allows coping with EURO IV emissions level with optimised catalyst light-off, transient running conditions, drive-ability, etc.

INTRODUCTION

Reduction of greenhouse gases emissions remains one of the most important challenges stakes for the next decades involving all energy consumers and especially the industrial actors, although the need to make better use of energy is not appreciated to the same extent by all industrial countries.

For the automotive sector, the European car manufacturers' commitment (reduction of the average fuel consumption for new vehicles to 140 grams of CO_2 produced per kilometre by 2008) appears to be a valuable contribution towards meeting the recommendations of the European Commission.

A sharp CO_2 emission reduction requires efforts both to reduce vehicle fuel consumption, and to widen the use of fuel with advantageous molecular hydrogen to carbon (H/C) ratio.

Alongside vehicle optimisation itself, improving engine efficiency remains the main target. Reduction of engine displacement volume (commonly named downsizing) is an efficient way to achieve this target. Due to its interesting chemical properties, natural gas can be used to achieve low carbon dioxide levels (H/C ratio close to 4), while at the same time maintaining high thermal efficiency through dedicated engine development.

The main objective of this project is to confirm (with a dedicated demonstrator vehicle) the high efficiency of CNG fuel when used in urban conditions. The target is to achieve low CO_2 emissions with a maximum level of 100 g/km on the MVEG driving cycle, while keeping pollutant emissions below EURO IV level.

First, this paper describes the optimisation work done on the Smart 600 cc engine for the gaseous version, and the Smart vehicle itself (adaptation of driveline transmission and implementation of natural gas storage tanks is not presented in this paper). Then, results obtained on test bench with specific electronic control unit developed by IFP have been compared to gasoline ones. Finally, the performances on chassis dynamometer bench are introduced.

NATURAL GAS AS ALTERNATIVE FUEL

As one of the most promising fossil energy resources, Natural Gas is widely used for transportation and more especially for automotive applications.

Some chemical properties of gaseous fuel are advantageous for emissions and consumption trade-off; others require specific adaptations due to gas phase in normal condition of use. Those that distinguish the most Natural Gas are highlighted.

• High Research Octane Number (RON), higher than 130 as compared to 95 for European Eurosuper premium gasoline. For spark ignited combustion this results in a reduced endency towards engine knock. Thanks to this characteristic, a NG engine can be designed with a higher compression ratio than for common gasoline combustion, resulting in better fuel efficiency. The following relation directly links the compression ratio [ε] to the thermal efficiency [η]. Symbol [γ] represents the specific heat capacity of mixture.

$$\eta = 1 - \varepsilon^{\gamma}$$

The possibility to run with Maximum Best Torque [MBT] advance limits the exhaust gas temperature and avoids the use of strategy applied to gasoline engine, which consists to cool down the exhaust gases by increasing the air to fuel equivalence ratio. Continuous stoichiometric conditions, best torque spark advance and high compression ratio are key factors to enhance thermal efficiency and drastically reduce exhaust emissions, especially for turbocharged engine.

• The main component of natural gas is methane (between 80 to 99%). Thus, H/C molecular ratio from natural gas is close to 4, higher than gasoline (around 1,85). This property generates a reduction of CO_2 emissions up to 23% for the same engine efficiency (of course this value depends on natural gas composition), **Table 1**.

	CO_2 [g/g fuel]		LHV [kJ/kg]	CO_2 [g/kJ] x 10^3	
NG	2,75	-13,5%	48450	56,8	- 23,7%
Gasoline	3,18	0	42690	74,5	0
Diesel fuel	3,17	- 0,3%	42770	74,1	- 0,5 %

Table 1. Theoretical CO_2 emissions for stoichiometric combustion

Taking into account the necessary improvements on engine and vehicle to fulfil the 2008 or 2012 commitments for CO_2 emissions (European cars on the road should emit 140 g/km then 120 g/km), natural gas reveals itself as an efficient solution suitable with today's technologies. From the today's European performance (~160 g CO_2/km), the gasoline consumption should be reduced by 14% in the next 4 years; it seems to be not easy taking into account gains already obtained, on small gasoline and Diesel vehicles on the market. More, recent surveys on well-to-wheel analysis confirm the positive impact of natural gas on CO_2 reduction (MIT, Eucar-Concawe, IFP, JCAE, etc).

• Natural gas engines shows a significant reduction of pollutant emissions compared to the corresponding gasoline baseline engine; the gaseous state of this fuel avoids wall wetting effects on intake manifold and cylinder liner especially at cold start conditions. Thus, it induces fuel consumption savings and above all, unburned hydrocarbons emissions reductions. Concerning unburned hydrocarbons, the reduction comes also from oil film adsorption-desorption phenomena, which is one of the main sources of unburned fuel. The percentage of HC emissions from this source is smaller for methane than for gasoline. Finally, the raw HC emissions should be 50% lower than those of the gasoline baseline engine are.

• Few intermediate operations are necessary to use Natural Gas on vehicle (no refining). Its composition, when delivered from the natural gas grid, depends on output well location and also on time (in accordance with suppliers' strategy). This variation affects its combustion characteristics. On

Figure 1, Air/Fuel ratio, Lower Heating Value and Knock Index are plotted for one calendar month Variations are important, but fortunately they do not affect the energy content for a given equivalence ratio as shown on **Figure 2**, only the gas flow sprayed out from injector varies.

Whatever the gas composition, rated power from engine equipped with Air/fuel ratio close-loop control system does not undergo gas variations.

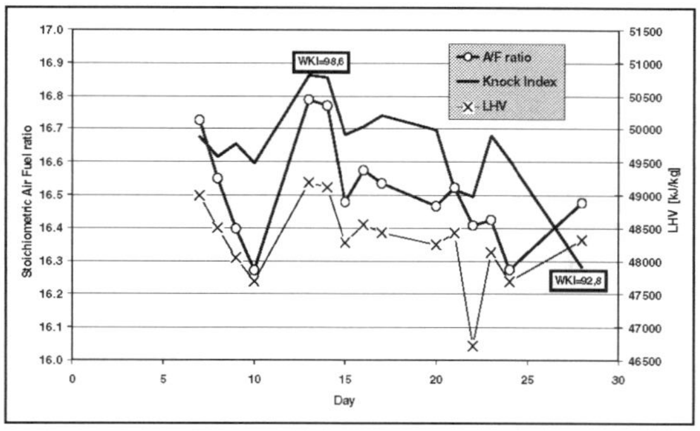

Fig. 1 Air/Fuel ratio, Low Heat Value and knock Index variation during one month

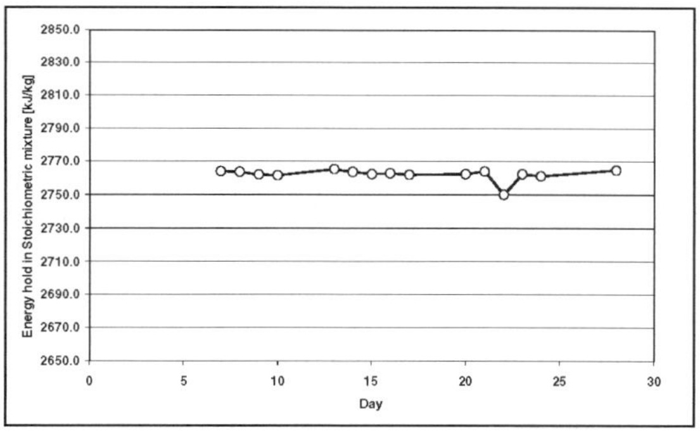

Fig. 2 Energy in a NG/air stoichiometric mixture

• Volumetric Lower Heating Value of Natural Gas is very low compared to liquid fuels, **Table 2**. For gaseous fuels a high capacity tank and high pressure are necessary to store the energy needed to achieve a competitive driving range.

For instance, at a storage pressure of 20 MPa, the volume necessary for same energy on board is five times as big compared to gasoline; furthermore, tank implementation on chassis vehicle is not so simple.

	density kg/m3	LHV kJ/kg	Energy MJ/m3		volume (same energy)
Gasoline	750	42690	32 020		
Diesel fuel	835	42770	35 710	+ 11 %	x 0,9
gaseous Methane 1013 hPa, 273 K	0,716	50010	36	- 100 %	x 889
gaseous Methane 20 Mpa, 293 K	173 [AGA8]	50010	8 652	- 73 %	x 3,7

Table 2. energy per storage volume for common fuel

• Gaseous fuel affects volumetric efficiency of a PFI engine, between 10 and 15% is lost depending on engine optimisations. The use of supercharger is necessary to keep output Power and Torque comparable to that of the alternative gasoline version.

STOICHIOMETRIC APPROACH

At present, the selection of combustion mode, stoichiometric or lean burn corresponds to a choice between fuel efficiency and low exhaust emissions.

The stoichiometric approach is more suitable to cope with stringent emission regulation, using proven three-ways catalyst technology. More, for a dedicated natural gas engine, the penalty in terms of consumption is acceptable, with higher exhaust temperature helpful to reduce unburned hydrocarbon during the cold start-up running.

Figure 3 gives a comparison of unburned hydrocarbon emissions for the same engine, running on gasoline and on natural gas. Notice that for natural gas application, unburned hydrocarbon are mainly composed of methane which has very high stability. Thus, in order to limit the unburned hydrocarbons, it is necessary to apply some management strategies which all lead to fuel consumption increasing. This would not be necessary if standard emissions regulations would impose different limits for methane and non-methane hydrocarbons.

In case of downsizing, the drop of performance due to displacement reduction can be partially (not all over the speed range) removed with stoichiometric turbocharged engine, with fewer penalties to control transien: conditions.

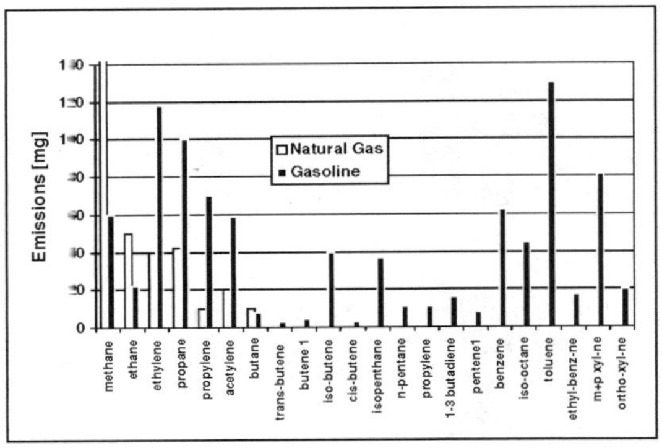

Fig. 3 comparison of unburned hydrocarbon emissions for the same engine running on gasoline and on natural gas (small BI-fuel vehicle)

DOWNSIZING AND SUPERCHARGING

Looking at running points in demand to propel a vehicle at constant speed, part loads are mainly used. For instance, a medium size vehicle at constant speed of 70 km/h needs an output power of ~7 kW. Nevertheless, specific consumption is all the more high as the load is low. For a 2 Litre displacement engine at 2000 rpm, 7 kW represents a brake mean effective pressure (BMEP) of 2,1 bar and a specific consumption over 400 g/kWh for a conventional gasoline FPI engine, **Figure 4.** For a smaller engine with displacement of 1 Litre, the BMEP is twice and the consumption is reduced to 300 g/kWh representing a gain of 25%.

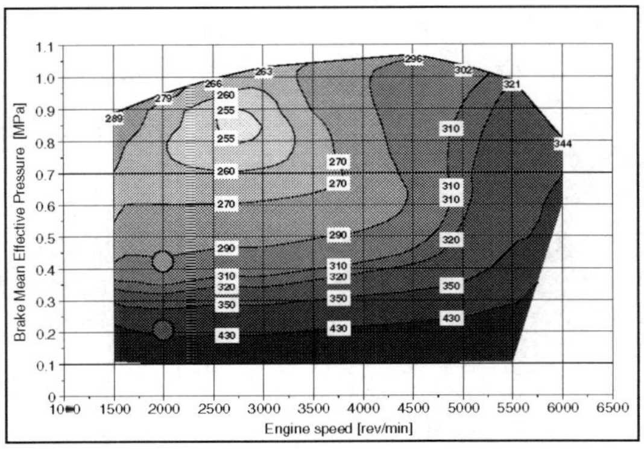

Fig. 4 Specific consumption for a conventional medium size SI-FPI gasoline engine

This example highlights the benefits when reducing engine size: it enables high engine loads with better fuel consumption reduces friction losses. Obviously, size reduction has to be limited in order to avoid to high BMEP (especially for gasoline application: knock sensitivity and pollution).

Nevertheless, reducing swept volume leads to affect performances for naturally aspirated engine in terms of output power and maximum torque (because of CNG volume needed). The use of supercharger could improve natural gas performances. For Natural Gas injection, the penalty is more severe due to gaseous phase, which could affect volumetric efficiency by over 10%.

TECHNICAL FEATURES

The Smart vehicle offers characteristics suitable for urban transportation: small dimensions, light weight and low consumption helping to reduce CO_2 emissions.

Designed by the SUPREX Company on behalf of MCC Company, the mechanical features of this 3 cylinder-in-line engine are: 63,5 mm bore, 63 mm stroke, 599 cc swept volume, 9,5:1 gasoline compression ratio. The output performances mentioned below are achieved through use of a GT12 cooled turbocharger produced by AlliedSignal Automotive [GARRETT], allowing over-boost conditions. In this case, the intake manifold pressure is regulated to 2 bar for NG instead of 1,8 bar for gasoline.

- Output Power: 40 kW at 5250 rev/min
- Maximum Torque: 80 Nm from 2300 rev /min to 4500 rev /min

The chain drive camshaft in the cylinder head (with hydraulic clearance compensation), actuates the two valves per cylinder, **Picture 1**.

Two spark plugs per cylinder, the electrodes of which are visible on **Picture 2,** help to stabilise combustion at delayed spark advance (at low load and low engine speed to light off catalyst during cold condition). The air coming from the compressor is cooled with a dedicated air/air cooler. No EGR circuit is available.

Picture 1 Camshaft in cylinder head

Picture 2 Spark plug and valve location

The exhaust manifold has an integrated turbine house as showed on **Picture 3**. This design reduces thermal losses and makes better use of the exhaust gas energy.

VEHICLE GASOLINE REFERENCE TESTS ON CHASSIS DYNAMOMETER

The smart vehicle has been characterised by the UTAC Company: aerodynamic and friction losses were determined. Many tests have been carried out by IFP-Lyon in order to establish reference data according to UTAC results.

Tests confirm that our gasoline vehicle pollutant emissions meet EURO IV regulation, but fuel consumption is higher than official figures. For CO_2 emissions, **Table 3**, the gasoline Smart with a mileage of 3150 km is quoted 140 g/km. Checking tests at IFP-Lyon confirm that CO_2 performances are quiet surprising, with a noticeable reduction for high mileage, 132 g/km for an aged vehicle (52000 km).

Picture 3 Exhaust manifold

This reference step is very important to be able to quantify the real impact of using natural gas compared to gasoline.

	Mileage km	CO g/km	UHC g/km	NOx g/km	CO_2 g/km
Euro IV - 2005		1,00	0,10	0,08	
UTAC results	3150	0,26	0,08	0,04	140,5
IFP results	3270	0,30	0,09	0,04	136,4
IFP results	52220	0,38	0,06	0,05	131,7

Table 3 UTAC and IFP MVEG results

Deep analyses of regulated exhaust emissions confirm a large margin for EURO IV. More than 90% of unburned hydrocarbon are produced during the first ECE cycle, 70% of CO emissions and 60% of NOx. For fuel consumption, 13% of total fuel burned during the MVEG cycle is for the first ECE cycle, which weights the most for emissions results.

DEDICATED NATURAL GAS VERSION

The first development work focused on the engine itself for an optimal use of natural gas characteristics (compression ratio, gas rail implementation, gas injectors). Then tests were performed on a test bench with in-house electronic management system ACEbox™ (Automotive Control Engine based on XPCtarget™) developed by IFP. Alongside engine optimisation, a vehicle was prepared by modifying the original gearbox and by implementation of gas tanks and a high-pressure fuelling line. Finally, the gas engine was installed in the vehicle with the in-board electronic management system based on the ACEbox™ device.

NATURAL GAS TANK IMPLEMENTATION

Three tanks build with composite material have been designed and machined by ULLIT Company. The total storage capacity is 35 litres, and the driving range for fuel consumption representative of MVEG cycle is approximately 150 km.

The tank size is compatible with the space available under floor when the gasoline tank is removed. The ground clearance and volume of the boot are not modified **pictures 4** and **5**.

Picture 4 Ullit gas tanks

Picture 5 under floor view

GAS ENGINE ADAPTATIONS

The compression ratio was increased from 9,5:1 to 12,0:1. New pistons replace the gasoline ones, able to resist higher in-cylinder pressure, which increases with MBT Advance settings. Connecting rods are strengthened; then, all the mobile parts for gas engine are redesigned **picture 6**.

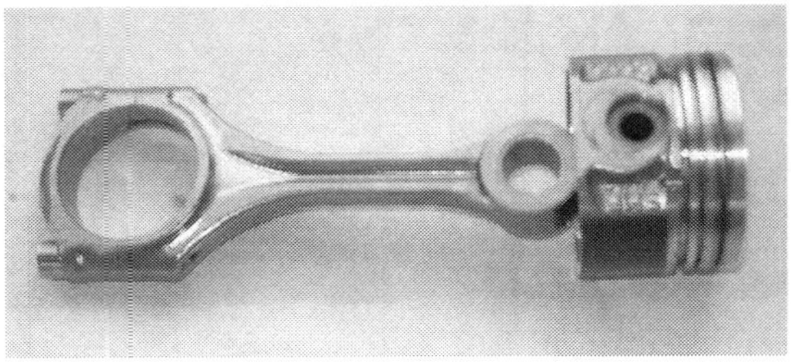

Picture 6 connecting rod and gas piston

Gas injectors developed by Keihin for the Honda Civic vehicle were selected for the low rail pressure needed useful to reduce leakage after engine switch-off (avoiding flash-back and HC peak when starting). The gas pressure reducer Sherex reduces the pressure from over 200 bar (corresponding to filling condition) to 4 bar At 4 bar sonic flow conditions are maintained under turbocharged conditions.

The main challenge for the gas rail design is its implementation within the limited space available on the gasoline engine. **Picture 7** shows the gas rail in connection with the gas injectors bolted on the intake manifold.

Picture 7 gas rail in connection with gas injectors

As mentioned, exhaust aftertreatment remains an essential stage to succeed in reducing exhaust emissions. In addition to light-off strategies (as for instance delayed spark advance) main catalyst formulation and loading have been deeply optimised in order to overcome the high stability of methane. In addition, three small metallic converters located between cylinder-head duct and exhaust manifold (loaded with the same formulation) improve post-treatment efficiency especially during the critical first seconds period after start-up, **Picture 8.**

Picture 8 metallic light-off implementation

ON-BOARD ENGINE MANAGEMENT SYSTEM

To control engine performance and to be free in from any engine management strategy implementation, IFP has decided in accordance with the partners involved in the project to manage the gas engine using its own device.

At first the ACEbox™ system designed by IFP has been used for engine bench tests, able to drive engine on bench as on vehicle. Mapping on steady state conditions is optimised before data implementation for vehicle tests, **picture 9**.

Picture 9 ACEbox™ system and engine on bench

The on-board system developed in collaboration with FH Electronics Company is based on ACEbox™ system and is composed of a Supervisor computer located in the boot and a Sequencer computer, located back the front right seat, in charge of timing and of input/output signal management, **figure 5**.

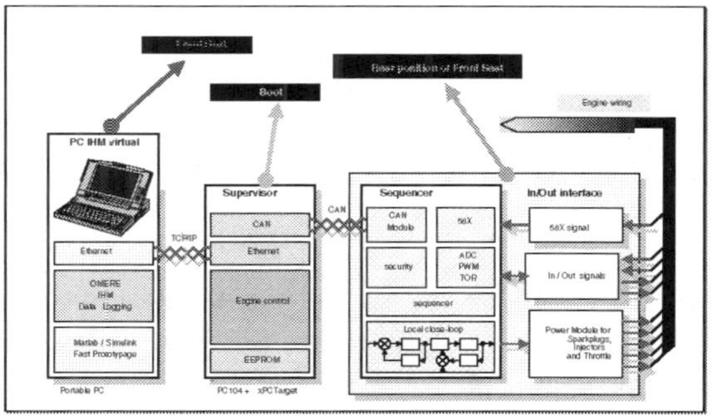

Fig. 5 Scheme of on-board system

As showed on **pictures 10** and **11**, gas tanks under floor and small EMS don't reduce luggage space available. For a small vehicle as the Smart one, maintaining acceptable luggage space is important. At the end of the optimisation tests, Supervisor computer (the biggest one) will be removed.

Picture 10 Boot of Demonstrator with on-board EMS

Picture 11 Supervisor and Sequencer computers

TESTS ON ENGINE BENCH: COMPARISON BETWEEN GASOLINE AND NATURAL GAS VERSIONS

Results obtained on gas are extremely promising compared to gasoline reference performances. At full load conditions, the potential of natural gas as a high octane number fuel is demonstrated: in **figure 6** spark advance timing and equivalence ratio are shown versus speed both for natural gas and gasoline settings.

From 2000 rev/min, gasoline spark advance must be reduced due to knock limit, and from 2500 rev/min enrichment is necessary to control exhaust temperature which drastically increase when spark advance delayed. The mean gap reaches 8 CA while equivalence ratio for Natural Gas operation is stoichiometric at all engine speeds.

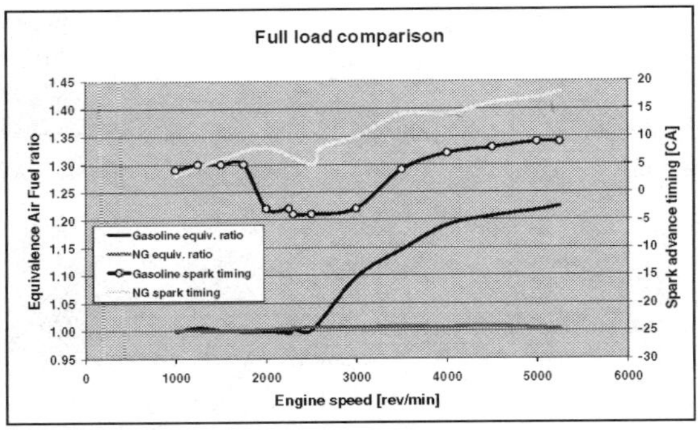

Fig. 6 Equivalence ratio and spark timing at full load (gasoline and Natural Gas)

One of the major drawbacks of gasoline settings is an increase of fuel consumption. **Figure 7** compares thermal efficiency at full load taking into account over-enrichment and spark timing for gasoline. At low engine speed, reduction of volumetric efficiency due to natural gas injection, affects brake means effective pressure. The turbocharger is able to compensate airflow reduction from 2600 rpm, so 300 rpm higher than for gasoline. New strategies under development for better management of the two wastegates should reduce this difference, but as of now, those performances with natural gas guarantee transient vehicle behaviour in accordance with client expectation.

One other solution to improve torque at low speed is electrical assistance. A light hybrid approach based on a 3 kW motor-alternator , should supply extra power for peak load condition and so totally compensates the penalty as showed on **figure 7**.

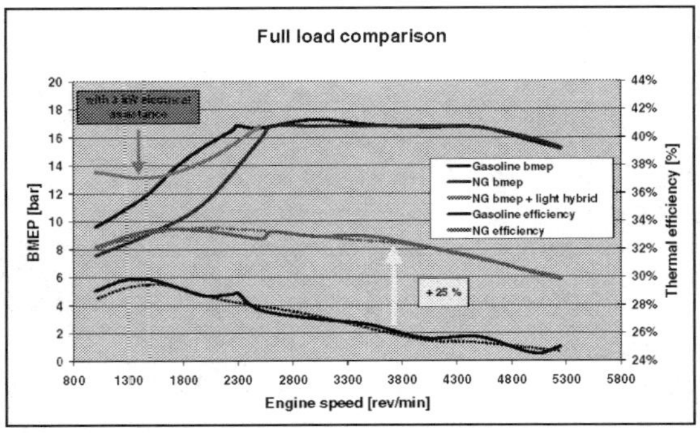

Fig. 7 BMEP and thermal efficiency at full load (gasoline and Natural Gas)

For the same load, thermal efficiency for natural gas is 6 points higher than gasoline ones so 25% more, which reveals the impact of settings and the favourable effect of the increased compression ratio. This benefit is noticeable for a wide engine speed range.

The second major effect is on exhaust emissions, both for NOx and Unburned Hydrocarbon as showed on **figure 8**. For stoichiometric mixture, NOx emissions with natural gas are lower due to combustion velocity and although spark advances timing delayed for gasoline. Unburned hydrocarbon emissions increase for gasoline as soon as mixture is rich, whereas stay low for natural gas.

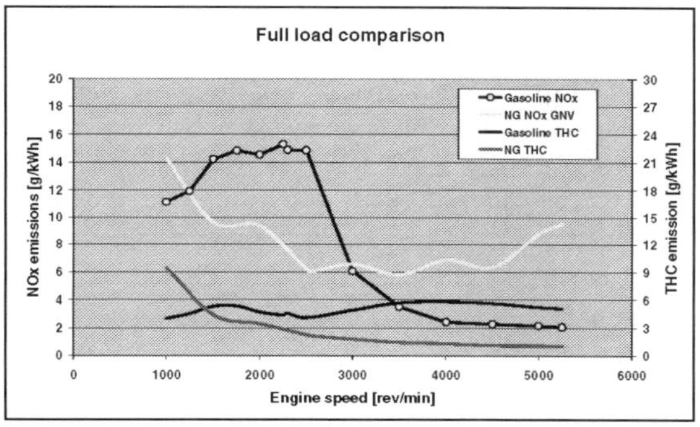

Fig. 8 NOx and UHC raw emissions at full load (gasoline and Natural Gas)

Results at part load conditions also distinguish natural gas engine. At 2000 rpm, **figure 9**, thermal efficiency for natural gas exceeds that of gasoline with a large margin from 2 bar to 8 bar (BMEP). For higher load, the variation is reduced as expected: indeed, for this engine speed, stoichiometric condition and spark advances settings close to MBT are used both for natural gas and gasoline.

At 4000 rpm, thermal efficiency is higher for natural gas with a difference that is enhanced as the load becomes higher (over enrichment and advance strategies for gasoline).

The difference between gasoline and natural gas specific fuel consumption calculated for the same fuel is plotted on **figure 10**. As seen, the gasoline engine consumption is noticeable higher at low and high loads. The lowest difference calculated is 7 g/kWh at 1500 rev/min and 7 bar (star mark). For MVEG cycle, some steady state running points are plotted (circle marks), mainly located in the highest difference area.

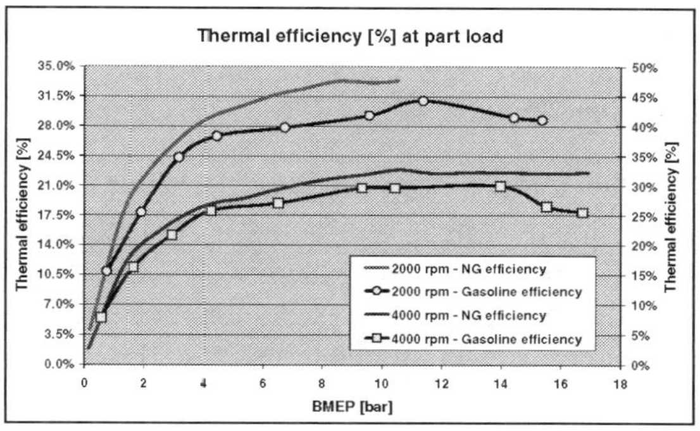

Fig. 9 Thermal efficiency at part load

Based on those results obtained on engine test bench, simulations performed with AMESim software reveal that the target in terms of CO_2 emissions (lower than 100 g/km on MVEG cycle) is attainable with confidence.

Calibration tests on a chassis dynamometer bench are in progress andreveal a big potential of this demonstrator vehicle to cope with the CO_2 target.

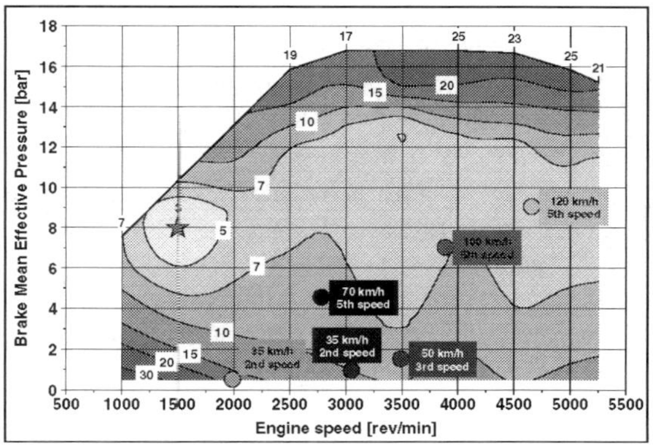

Fig. 10 BSFC difference between gasoline en NG engine

CONCLUSION

Partnership between Gaz de France and IFP with funding from ADEME confirms the great potential of natural gas when used as a dedicated fuel on a downsized engine. Simulations performed with engine bench results are promising; targets of CO_2 emissions less than 100 g/km are realistic but must be confirmed with chassis dynamometer tests.

The new gas tanks made of composite material are implemented so that the bootspace inside the vehicle is not affected. The extra weight is limited and no more than 45 kg.

The development of in-house on-board electronic management system represents a strategic stage to be able to assess the potential of our approach focusing an optimisation of vehicle behaviour while meeting EURO IV regulations. For this challenge, the installation of extra metallic light off catalysts should be effective.

Finally, the development of the NG Smart demonstrator is in line with the IFP strategy that considers that the natural gas pathway is one of the most efficient ways to reduce both greenhouse gas and pollutants emissions from transportation.

Acknowledgements. On behalf of that study, IFP would like to acknowledge GDF for its co-operation and ADEME for its interest and confidence in this project and for its financial support

REFERENCES

1. M. Heitzer, P. Langen, F. Kramer, O. Meyer, "Achieving low emissions and high engine efficiency with natural gas", IANGV Yokohama 2000
2. X. Sun, T. Wiedmann, S. Hussain, "Fuel management and exhaust emissions of light-and heavy-duty trucks operating on CNG and LPG ", SAE paper 971661
3. T. Suga, T. Muraishi, R. Bienenfeld, "Potential of a low-emission natural gas vehicle for the 21st century", IANGV Yokohama 2000
4. R. Tilagone, G. Monnier, A. Satre, Y. Lendresse, J. Personnaz, D. Cullerier, F. Turin, "Development of a lean-burn natural gas powered vehicle based on a direct-injection diesel engine", SAE paper 2000-01-1950.
5. Johansson, B., and Olsson, K., "Combustion Chambers for Natural Gas SI Engines Part I: Fluid Flow and Combustion", SAE paper 950469, 1995
6. Olsson, K., and Johansson, B., "Combustion Chambers for Natural Gas SI Engines Part I: Combustion and Emissions", SAE paper 950517, 1995
7. Tilagone R., Monnier G., Chaouche A., Baguelin Y., and De Chauveron S., "Development of a high efficiency, low emission SI-CNG bus engine", SAE paper 961080.- Detroit (USA) - 1996.
8. Gaz de France: « Combustibles gazeux et principes de la combustion (BT104) », January 1993
9. Union européenne: « bilan Auto-oil", published in 2000
10. Cornetti G: « Alternative fuels: the natural gas », ATA, vol. 52, Nov.-Dec. 1999
11. Cedigaz: « Natural gas in the world », 2002
12. J-C. Guibet: « Carburants et Moteurs, vol. 1et 2 » ; 1997
13. "Journal Officiel des Communautés Européennes", L107/10. « Directive 2001/27/CE de la Commission du 10 April 2001 »
14. Norm ISO 15403: « Gaz naturel - Désignation de la qualité de gaz naturel pour usage comme carburant comprimé pour véhicules », 2000
15. Norm ISO 6976: « Gaz naturel - Calcul du pouvoir calorifique, de la masse volumique, de la densité relative et de l'indice de Wobbe à partir de la composition », 1997
16. IANGV (International Association for Natural Gas Vehicles): « Emissions report, part 2: fuel properties », 31 Mars 2000
17. Buthker E.: « World Wide NGV standards and Regulations. Session 2A », IANGV, October 17 - 19, 2000
18. AFGNV (Association Française du gaz naturel pour véhicules.): « Rapport d'activité », 2001
19. P. Zanetta: « GNV: une première étape est franchie. », Pétrole et gaz informations , n° 1755, December 2001
20. Plouchard G.: « Le gaz naturel véhicules: les enjeux énergétiques et environnementaux de son développement durable, Rapport du Groupe de travail Energie Environnement de l'AFGNV », October 2002
21. Nylund N.O., Laurikko J., Ikonen M,: « Pathways for natural gas into advanced vehicles », 30.08.2002
22. Argonne National Lab. vol 1-2: « Well to Wheel Energy use and Green Gas Emissions of Advanced Fuel/Vehicle Systems - North American Analysis », June 2001
23. Weiss M.A., Heywood J.B., Drake E.M., Schafer A., AuYeung F.F.: « On the Road in 2020. A life -cycle analysis of new automobile technologies », MIT EL 00-003, October 2000
24. Choudhury R.: « Well to Wheel analysis of energy use and greenhouse gas emissions of advanced fuel/vehicle systems », May 2002
25. European commission: « Livre Blanc. La politique européenne des transports à l'horizon 2010. L'heure des choix », COM(2001)370, 20 September 2001

26. Ly H.: « Effects of Natural Gas Composition Variations on the Operation, Performance and Exhaust Emissions of natural Gas-Powered Vehicles », IANGV, December 2002
27. European commission: « Livre Vert: vers une stratégie européenne de sécurité d'approvisionnement énergétique », COM(2002)321 du 26.6.2002
28. J. ORSELLI: « Energies nouvelles pour l'automobile », 1992
29. P. ROUVEIROLLES: « Les spécifications des carburants GPL et GNV », EUROFORUM June 2000.
30. A. DOUAUD: « Quelles énergies pour l'automobile de demain », SIA November 2001.

INTRODUCTION OF THE USE OF HYDROGEN IN IC ENGINES

Which Fuels For Low CO_2 Engines?
© P. Duret, X. Montagne (Eds.) and Editions Technip, Paris, 2004, pp. 159 – 169
27 rue Ginoux, 75015 Paris

EMISSIONS REDUCTION THROUGH HYDROGEN ENRICHMENT

Jean J. Botti, M. James Grieve, Carlton E. Speck

Delphi Corporation, USA

ABSTRACT – *Hydrogen has unique properties in dilute combustion and catalytic reactions compared to other fuels. While the long-term vision for production and use of renewable hydrogen in transportation is theoretically attractive, significant economic and technical barriers remain in all areas. This paper will focus on two technologies which Delphi is developing to allow hydrogen to be produced and used on-board vehicles to capture the efficiency and emission control benefits of hydrogen in the short to mid-term. These technologies are:*

On-board reforming for emission control with internal combustion engines (ICE)

Solid Oxide Fuel Cell (SOFC) for auxiliary power, heat and hydrogen generation.

INTRODUCTION

Hydrogen as a fuel for automotive internal combustion engines has had a long history of study in academic and industry circles. Hydrogen has unique properties with respect to combustion and catalysis. Hydrogen has very wide flammability limits compared to gasoline. Thus, hydrogen-fueled engines can operate under extremely high levels of dilution (either ultra-lean or with high levels of EGR) – allowing engine operation with high efficiency and low emissions. Hydrogen also greatly reduces exhaust catalyst light-off temperatures - enabling enhanced aftertreatment performance. While direct fueling of vehicles with hydrogen can offer extremely high efficiency with internal combustion engine hybrids [1], Delphi sees this as a relatively long-term option due to economic, technical and political barriers. On-board reformation can be employed to achieve much of the benefit in emissions and efficiency without the short-term need for production, distribution and storage infrastructure for hydrogen. On-board reforming in emission control, fuel efficiency improvement and consumer functions would allow the sales volume and technical maturity of these hydrogen-utilizing systems to increase. This may result in an accelerated transition to direct hydrogen if, and when, the barriers are overcome.

Delphi Corporation has published extensively on its on-board reforming and SOFC programs. Integrating H_2 enrichment into an engine management system (EMS) enables SULEV emissions for larger / heavier passenger cars and trucks [2 – 8]. Compared to other SULEV approaches, this architecture permits a reduction in the size, precious metal loading and backpressure of the exhaust catalysts (which otherwise would be required for this extremely low emission level). Integration of an SOFC APU system allows notably higher efficiency for electrical generation compared to traditional on-board systems [9 – 12]. The SOFC APU will allow full electrification of accessories such as fans, pumps, blowers and compressors. This permits HVAC, lighting, entertainment, communication and battery charging functions to be independent of the ICE. While the use of reforming to enhance engine combustion and catalytic aftertreatment and the application of SOFC for efficient, on-board, engine independent electric power generation may be pursued as separate strategies, a substantial synergy may be exploited where the SOFC takes on the reforming function and becomes the source of syngas for the engine [13 – 15]. The very high marginal efficiency of syngas generation (due both to its generation as a co-product with electricity and heat and to the use of waste heat in the stack to drive a substantially endothermic reforming process) allows integration of the SOFC into the propulsion system. This allows higher efficiency engine combustion – with ultra-lean combustion modes at light loads and extreme EGR modes heavy loads.

The deployment of these technologies into mass production would be important steps toward an economic and sustainable low CO_2 transportation - based on either renewable hydrogen or bio-alcohols. On the one hand, the co-fueling of reformed hydrogen in ICE and the use of SOFC for auxiliary power and in hybrid mechanizations, will build a fleet of vehicles, which could use direct hydrogen efficiently and economically. This could bring important economies of scale for hydrogen consumption as a renewable hydrogen infrastructure is created. On the other hand, the use of renewable alcohols (ethanol and methanol) is a viable pathway for reducing CO_2 in transportation. On-board reforming and SOFC technologies are expected to be very efficient and attractive ways of using alcohols – with greater benefits than the current practice of using alcohols directly (blended with conventional fuels) or an inputs to bio-diesel production.

This paper will first present observations of emissions reduction from a conventional gasoline engine when the engine is fed hydrogen-enriched fuel during cycle 1 of the EPA Federal Test Procedure. Results with both "synthetic" (bottled) reformate and reformate generated by a prototype on-board gasoline reformer will be included and compared. Next, the potential for additional efficiency improvements and CO_2 reduction will be presented for systems utilizing alternate, renewable fuels. Finally, an advanced vision for achieving ultra-low emissions in hybrid systems employing a SOFC auxiliary power unit (APU) with an ICE and/or an electric drive will be presented.

REFORMER-ASSISTED INTERNAL COMBUSTION ENGINE OPERATION

One of Delphi's SULEV strategies under development is H_2 enrichment. This approach uses an on-board gasoline reformer to produce H_2-rich reformate for enabling very low vehicle emissions from a conventional spark-ignition engine. In the following, the fundamentals and advantages of enrichment are reviewed, and experimental results presented.

FUEL REFORMING FUNDAMENTALS

Hydrogen may be made on-board a vehicle by combining fuel and an oxidant under very fuel rich conditions in a catalytic reformer. This produces reformate (also called syngas) a gaseous fuel containing primarily H_2 and carbon monoxide (CO). This is shown, schematically, in Figure 1.

Fig. 1 Catalytic fuel reforming schematic.

The oxidant described above is usually air, but it may also be combustion products (water and carbon dioxide) as shown in Figure 2. Using air, the reaction is called catalytic partial oxidation reforming (CPOx). It is highly exothermic. This is convenient for small fast start-up devices – because the inputs can be delivered at low temperature and the heat released in the reforming reaction provides the sensible heat to keep the catalyst in an appropriate temperature range (about 1000 °C). Of course the downside to this is that the lower heating value of the syngas produced is, at best, 83% of the fuel delivered to the reformer – so there is an inherent loss in efficiency.

CPOx (Exothermic) Reforming (O/C = 1.000, 1000 °C, 83%)

$$C_7H_{12} + 3.5\,(O_2 + 3.776\,N_2) \rightarrow 6\,H_2 + 7\,CO + 13.22\,N_2 + Heat$$

Exhaust Gas (Endothermic) Recycle (O/C = 1.715, 750 °C, 126%)

$$C_7H_{12} + 9\,H_2O + 10.5\,CO_2 + Heat \rightarrow 10\,H_2 + 10\,CO + 5\,H_2O + 7.5\,CO_2$$

Fig. 2 Nominal Reactions for Gasoline for CPOx and Exhaust Gas Reforming.

On the other hand, using stoichiometric exhaust gas as the oxidant requires a large input of heat into the process (to preheat the reactants and to supply the endotherm of reaction). It also requires a significantly larger catalyst volume and excess oxidant in order to work at practical temperatures. The upside is that this endothermic process delivers syngas with a higher lower heating value, which is about 126% of the fuel delivered to the reformer. This increase is due to the input of waste heat available in the system. So, waste heat must be available (in both quantity and at the right temperature) in order to realize this gain in efficiency at the system level.

Practical reformer devices may use a combination of these processes to realize an appropriate efficiency, size and dynamic range for a real application. For example, using lean engine exhaust as an input to the reformer is a method to deliver heated air + stoichiometric exhaust together.

Fuel Properties

Table 1 compares selected combustion related properties of H_2 and CO to those of gasoline. It can been seen that properties of CO are often between those of hydrogen and gasoline – so reformate, which contains roughly equal amounts of H_2 and CO, will retain attractive properties compared to gasoline. These properties may be viewed for their impact on engine efficiency and exhaust emissions and aftertreatment. First, let us consider the impact on engine efficiency. The high flame speed and low ignition energy of H_2 and reformate relative to gasoline, permit the use of very lean or high EGR operation. With lean operation, the engine can have high thermal efficiency because of an increase in the ratio of specific heats of the charge closer to that for air (k=1.4). The higher flame speed of H_2 and reformate helps to maintain combustion as near to constant volume as possible even with lean mixtures. Pumping losses are lower with H_2 or reformate relative to gasoline due to the lower energy density of H_2 and CO (in addition to the high dilution tolerance). This contributes to better efficiency at part load. The high spontaneous ignition temperature of H_2 and CO (coupled with the high flame speed) permit the use of higher compression ratios to boost the thermal efficiency further [7]. Second, let us consider the impact on emissions. Ultra lean or highly dilute combustion results in low peak cycle temperatures and very low NOx emissions. Lean operation can also result in low CO emissions. H_2 or reformate fueled engines have low and non-reactive HC emissions (related to light hydrocarbons that survive the reforming and to lubricating oil).

Property	Gasoline	H2	Carbon Monoxide
Lean limit equivalence ratio in air	0.58	0.1	0.34
Flammability limits (volume %)	1.2-6	4-75	12-74
Laminar flame speed (m/s)	0.37-0.43	1.9 – 2.7	~ 0.4
Spontaneous ignition Temperature (°C)	280-400	574	609
Net energy density (MJ/m3)	202	10.3	12.6
Quench distance (mm)	~ 2.0	0.64	-
Minimum ignition energy (mJ)	0.24	0.02	-
Light off Temperature (°C) over platinum based catalyst	~ 270	~ 30	~ 220

Table 1. Comparative properties of gasoline, hydrogen and carbon monoxide.

Furthermore, smoke (particulate matter) should also be extremely low. When blended with gasoline, H_2 and reformate can help reduce the HC emissions by decreasing the quench distance for better flame propagation in crevices. In terms of aftertreatment, H_2 and reformate are attractive for catalyst heating and light off, because they can be more easily and complete reacted on a catalyst. Thus, a rapid catalyst light off with minimal emission breakthrough should be possible with reformate added to the exhaust during startup.

HYDROGEN ENRICHMENT STATEGY

Delphi's hydrogen enrichment strategy with an on-board reformer (see Figure 3) uses attributes of reformate to augment (not replace) a 3-way catalytic aftertreatment system. Briefly, the hydrogen enrichment strategy during cold start consists of:

- fueling with 100% reformate during cold start for near-zero engine-out HC and NOx;
- reacting reformate and air in the exhaust during cold start for rapid exhaust catalyst light off.

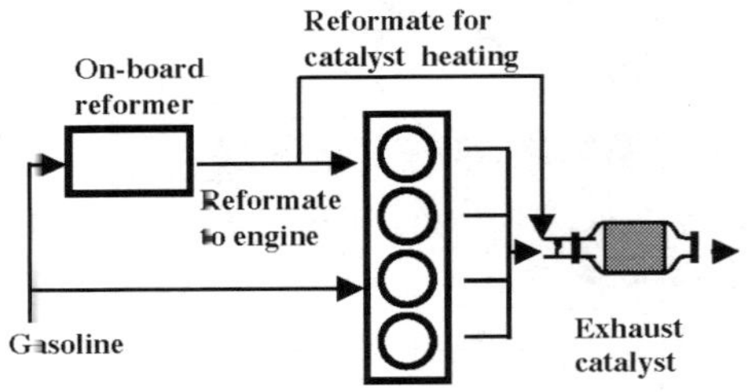

Fig 3 Schematic of the on-board reformer system.

EXTENDED HYDROGEN ENRICHMENT TESTS

Encouraging results in the emission reduction capability of H2 enrichment have been reported in earlier Delphi publications. More recently, Delphi has integrated a prototype reformer into a vehicle and demonstrated the benefits on a full US FTP emission test. The results are described in the following:

Baseline Testing

The test bed for the extended tests was a 2002 Chevrolet Trailblazer. Baseline emissions were evaluated using the Environmental Protection Agency (EPA) Federal Test Procedure (FTP) with gasoline fueling. For baseline testing the vehicle had its original engine management system (EMS) and its original equipment (OE) exhaust after-treatment catalytic converter. The converter had a relatively fresh catalyst with approximately 3000 miles. Figure 4 shows the cumulative (HC) and (NOx) emissions for the first 18 cycles of the FTP test. As indicated in Figure 4, the major fraction of the emissions occurs during the first cycle. The total HC emissions in this test were 63 mg/mi and the first cycle emissions were 54 mg/mi. This represents 86% of the total HC emissions emitted in the first cycle. The total NOx emissions in this test were 95 mg/mi and the first cycle emissions were 65 mg/mi. This represents 68% of the NOx emissions emitted in the first cycle. These HC and NOx emissions levels are well in excess of the SULEV standard of 10 mg/mi and 20 mg/mi for HC and NOx respectively. Since reducing emissions in cycle 1 has a disproportionate impact on total pollutants, the focus of this work was on reducing cycle 1 HC emissions using reformate fueling.

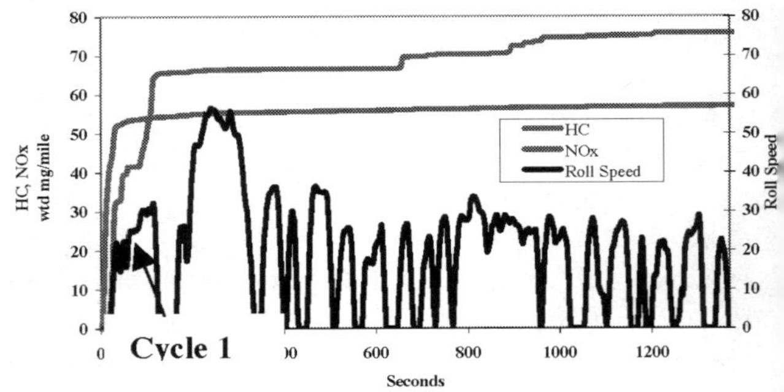

Fig. 4 Cumulative HC and NOx Tailpipe Emissions FTP.

Reformate Fueling

Reformate testing was conducted in two parts. Firstly, the system was set-up so that the vehicle could be fueled with bottled reformate. The bottled reformate was a blend of 21% hydrogen, 23.5% carbon monoxide and the balance was nitrogen. This simple blend simulates the mole fractions of the main components of reformate made from the fuel reformer system. In order to set-up for reformate fueling the vehicle engine management system (EMS) had to be modified so that there was communication between the vehicle and the reformate controller. Figure 5 shows a plot of the roll speed during the first cycle of the FTP test. During reformate tests, reformate is fueled to the engine at crank and through approximately 120 seconds of the test. Reformate fueling stops at the beginning of the final deceleration in the first cycle of the test. There are portions of the first cycle when the reformate flow is unable to supply the required power. During those portions, gasoline fuel supplements the reformate fueling. After the first cycle the vehicle is fueled with gasoline alone for the rest of the test.

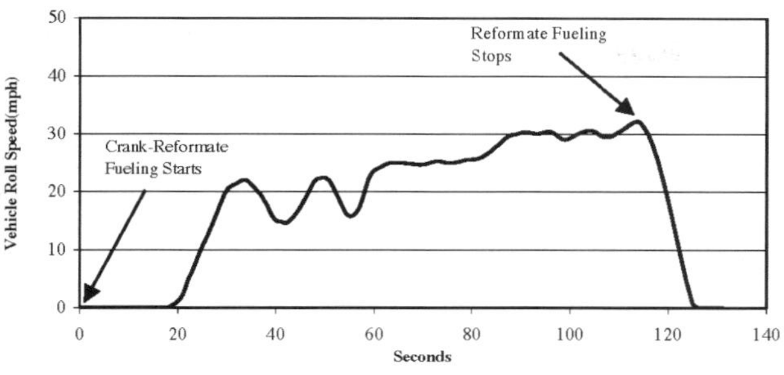

Fig. 5 FTP First Cycle Reformate Fueling Strategy.

The second part of the reformate testing was conducted with a gasoline fuel reformer system making reformate on-line using partial oxidation catalytic reforming. In this system gasoline and air are supplied to the reformer system where a partial oxidation catalytic reformer converts the fuel to reformate. During steady state conditions the hydrogen and carbon monoxide concentration exiting the reformer were approximately 22 and 23%, respectively. Figure 6 shows a schematic of the components associated with the fuel reformer system.

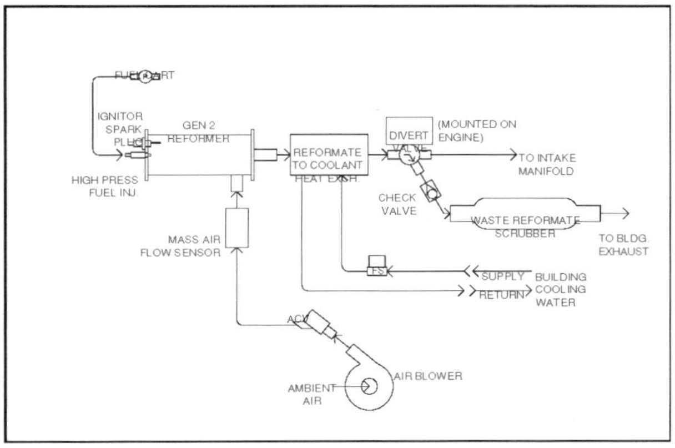

Fig. 6 Prototype Gasoline Fuel Reformer and Test Configuration.

When the system was tested with the gasoline fuel reformer system the reformer was started and allowed to reach steady state reformate production prior to starting the emissions test. During this stabilization period, reformate is diverted to the waste reformate scrubber. At the start of the test the divert valve is activated and reformate flows to the intake during engine crank. Figure 7 shows a picture of the vehicle and associated reformer system hardware.

Fig. 7 Vehicle and Reformer System Hardware.

After the baseline testing was completed emissions tests from three configurations were evaluated as part of the confirmation testing. Table 2 gives a brief description of these test conditions. For these confirmation tests the EMS was modified to enable reformate fueling and the exhaust after treatment system was replaced with a catalytic converter that had been aged for the equivalent of 100,000 miles. This converter was moved approximately twelve inches closer to the engine to allow for faster warm-up.

Configuration #	Description
1	Delphi EMS and Calibration with close coupled aged converter (100k mi equivalent), gasoline fueling
2	Delphi EMS and Calibration with close coupled aged converter (100k), bottled reformate fuel
3	Delphi EMS and Calibration, close coupled aged converter (100k), reformate fuel made from reformer

Table 2. Confirmation Test Conditions

Figure 8 shows plots of the total HC emissions versus time for the first cycle of the FTP test using the 2002 Chevrolet Trailblazer. Total HC emissions during the first cycle of the test were reduced from 54 mg/mile to 8 mg/mile using bottled reformate fueling during this cycle. HC's are emitted during the first cycle even during reformate fueling. As indicated previously, gasoline fuel supplements the reformate fueling starting at the first acceleration in the test cycle. This accounts for some of the additional HC's emitted in this cycle with reformate fueling. When using reformate generated by the reformer the emissions were reduced to 9 mg/mi. This is higher than the result with bottled reformate. This is due to the fact that the reformer produces other constituents including some HC's that are not present in the bottled reformate. As a result, bottled reformate is more effective at reducing overall HC emissions.

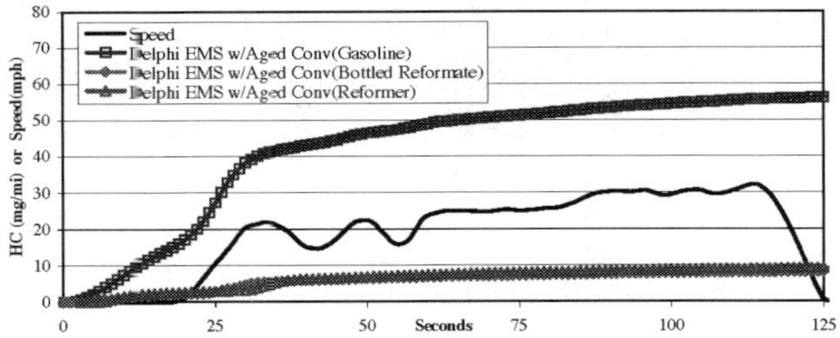

Fig. 8 Weighted Tailpipe HC Emissions – First Cycle of the FTP Test

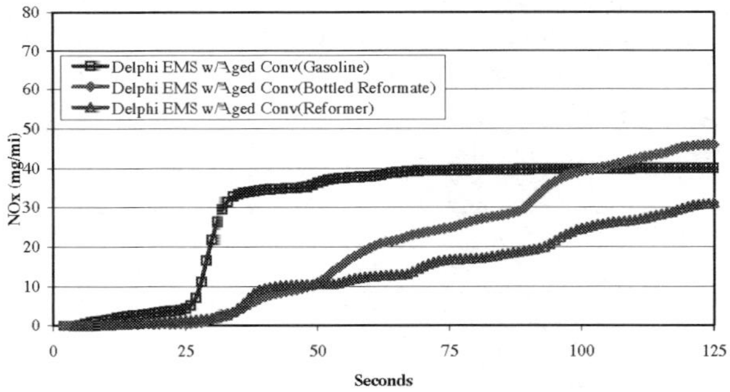

Fig. 9 Weighted Tailpipe NOx Emissions - First Cycle ofthe FTP Test

As shown in Figure 9, reformate fueling does not have as large an impact on NOx emissions in the first cycle. Total NOx emissions during the first cycle of the test were reduced from 38 mg/mi to 30 mg/mi using reformate supplied by the reformer. With bottled reformate the NOx emissions at the end of cycle 1 was 45 g/mi. This is higher than the test with the reformer and even higher than the test with gasoline fueling. Using bottled reformate, the NOx emissions showed lower NOx early into the cycle; however, there was a continuous increase in NOx emissions through the remainder of the cycle resulting in 45 mg/mi at the end of cycle 1. For the reformer test, the tailpipe NOx emissions at the end of the first cycle were the lowest. Unlike bottled reformate, reformate produced by the reformer contains some CO_2 and water as by-products. These serve as added diluents in the engine. Figure 10 shows the engine out NOx emissions for the tests. These added diluents might explain why the engine-out NOx levels were lowest with the reformer. Like the tests with bottled reformate, the tests with the reformer show a positive slope in tailpipe NOx throughout the first cycle. This indicates lean fueling for the reformer tests leading to incomplete NOx conversion across the exhaust catalyst. With reformate fueling, especially when the fueling is supplemented with gasoline, there is considerable work yet to do to improve air/fuel control.

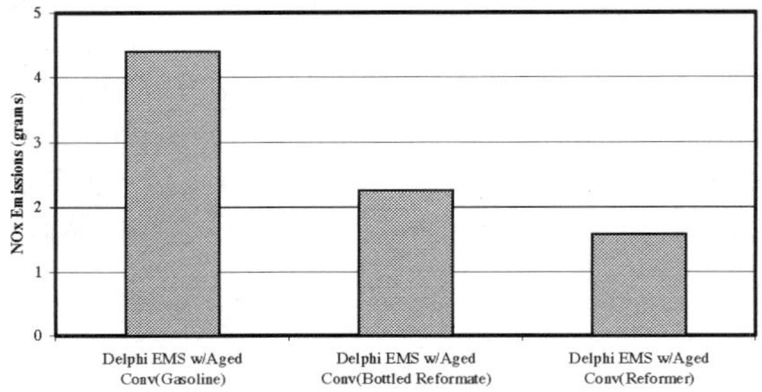

Fig.10 Engine Out NOx Emissions - First Cycle of the FTP Test

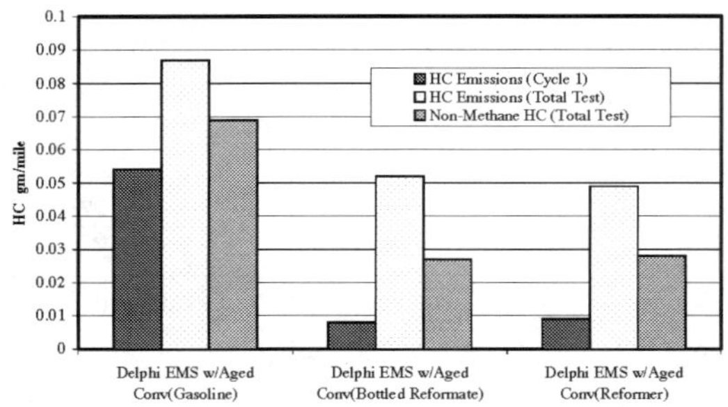

Fig. 11 Tailpipe HC Emissions Summary

Figures 11 and 12 show the HC and NOx emissions summaries for the test conditions that were evaluated. Cycle one HC emissions are reduced from 54 mg/mi with gasoline fueling to 8 mg/mi with reformate fueling with bottled reformate. With the reformer the first cycle HC emissions were 9 mg/mi. HC emissions for the total test are reduced from

87 mg/mi to 52 mg/mi using bottled reformate. With the reformer the total test HC emissions were 49 mg/mi. Non-methane HC emissions for the total test were 27 and 28 mg/mi for the bottled reformate and the reformer tests respectively.

The NOx reductions observed in Figure 12 for the total test for the reformate fueling conditions are clearly related to conditions beyond cycle 1 and are not related to reformate fueling. Additional work, focusing on calibration and air/fuel control for NOx emissions will be required to address these issues.

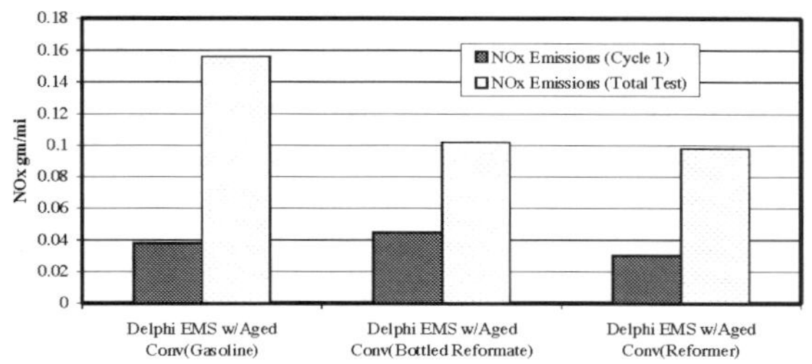

Fig. 12 Tailpipe NOx Emissions Summary

EFFICIENCY AND EMISSIONS IMPROVEMENTS WITH RENEWABLE FUELS

The primary focus of a CPOx-based H_2-enrichment system is start-up emission control and accelerated catalyst and engine warm-up. The loss in efficiency of CPOx reforming is mitigated by the ability to run the engine lean and essentially unthrottled and by the ability to recover by-product heat (to speed the passenger compartment and engine warm-up). In comparison, traditional methods of accelerated catalyst heating (for example spark retard and rich fueling with secondary air) tend to cause a small loss in fuel efficiency over the drive cycle – so some compromise to fuel economy is expected to meet emissions during a cold start. H_2-enrichment is likely to be comparable or slightly better than traditional methods.

During warmed-up conditions, where the engine is normally constrained to stoichiometric operation (and EGR is the diluent), the small gain in efficiency allowed by higher EGR rates (enabled by H2 enrichment) is normally less than the loss in fuel lower heating value due to CPOx reforming. Thus, it would be most desirable to shift towards a more endothermic reforming process under warmed-up conditions. If fully endothermic reforming with exhaust gas is used (as showed in the second equation of Figure 2), the gains in engine efficiency (with higher EGR) would be compounded by the gains in fuel lower heating value for the fraction of engine fuel which is reformed. Unfortunately, with gasoline (and even more so with diesel engines) the exhaust gas is often too low in temperature to allow endothermic reforming with a practical reformer size. So fully endothermic exhaust gas reforming will be limited to very high engine loads and stoichiometric or near-stoichiometric engine operation for these fuels.

Substitution of alcohol fuels (either methanol or ethanol) as the reformer input allows fully endothermic reforming at as little as 250°C. This temperature is almost always available in the exhaust of a gasoline engine. A flex-fueled (alcohol/gasoline) system could provide the SULEV cold-start advantages with either fuel and enhanced fuel efficiency when using alcohol blends. As alcohols can be made economically from renewable (biomass) sources, the CO_2 benefits of blending of alcohol fuels into conventional road fuels is well established. But the ability of efficiently reforming the alcohol to enhance the engine efficiency compounds the CO_2 reduction benefit.

DELPHI'S POWERTRAIN VISION BACKGROUND

Delphi has been doing R&D on fuel cells since 1990. A solid oxide fuel cell (SOFC) program began in 1999 with a customer-linked program to develop a 3-5 kW auxiliary power unit (APU) product for luxury passenger cars running on gasoline. The first proof of concept was completed in late 2000. While other developers have shown Hydrogen PEM APU concepts for heavy-duty trucks, SOFC has significant advantages by being able to operate on gasoline, diesel or natural gas as well as (in the long term) hydrogen. Fuel is still reformed to syngas (a mixture of hydrogen and carbon monoxide) but the reformate requirements are much less severe than for PEM. The simple, 1-step reforming process previously described for H_2 enrichment can be used. For system start-up the CPOx reforming process is attractive, but it is possible to migrate to a fully endothermic reforming process during warmed-up operation of the SOFC. This is because the fuel cell runs at constant temperature of about 800 °C. This is above the temperature required for endothermic reforming of conventional fuels.

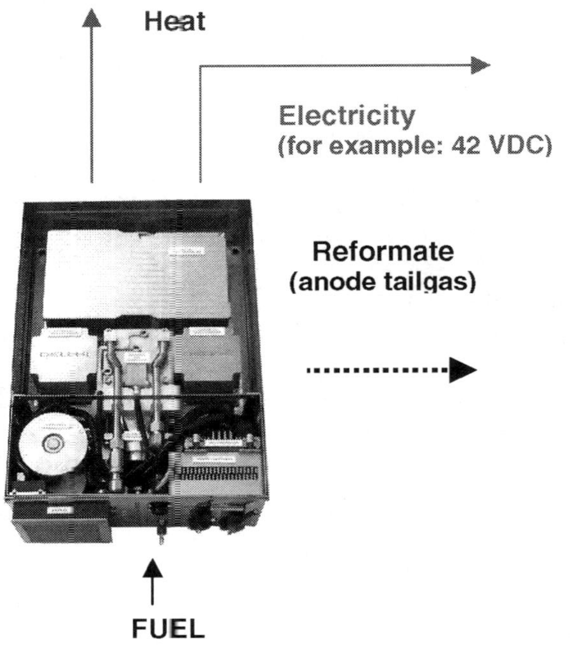

Fig. 13 Prototype SOFC APU, showing 3 potential products

Thus, it is possible to use waste heat from the stack to drive the endotherm of reaction for the fuel reformer. Key to SOFC's efficiency and future applications is the fact that the SOFC inherently makes 3 products: electricity, hydrogen-rich syngas and high grade waste heat (see Figure 13). This has been part of Delphi's vision for SOFC since the beginning of our program. A well thermally integrated system can make some syngas as a by-product at high electrical loads at essentially zero marginal cost (because there is extra waste heat generated at stack temperature compared to the internal needs for reformate in the SOFC system). Additional syngas can be easily made (to match the requirements of the engine or aftertreatment system) by modulating the instantaneous power of the SOFC or by increasing the throughput of the reformer. Looking only at electricity, the efficiency of SOFC has clear advantages over conventional engine driven electric generators. But taking credit for the by-product syngas and, again, for its compounding benefit to engine efficiency makes SOFC an even more compelling alternative. In addition to the electric power and by-product syngas the exhaust of the SOFC still is relatively hot and may be very valuable for premium vehicle heating fucntions. Thus, while today's research focuses on the SOFC as an APU for powering electrical accessories, our vision is to utilize the SOFC APU in a new family of hybrids that achieve additional gains in performance, fuel economy, and emissions.

The SOFC APU is a practical first step for introduction of fuel cells in transportation. But its linkage to luxury functions, (like electric air conditioning) instead of propulsion and its use of conventional fuels, instead of direct hydrogen may make it seem like a timid first step. However, a variety of future integration strategies for combined cycle SOFC/ICE, SOFC hybrid vehicles and, of course, the SOFC/Li-Ion range extender EV offer a wide spectrum of future "green" applications. In addition, the SOFC is capable of burning hydrogen or other renewable fuels efficiently. So low or zero net CO_2 vehicle systems can be conceived using SOFC with hydrogen or bio-alcohols – backed up by compatibility with conventional transportation fuels.

SYSTEM MODEL AND PREDICTIONS

<u>SOFC / ICE hybrid vehicles</u>

Figure 14 is a schematic that highlights Delphi's vision for an advanced propulsion system with inherently low emissions. As drawn it models a full series/parallel hybrid, but by deleting or scaling the emphasis of various components it can represent a soft or mild parallel hybrid or an electric vehicle in either a full series or range extender mode. The

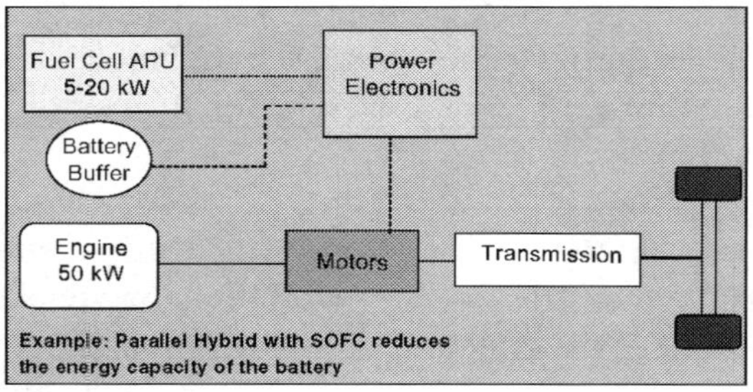

Fig. 14 SOFC APU Plus Full Hybrid

example hints at the predictions of modeling that replacing or substantially downsizing the battery in a hybrid vehicle with a SOFC can significantly reduce weight and space resulting in improved performance and fuel economy. A two-seater vehicle could become a four-seat one. Some typical predictions for specific configurations follow.

Preliminary simulation results predict that there is potential for excellent improvement in fuel economy when a SOFC APU is added to a full hybrid (see Figure 15). For example, gains of up to 60% can be expected if the unit is sized to allow full electric operation over the NYCC drive cycle (a severe, very low speed cycle). Optimization of the powertrain control strategy will also reduce emissions and fuel consumption in at least two ways:

(1) an EV mode can be used where the ICE would otherwise be inefficient

(2) by-product syngas from the SOFC could be used to fuel the ICE.

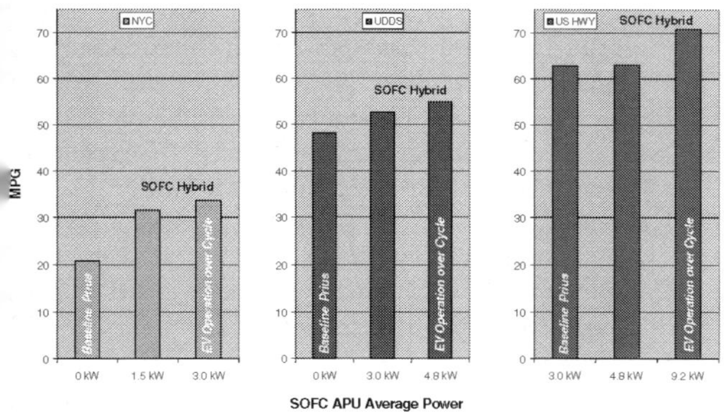

Fig. 15 Results of simulation of SOFC hybrids applied to a hybrid vehicle similar to the first generation Toyota Prius

Furthermore, this SOFC reformate could significantly reduce cold start emissions of the engine as demonstrated earlier in this paper. Rather than completely replacing the battery, near term gains can be made by employing a small (3 – 5 kW) SOFC in conjunction with a higher power (35 kW) lithium battery or supercapacitor. This would reduce both the mass and volume of the battery and permit the elimination of the alternator. Finally, the high-temperature exhaust of the SOFC could provide high-efficiency heat to the vehicle's cabin during periods of cold weather, thereby eliminating the need to periodically run the engine to provide this function.

SOFC EV Range Extender

By adding an SOFC as an on-board battery charger to a standard plug-in electric vehicle, one can substantially increase the range and utility of the EV. Figure 16 shows typical predictions when it is assumed that a 10-kW SOFC APU is employed and a 100-kg lithium battery replaces the NiMH battery of the GM EV1. The model also assumes that the majority of the trips (85% of the trips and 60% of the miles) are in a pure EV mode where the energy was derived from the power grid, thereby being petroleum free (low net CO_2 emissions). It is predicted that through substantially reduced weight and increased on-board energy storage, the range is more than tripled. Fuel economy is improved by a factor of 11, compared with the PNGV goal of 3. In addition, 9500 W of high-quality heat is available for cabin heating or cooling. Another advantage is that the fuel cell can be run-on after the trip, thereby allowing the vehicle batteries to be recharged independently of the grid if required. In addition, the batteries can be kept warm in cold weather, thereby insuring power density and range. This configuration is one that could compete with a PEM fuel cell EV as a long-term, environmentally sustainable powertrain system. It is inherently near zero emissions, and is predicted to be 50% more efficient than the next best competitive system, a diesel engine APU series EV.

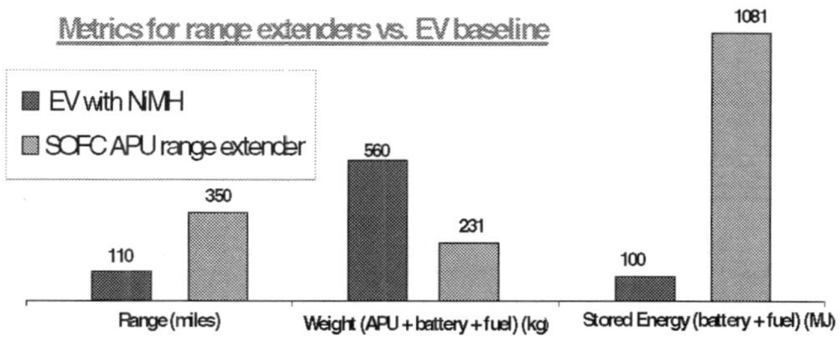

Fig. 16 SOFC EV Range Extender Modeling Results

CONCLUSION

1. Hydrogen has well known efficiency and emissions advantages as a fuel for fuel cell and ICE powertrains. While renewable hydrogen is

an attractive long-term scenario as a significant transportation fuel, the generation and use of hydrogen on-board the vehicle brings numerous advantages for emissions, efficiency, economics and utility.

2. Catalytic partial oxidation reforming is attractive for fast start-up systems used for SULEV emission control functions for gasoline engines.

3. Exhaust gas reforming can be used to increase the efficiency of a super low emissions gasoline powertrain by recovering exhaust waste heat (in an endothermic reforming process) and by increasing the dilute limit of the engine.

4. Renewable alcohol fuels have significant advantages in their ability to be endothermically reformed at low temperature. Thus, a strong synergy exists between the use of on-board reforming technology for SULEV ICE and the deployment of alcohol fuel blends.

5. Solid oxide fuel cell systems are another user of on-board reformer technology. In addition to the high fuel-to-electric efficiency of the fuel cell itself, waste heat from the SOFC stack can be used in an endothermic reforming process. This allows the export of by-product syngas for use in ICE combustion and aftertreatment.

6. Relatively small SOFC systems can be used in a variety of hybrid electric vehicle configurations to improve efficiency, emissions and electric propulsion capability of the vehicle.

7. In addition to the ability of SOFC hybrids to use conventional fuels (gasoline and diesel), they are well suited to run on hydrogen or alcohols when these become economically available from renewable sources.

8. SOFC Range extender EVs also have the ability to use renewable or low CO_2 electricity from the grid to supply a substantial fraction of vehicle miles traveled for personal transportation. This type of vehicle is also compatible with conventional or renewable fuels. The economic incentive for reducing transportation fuel consumption will be the relative retail cost of electricity compared to the fuel, which would otherwise be used.

REFERENCES

1. Aceves, S. Smith, R., "Lean-Burn Hydrogen Spark Ignited Engines: The Mechanical Equivalent to the Fuel Cell," ASME Fall 1996 ICE Conference.
2. Grieve, M. J., "Hydrogen Leveraging for Near Zero-Emission Vehicles with Conventional or Mild Hybrid Powertrains and Gasoline Fuel," Global Powertrain Congress, Detroit, MI, October, 1998.
3. Kirwan, J. E., Quader, A. A., and Grieve, M. J. "Advanced Engine Management Using On-Board Gasoline Partial Oxidation Reforming for Meeting Super-ULEV (SULEV) Emissions Standards," SAE Paper 1999-01-2927, August 1999.
4. Grieve, M. J., Kirwan, J. E. and Quader, A. A., "Integration of a Small On-board Reformer to a Conventional Gasoline Internal Combustion Engine System to Enable a Practical and Robust Nearly-zero Emission Vehicle," Global Powertrain Congress, Stuttgart, Germany, October 1999.
5. Kirwan, J. E., Quader, A. A., and Grieve, M. J., "An On-Board Gasoline Reforming System for Meeting SULEV Emissions Requirements in a Spark-Ignition Engine," Global Powertrain Congress, Detroit, MI, June 2000.
6. Kirwan, J. E., Quader, A. A., and Grieve, M. J., "Development of a Fast Start-up On-Board Gasoline Reformer for Near Zero Emissions in Spark-Ignition Engines," Aachen Colloquium on Automobile and Engine Technology (Aachener Kolloquium Fahrzeung- und Motorentechnik), October, 2001.
7. Kirwan, J. E., Quader, A. A., and Grieve, M. J., "Fast Start-up On-Board Gasoline Reformer for Near Zero Emissions in Spark-Ignition Engines," SAE Paper 2002-01-1011, March 2002.
8. Kirwan, J. E.; Quader, A. A.; Grieve, M. J.; Rahmoeller, K. M.; Wadumesthrige, K.; and Fisher, G. B.; "On-Board Reformer Development for Low Emissions in Spark-Ignition Engines," Global Powertrain Congress, Ann Arbor, MI, September, 2002.
9. Zizelman, J., Botti, J., Tachtler, J., Stobl, W., "Solid Oxide Fuel Cell Auxiliary Power Unit – A Paradigm Shift in Electric Supply for Transportation," Convergence 2000, SAE 2000-01-C070, Detroit, MI October 2000
10. Zizelman, J., Mukerjee, S., Shaffer, S., "Solid Oxide Fuel Cell Auxiliary Power Unit – A Development Update," SAE 2002-01-0411
11. Keegan, K. Khaleel, et al., "Analysis of a Planar Solid Oxide Fuel Cell based Automotive Auxiliary Power Unit," SAE 2002-01-0413
12. Mukerjee, S., Chick, L. et al., "Development of a Solid Oxide Fuel Cell Stack by Delphi and Battelle,"based Automotive Auxiliary Power Unit,"

8th International SOFC Symposium, Paris, France 2003
13. Botti, J. J., "The Revolution Through Evolution: Delphi Solid Oxide Fuel Cell for APU and Hydrogen Reformation," Powertrain International Magazine, November, 2002.
14. Quader, A. A., Kirwan, J. E., Grieve, M. J., "Engine Performance and Emissions near the Dilute Limit with Hydrogen Enrichment using an On-Board Reforming Strategy," SAE Paper 2003-01-1356
15. Botti, J. J., Grieve, M. J., Miller, C. E., "Power Generation System and Method," US Patent 6,230,494 B1

Contact.
Jean J. Botti, jean.botti@delphi.com

Which Fuels For Low CO_2 Engines?
© P. Duret, X. Montagne (Eds.) and Editions Technip, Paris, 2004, pp. 171 – 180
27 rue Ginoux, 75015 Paris

REDUCING HYDROCARBONS EMISSIONS WITH DIRECT INJECTION OF NATURAL GAS IN A HYDROGEN ASSISTED JET IGNITION ENGINE

Deanna E. Wang, Harry C. Watson
The University of Melbourne

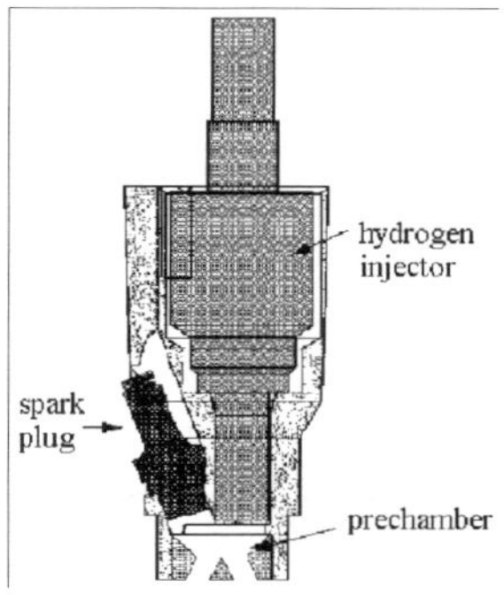

Fig. 1 Hydrogen assisted jet ignition unit

ABSTRACT – *This paper presents an investigation of gaseous direct injection strategies on engine performance and exhaust emissions (in particular, hydrocarbons emissions). The work was undertaken on a natural gas fuelled engine. Hydrogen assisted jet ignition was employed to allow ultra-lean (λ=2) operation. The use of natural gas resulted in low CO_2 output and the ultra-lean air/fuel ratio reduced the CO and NO exhaust emissions to negligible levels. Experiments showed that the engine-out HC level can be reduced by DI to as low as 67% of the baseline homogeneous value with mid-range end of injection times. Modelling of the injection, mixing and combustion processes was also carried out. These results indicate that the decrease in HC when DI is used is mainly due to a decrease in wall layer hydrocarbons.*

INTRODUCTION

Currently there is much interest in natural gas (NG) due to its clean combustion characteristics and world-wide availability. The low sulphur content of NG results in negligible exhaust SO_x, and the high H:C molecular ratio is responsible for minimal particulate emissions and less CO_2 than petrol or diesel engines. NG is also highly resistant to autoignition allowing increased compression ratios and turbocharging boost pressures. However, NG also possesses several disadvantages. Aside from practical issues associated with the low volumetric energy density, NG engines may also be significant greenhouse gas contributors even accounting for the low CO_2 exhaust levels. This is due to the fact that methane (50-97% by volume of NG) is a greenhouse gas 21 times as potent as CO_2 as determined by the International Climate Change Protocol. Consequently, the HC emission level must be carefully controlled for a NG engine to be viable. Furthermore, methane is quite unreactive in comparison to other hydrocarbon fuels. This results in a higher level of unburned hydrocarbons and a lower laminar flame speed (which increases the combustion duration and therefore decreases efficiency).

Hydrogen assisted jet ignition (HAJI) is a combustion enhancement ignition strategy developed by The University of Melbourne. It replaces the spark plug in a conventional SI engine, and consists of a hydrogen injector, a small prechamber and a spark plug, see Fig. 1. Prior to ignition, hydrogen is injected into the prechamber to form a rich mixture. The wide flammability limits of

H_2 guarantee stable and repeatable ignition. Increasing pressure forces the burning hydrogen gas through a number of orifices into the main chamber where it ignites the main chamber fuel (in this case, natural gas). Approximately 1.5% of the total energy is supplied by hydrogen. The presence of exceedingly reactive partial products and high levels of turbulence result in an ignition energy at least two orders of magnitude higher than that of a spark [1].

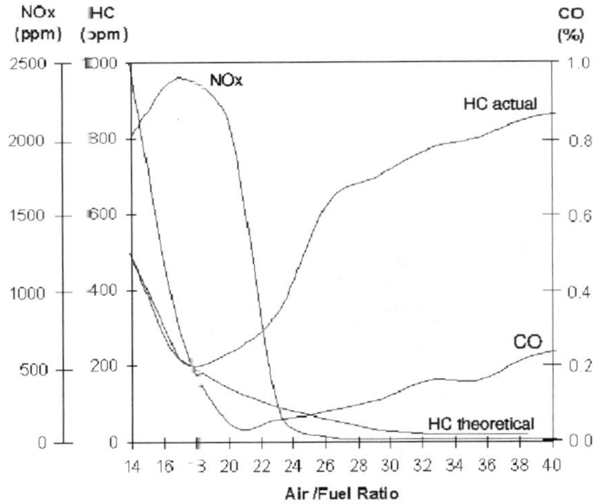

Fig. 2 Variation of emissions with relative air/fuel ratio in a second generation HAJI-equipped four-stroke engine

It has been shown that a HAJI-equipped engine may run as lean as λ=5 [2]. Operating an engine in the ultra-lean region results in ultra-low NO_x and CO levels without the use of complex and expensive aftertreatment. Furthermore, the work produced may be reduced to no-load quantities by simply increasing the air/fuel ratio as there is effectively no lean limit nor the need for a 3-way catalytic converter. This results in reduced pumping losses estimated to improve efficiency by around 15-20% [1]. Figure 2 shows some baseline results for a HAJI-equipped petrol engine. Most notable is the strong positive trend of HC emissions with air/fuel ratio for λ>1.6. Research by Lawrence indicates that quench and crevice are the main sources of unburned HC in lean burn HAJI-assisted combustion. Furthermore, Lawrence determined that the contribution of quench sources to engine-out HC increases in proportion as the mixture becomes leaner [3]. This is attributed to the strong influence of peak temperature on wall quench thickness and consequently quench HC contribution.

Direct injection (DI) has emerged as the most promising new engine technology with the ability to improve fuel economy and decrease emissions (as evidenced by an open debate on the future of combustion in piston engines held in 1997 at the SAE Congress [4]). This is due primarily to the ability of DI to stratify the charge. Developers aim to keep fuel away from regions which do not burn well (e.g. near walls) whilst increasing the amount of fuel near selected sites such as the spark plug. This facilitates a reduced heat loss to the wall as well as a decrease in wall wetting, wall quench and crevice HC sources. Additionally, a higher proportion of fuel near the spark can significantly extend the lean limit of an engine. Direct injection also eliminates the transport lag of the fuel, and can bring about incylinder charge cooling for liquid fuels. However, a richer region near the spark will result in localised high peak temperatures and higher levels of NO_x.

As discussed earlier, natural gas holds many emissions advantages over other hydrocarbon fuels. However, there are also significant drawbacks including the low flame speed and high global warming factor of methane. Since HAJI has been shown to greatly increase the speed, stability and reliability of combustion, it was an obvious progression to use natural gas as the main fuel in conjunction with HAJI ignition. By operating in the ultra-lean region, low levels of NO_x and CO were expected. However, it was imperative that engine-out HC levels be controlled for the system to be viable. Hence, direct injection was introduced with the intention of reducing wall layer HC (ascertained by Lawrence to be the main source of unburned hydrocarbons). It was anticipated that the combination of natural gas, direct injection and ultra-lean burn facilitated by HAJI would result in a spark ignition engine with ultra low engine-out emissions. Some preliminary work involving emissions measurements and incylinder flame visualisation was carried out (reported in [5]). The results from this research were positive enough to justify a more detailed investigation of direct injection of natural gas with HAJI.

OBJECTIVES

The purpose of this study was to investigate the effects of various natural gas direct injection strategies on engine efficiency and NO, CO and HC emissions. In particular, the influence of three injection parameters; start of injection time, end of injection time and injection duration, on engine performance was explored.

EXPERIMENTATION

The long term aim of this study is to achieve a 'top hat' fuel stratification pattern in which the fuel/air ratio is zero near the walls, and uniform in the middle of the cylinder (see Fig. 3). It is expected that a centralised air/fuel mixture will increase engine efficiency and eliminate the contribution of wall sources to the overall exhaust hydrocarbons level. This paper presents work undertaken with a single injector. It is envisaged that dual opposed gaseous injection will be investigated in the future.

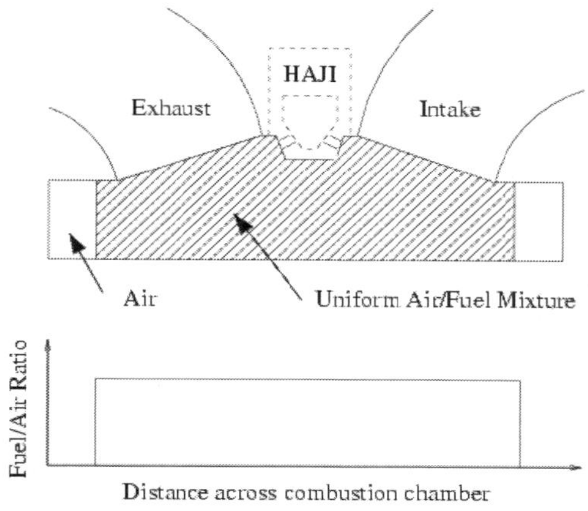

Fig. 3 Schematic of 'top-hat' fuel stratification

APPARATUS

A summary of the engine test rig is given in Table 1. Two Siemens DEKA DI (preproduction) injectors were used — one in the HAJI unit and a second as the natural gas injector. The injectors are single hole, inward opening swirl injectors which produce a hollow cone spray. The NG injector was located in the cylinder head next to the intake valve (see Fig. 4). Previous investigations by Dober determined that an improvement in efficiency and emissions could be obtained by increasing the swirl ratio to 1.2 in a HAJI direct injection butane engine [6]. Hence, this level of swirl was used in the experiments reported in this paper, in the anticipation that the trend of performance with swirl would be similar in a gaseous fuelled engine. The increase in swirl level was achieved by blocking the intake port next to the injector (shown hatched in Fig. 4).

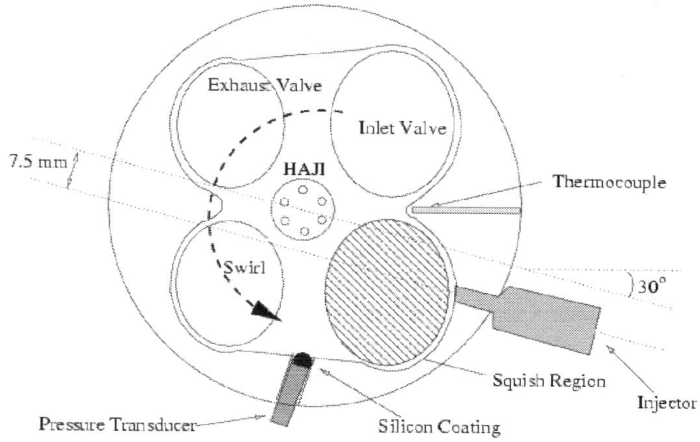

Fig. 4 Schematic of the injector location in the head

Table 2 details the control/data acquisition system and the instrumentation used. It should be noted that a Cambustion HFR400 Fast Flame Ionisation Detector (Fast FID) was used to calibrate the sensitivity of the NDIR exhaust gas analyser to natural gas exhaust hydrocarbons.

Table 1 Specifications of the engine test rig

Short engine:	
Manufacturer	University of Melbourne
Capacity	410 cm^3
Bore	89 mm
Stroke	66 mm
Cylinder head:	
Manufacturer	Toyota Australia
Designation	Corolla 4A/7A
Intended bore	83 mm
Cam/valve	4 valve/cylinder, DOHC
Clearance volume	44.1 cm^3
IVO	0 °ATDC
IVC	225 °ATDC
EVO	495 °ATDC
EVC	0 °ATDC
HAJI unit:	
H_2 start of injection	264 °ATDC
H_2 injection duration	4.0 ms
H_2 flow rate	2 mg/s
Prechamber volume	1.3 cm^3
Orifices	6, 1.25 mm diameter
System:	
Compression ratio	10.3:1
Swirl ratio	1.2:1
Cooling	none
Dynamometer	CD, variable speed

Table 2 Control/data acquisition system and instrumentation

Control/DA hardware	486 DX2 66 MHz PC PC30D A/D card
Control/DA software	Custom program written in Turbo Pascal 6.0 (DOS)
Signal duration resolution	25 ms
Timing resolution	1 °CA
Temperature sensor	K-type thermocouple
Pressure sensor	Kistler 601B1 transducer
Exhaust gas analyser	Autodiagnostics NDIR 5-gas system (HC, CO, CO_2, NO, O_2)

EXPERIMENT DESIGN

The aim of this study was to investigate the relationship between various fuel injection strategies and engine performance. Engine operation parameters were therefore held constant, and the injection parameters were varied. The engine speed was fixed at 1500 rev/min and the load at 300 kPa IMEP to replicate as closely as possible a mid-speed mid-load mapping point commonly reported in literature. The swirl ratio, HAJI operation variables and injector location were also fixed. The natural gas start of injection time (SOI), upstream pressure (Pup), end of injection time (EOI), injection duration (IDUR) and spark timing were classified as free variables. Previous work has shown that engine-out emissions levels in a HAJI-assisted engine are strongly influenced by air/fuel ratio [7]. To avoid confounding the results, it was decided that the experiments would be carried out at a single air/fuel ratio. An ultra-lean air/fuel ratio of λ=2 was chosen in order to benefit from low exhaust emissions without paying too high a penalty in performance. A fixed air/fuel ratio, speed and load means that any two injection parameters determine the remaining two parameters. For instance, for a selected start of injection time and injection pressure, only one injection duration (and therefore end of injection time) will result in the desired air/fuel ratio and load. The chosen injection times resulted in upstream gas pressures ranging from 1.2 MPa to 6 MPa. The throttle was adjusted throughout the experimental regime to maintain the air/fuel ratio and load since thermal efficiency varies with injection strategy and spark timing.

A baseline case was run with port induction of natural gas. This results in a homogeneous incylinder mixture, allowing an analysis of the effects of mixture stratification to be carried out.

As the engine rig has optical access it is uncooled. Furthermore, inserting the injector into the head required breaking into the water cooling jacket, and for this reason, head cooling was also impossible. Therefore an experimental procedure was developed to ensure consistent and repeatable results in the absence of engine cooling. The intention was to obtain data indicative of a steady-state situation. Accordingly, data was taken when the cylinder head temperature 2 mm from the combustion chamber surface reached 95°C. Once the data was collected, the engine was turned off to prevent overheating. To ensure each run experienced the same warm-up procedure, the engine was cooled to 70°C between runs. Repeatability was checked several times throughout the course of the experiments and found to be acceptable. For further detail regarding the experimental set-up, refer to [8].

RESULTS

The performance parameters (efficiency, CO, NO, HC) are shown in contour plots with end of injection time on the horizontal axis, and injection duration on the vertical axis. Consequently, contours of constant start of injection (SOI) times take the form of straight lines with a positive gradient. It can be seen from the contour plots that none of the performance parameters exhibit an overall positive gradient. Therefore, it can be concluded that engine performance is much more strongly influenced by EOI and IDUR than SOI times. Plots of the performance parameters against SOI were also compiled, verifying the low level of influence of SOI. For the sake of brevity, the SOI contour plots are not shown in this paper, but may be found in [8].

Figure 5 shows a contour plot of indicated thermal efficiency against EOI and IDUR. The dashed line indicates the baseline (homogeneous) MBT thermal efficiency for comparison. In general, engine efficiency increases as the injection duration is reduced. The highest thermal efficiencies were found for short injection durations (high injection pressures) and mid-range end of injection (EOI) times. A substantial region exists at the right hand

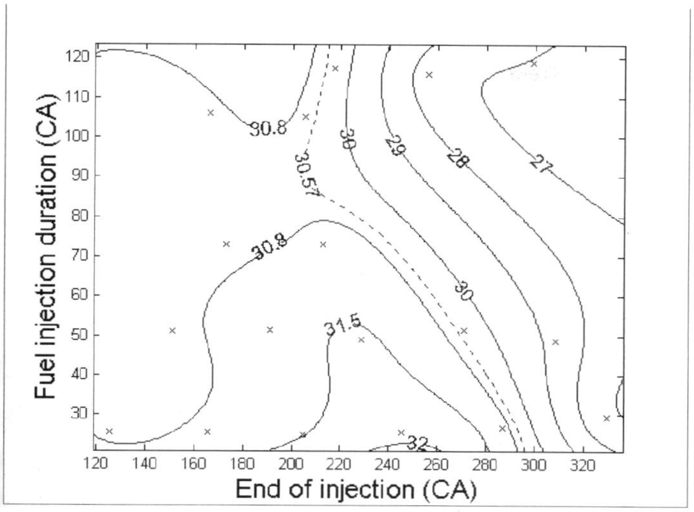

Fig. 5 Indicated thermal efficiency (%) as a function of end of injection time and fuel injection duration. The dashed line indicates the homogeneous MBT efficiency level.

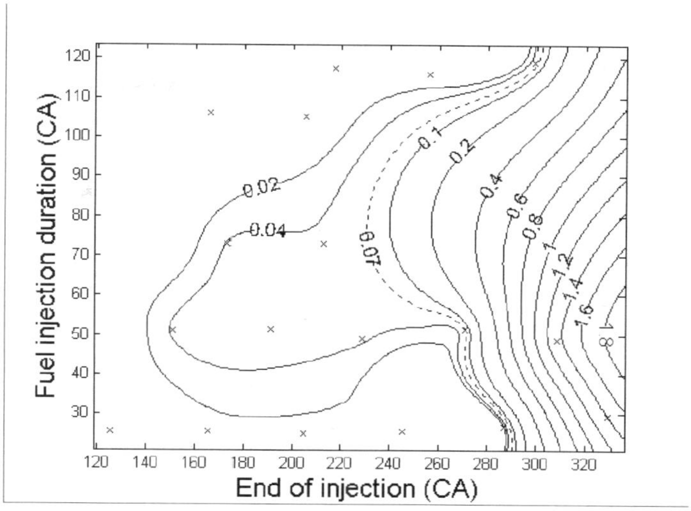

Fig. 6 NO (g/kWh) as a function of end of injection time and fuel injection duration. The dashed line indicates the homogeneous MBT NO level.

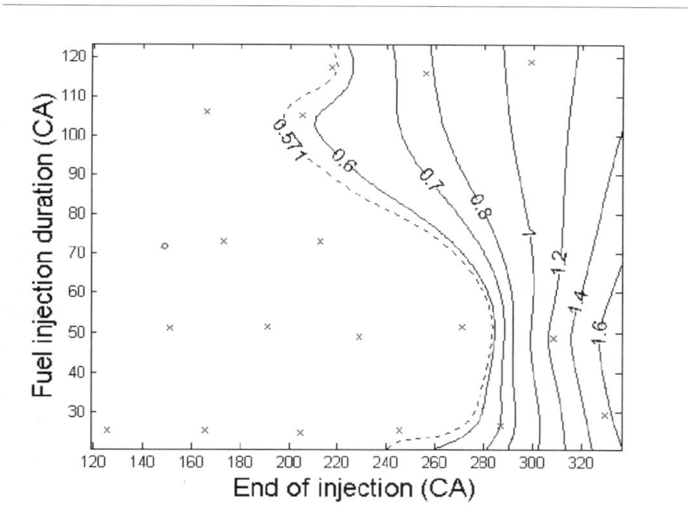

Fig. 7 CO (g/kWh) as a function of end of injection time and fuel injection duration. The dashed line indicates the homogeneous MBT CO level.

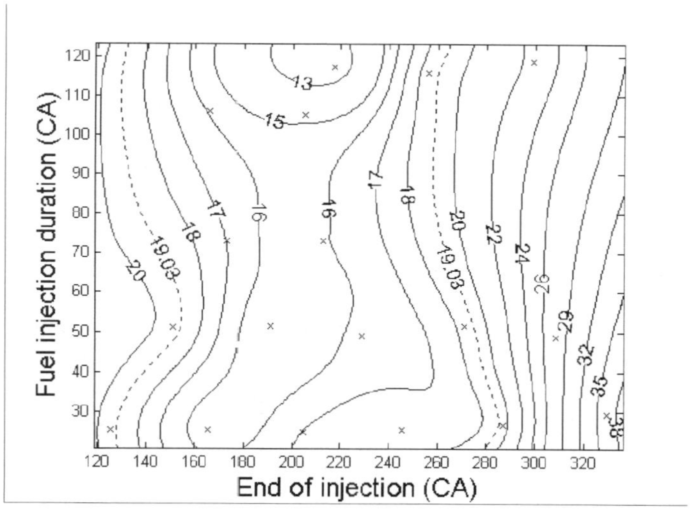

Fig. 8 HC (g/kWh) as a function of end of injection time and fuel injection duration The dashed line indicates the homogeneous MBT HC level.

side of the plot where the efficiency obtained with DI is lower than the homogeneous case. This undesirable outcome occurs for late EOI times, especially when coupled with long injection durations. The highest efficiency achieved by DI was 32.0%, corresponding to a relative increase of 4.6% over the homogeneous efficiency (30.6%). It should also be noted that the indicated thermal efficiency for the stoichiometric case is approximately 23%, resulting in a 33% relative improvement between the homogeneous ultra-lean and homogeneous stoichiometric operation points.

Figure 6 presents a plot of NO as a function of end of injection time and injection duration. The highest levels of NO were obtained at late EOI times, in particular for short IDUR's. The homogeneous NO level was recorded as 0.07 g/kWh. This compares favourably with the Euro 4 standards of approximately 1 g/kWh for diesel engines, and approximately 0.35 g/kWh for petrol engines. Earlier EOI times resulted in NO levels equal or lower than the homogeneous case, with the lowest emission levels measured as less than 0.01 g/kWh (around 1 ppm). It should be noted that due to the resolution of the exhaust gas analyser, NO levels less than approximately 0.04 g/kWh (around 5 ppm) should be considered as indicative only. Medium IDUR times appear to produce slightly higher NO levels than very short or long injection durations. Reasons for this are unclear, however it is evident that the experiments should be re-run with higher resolution NO analysis equipment in order to verify this trend.

Figure 7 shows a plot of CO against end of injection time and injection duration. Similar to NO, the highest levels of CO were obtained for very late EOI times, especially for short injection durations. This suggests that these operating points produce extremely stratified mixtures with significant regions of rich air/fuel ratio. Once again, the lowest emission levels were found at early to medium EOI, although the improvement over the homogeneous case achieved with DI at such EOI times was only small.

A plot of hydrocarbons emissions as a function of end of injection time and injection during may be seen in Fig. 8. Similar to CO and NO, very late EOI times produce high levels of HC emissions. However, very early EOI times also produce higher levels of HC than the baseline case. The lowest HC emissions were obtained for mid-range EOI times, especially for long fuel injection durations. Particularly interesting is the observation of the extremely vertical nature of the HC contour plot, indicating that HC emissions are strongly influenced by EOI time. It should be noted that the levels of HC emissions obtained experimentally are high in comparison to current legislation levels. However, the crevices in the research engine used for this work are approximately twice the volume as those in current production engines. Furthermore, although conventional 3-way catalytic aftertreatment is inappropriate for this system (due to the ultra-lean nature of the combustion), exhaust temperatures are hot enough for oxidation catalysts to operate. Finally, two investigations by Stone et. al. and Takagaki et. al. have shown that increasing the compression ratio in a lean burn natural gas engine can lower the engine-out hydrocarbons and NO, and also improve efficiency [9,10]. The compression ratio of the experimental rig used for this work is only 10.3:1 — much lower than that at which a natural gas engine can operate without knock occurring. Hence, a further reduction in exhaust HC emissions may be achieved by increasing the compression ratio of the engine.

From the performance parameters presented, two operating points may be identified as the most promising:

The first point has a mid-range EOI (around 220°ATDC) and a long IDUR. Operation at this point results in a similar efficiency to the homogeneous case, but lower NO, CO emissions and an HC level of 12.8 g/kWh (67% of the homogeneous case).

The second case has a mid-range EOI (around 240°ATDC) and a short IDUR. Operation at this point results in a 4.6% improvement in efficiency, lower NO, lower CO and a 10% reduction in HC.

MODELLING

The modelling in this project was carried out using the KIVA-3V code developed by the Los Alamos Laboratories. A detailed discussion of the code may be found in [11-13]. KIVA-3V enables the "numerical calculation of transient, two- and three-dimensional chemically reactive fluid flows with sprays". A discussion of the modelling undertaken is reported here only where it is relevant to the analysis of hydrocarbons sources. For a full report of the simulation work, see [8,14,15].

The simulation was broken into three stages. The first modelled the initial expansion phase of the jet as it leaves the injector nozzle. Results from this stage were then used as inflow boundary conditions for a full cylinder injection and mixing model. Finally, some of the stratified mixture patterns produced by the mixing model were used as initial conditions for the combustion model. To reduce computational time, methane was modelled as the fuel instead of natural gas. This was considered acceptable as methane comprises 93% (vol.) of the natural gas used in the experiments.

The combustion model was based on a coherent flamelet model developed by Boudier et. al. [16]. The combustion of methane was modelled as a

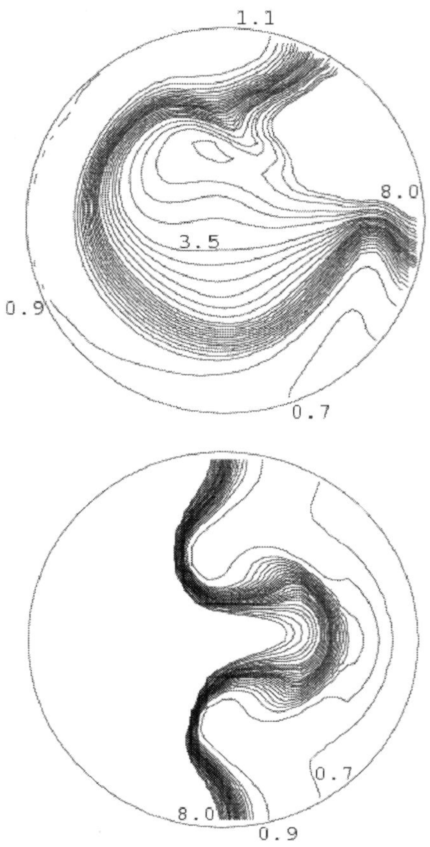

Fig. 9 Contour plots showing relative air/fuel ratio (λ) distribution at 352°ATDC for the stratified (top) and highly stratified (bottom) cases

single step forward kinetic reaction as presented by Dryer and Glassman [17]. Enhanced combustion was simulated by the modification of a flame growth term, allowing the modelled pressure traces to be matched with those obtained experimentally. An annular ignition region was used to replicate the ignition characteristics of HAJI in which the hydrogen jets quench at the HAJI orifices and re-ignite further out in the cylinder. A detailed hydrocarbons emissions model developed by Dober was used in this work [6]. This model accounts for HC sourced from the wall quench layer, the wall layer produced by gases exiting the top land piston ring crevice during expansion (termed wall crevice), the top land piston ring crevice, the head gasket crevice and the prechamber crevice. Unburned fuel originating away from the crevices and walls was labelled "free" HC. It should be noted that bulk quench was not modelled.

Matching the combustion simulations to the experimental results took many runs. As each

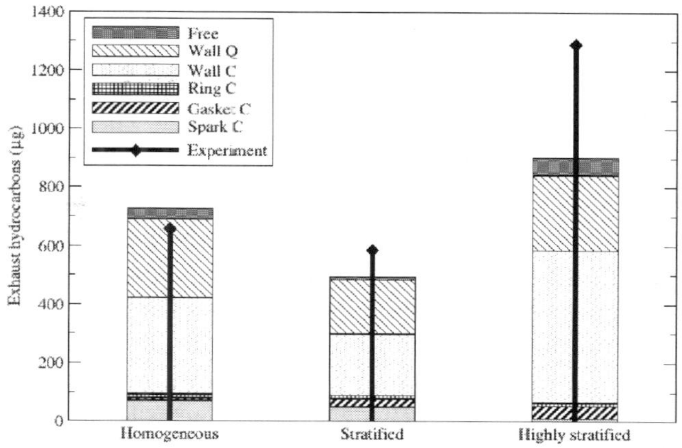

Fig. 10 Bar chart illustrating the modelled breakdown of exhaust HC (μg/cycle) for the three cases. The experimentally measured HC level is also shown

combustion simulation takes over a week to complete, only three stratification cases produced by the mixing simulations were carried on to the combustion model. The cases chosen were:

1. the baseline homogeneous case

2. the stratified case which delivered the best efficiency and acceptable emissions (SOI=220°ATDC, EOI=245°ATDC)

3. a highly stratified case which resulted in high emissions and low thermal efficiency (SOI=300°ATDC, EOI=329°ATDC)

Figure 9 shows the air/fuel ratio distribution in the plane 1 mm below the cylinder head at 352°ATDC (just before ignition) for the stratified and highly stratified cases. These stratification patterns were produced by the mixing model and used as initial conditions for the combustion model as explained earlier. It was observed that single injection DI of natural gas results in stratification patterns in which a bulk of the fuel lies around the outside of the cylinder. Note in particular that the stratified case is almost a complete inversion of the 'top hat' stratification pattern, which was one of the original aims of this project.

ENGINE-OUT HYDROCARBONS

Figure 10 shows the modelled breakdown in exhaust hydrocarbons for the homogeneous, stratified and highly stratified cases. The homogeneous case produced more HC than the stratified case and less than the highly stratified case both experimentally and numerically. The fuel remaining in the top land piston ring crevice at blowdown was shown to be unimportant compared with other sources of HC for all three cases. The other two crevice sources also contribute only small amounts of HC, although the gasket crevice HC's increase with high levels of stratification, and the prechamber crevice HC's decrease with high stratification. The mixing simulations show that the amount of fuel located near the cylinder walls increases with stratification, and the amount of fuel in the middle of the combustion chamber decreases. This explains why the gasket crevice (at the side of the cylinder) becomes more significant as a HC storage source as stratification increases, whilst the prechamber (in the middle of the cylinder head) becomes less significant.

In all cases, the dominant sources of unburned fuel are the wall layer hydrocarbons (wall quench and piston ring crevice wall hydrocarbons). In particular, the overall HC trend is determined mainly by the contribution from the top land piston ring crevice wall layer. The stratified case produces the least wall quench, however (unlike the wall crevice hydrocarbons) the homogeneous and highly stratified cases yield similar levels of wall quench HC. The reason for the lower levels of wall layer hydrocarbons in the stratified case becomes clear upon examination of the incylinder air/fuel ratio distribution (pictured Fig. 9). A majority of the mixture near the cylinder walls in the stratified case is close to stoichiometric. Hence, the wall quench layer in the stratified case will be thinner than that in the homogeneous case, where the mixture next to the wall is ultra-lean (λ=2). Furthermore, the close-to-stoichiometric mixture in the stratified case will result in higher temperatures near the walls. These higher temperatures encourage post-flame oxidation of the piston ring crevice wall layer and other crevice hydrocarbons located near the wall. This is supported by the combustion simulations which tracked the amount of post-flame oxidation that occurred. The model found that 1.7% of the hydrocarbons existing at exhaust valve open (EVO) underwent post-flame oxidation in the homogeneous case. In contrast, 4.7% of the hydrocarbons existing at EVO were later oxidised in the stratified case, but only 0.5% in the highly stratified case. It should be noted that the highly stratified case did not improve on the homogeneous case as the mixture near the walls was much richer than stoichiometric. In fact, a vast majority of the fuel near the walls had a relative air/fuel ratio of λ<0.7 (ϕ>1.43). Consequently, much of the mixture close to the walls does burn well, resulting in cooler temperatures near the wall and high levels of wall quench and piston ring crevice wall layer hydrocarbons.

One discrepancy between the modelled and experimentally measured HC levels is that the modelled stratified and highly stratified cases produced significantly less HC than the experiments. This is attributed to two mechanisms. Firstly, bulk quench was not included in the HC model. Hence, the modelled HC level will be underestimated in cases where bulk quench is significant. This is especially true of the highly stratified case, where the volume in which the mixture is very rich or very lean is large. Bulk quench will also occur to a lesser degree in the stratified case. Secondly, the top land piston ring crevice wall layer was modelled very simply. As the piston moves downwards, wall crevice gas is removed from the main grid and stored in concentric annular cells. At blowdown, the gas in these crevice wall layer cells is returned to the main grid. However, these wall layer cells are unresolved around the cylinder. Therefore, as the piston moves down one step, all the piston ring crevice gas laid along the wall in that step is removed and placed into a single wall layer cell. Consequently, the air/fuel ratio distribution of the wall layer gas is not preserved. Ideally, the wall crevice gas in the stratified mixtures would be returned to the main model with the same local air/fuel ratios that existed at the time each stage of the wall layer was created. However, in actual fact, the gas is returned to the main grid perfectly mixed. Hence the ultra-lean and ultra-rich regions along the wall are not preserved, leading to higher levels of post-flame oxidation of the wall crevice layer than would actually occur. The lack of resolution of the wall crevice model was not significant for the work originally carried out by Dober (the author of the code). This is because this original work dealt only with uniform incylinder mixtures or highly stratified mixtures where a thick air layer was assumed to insulate the fuel from the wall [6]. The shortcomings of the wall crevice model for this project were recognised early in the modelling process. However, due to run time and memory space limitations, extending the model to include higher wall layer resolution around the cylinder was unfeasible. It should be noted that this did not in any way affect wall quench, which was modelled directly in the main grid.

CONCLUSION

From the experiments and modelling reported in this paper, several conclusions can be drawn:

Direct injection of natural gas has the potential to increase thermal efficiency and decrease CO, NO and HC emissions for ultra-lean combustion if the injection strategy is carefully selected. In the research engine at the 1500 rev/min, 300 kPa IMEP world wide mapping point, two operating points in particular were isolated as advantageous over the baseline homogeneous case:

1. Mid-range end of injection time with a long injection duration (i.e. an early start of injection time and a low upstream fuel pressure). This operating point reduced the exhaust hydrocarbons level to 67% of the baseline value. Negligible NO was recorded, and the CO and thermal efficiency remained similar to the baseline case.

2. Mid-range end of injection time with a short injection duration (i.e. a medium start of injection time and a high upstream fuel pressure). This operating point increased the engine efficiency to 32%, corresponding to a relative increase of 4.6% over the baseline homogeneous case. Negligible NO was recorded, and small improvements in CO and HC were also obtained.

The performance parameters (efficiency, NO, CO and HC levels) were all found to be strongly influenced by end of injection time. The start of injection time seemed to have little affect on the performance of the engine. NO and CO levels were discovered to be reasonably invariant of injection duration, although long injection durations result in lower HC and short injection durations result in higher efficiencies.

The injection, mixing and combustion processes were simulated. A detailed HC model predicted levels of exhaust hydrocarbons similar to those recorded experimentally. In particular, the wall layer hydrocarbons (wall quench and the piston ring crevice wall layer) were found to be the dominant sources of hydrocarbons. A reduction in hydrocarbons in the stratified case was attributed to decreases in the wall quench and wall crevice sources. An increase in hydrocarbons in the highly stratified case was attributed to a substantial increase in hydrocarbons stored in the piston ring crevice wall layer.

Single injector direct injection of natural gas was found to produce stratification patterns dissimilar to the 'top hat' pattern originally desired. In fact, single injector gaseous direct injection encouraged fuel to accumulate around the cylinder walls instead of in the centre of the cylinder, resulting in an inverted 'top hat' pattern. However, the close-to-stoichiometric mixtures near the walls provided an unexpected bonus in the higher combustion temperatures. This led to thinner quench layers and more post-flame oxidation than in the homogeneous case. As a consequence, the level of exhaust hydrocarbons was significntly reduced from the homogeneous case. This suggests that an inverted 'top hat' stratification pattern may in fact be preferable to a poorly implemented 'top hat' pattern. This is because a 'top hat' distribution in which the mixture near the wall contains a small amount of fuel will engender thick quench layers and low levels of post-flame oxidation. Thus, it is possible for a poor-quality 'top hat' distribution to produce higher levels of hydrocarbons than an inverted 'top hat' distribution where the close-to-stoichiometric mixtures near the walls result in thin quench layers and high levels of post-flame oxidation.

Acknowledgments. This work was supported by an Australian Government Postgraduate Award (APA). Deanna Wang would also like to thank Prof. Harry Watson for his support as supervisor during this project

REFERENCES

1. Dober G., Watson H. (1999) Modelling the flame enhancement of a HAJI equipped spark ignition engine. *IPC-10 Proceedings*, No. 99091.
2. Lumsden G., Watson H. (1995) Optimum control of an SI engine with a λ=5 capability. SAE technical paper series, No. 950689.
3. Lawrence, J. (1999) *Hydrocarbon emissions from a HAJI equipped ultra-lean burn engine*. PhD Thesis, The University of Melbourne.
4. Beck N., Oppenheim A., Uyehara O. (1998) On the future of combustion in piston engines. SAE technical paper series, No. 980117.

5. Wang D., Watson H. (2000) Direct injector compressed natural gas combustion and visualisation, SAE technical paper series, No. 2000-01-1838.
6. Dober G. (2002) *Geometric control of hydrocarbon emissions*. PhD Thesis, The University of Melbourne.
7. Lumsden, G. (1995) *The phenomena of hydrogen assisted jet ignition*. PhD thesis, The University of Melbourne.
8. Wang, D. (2004) *An investigation of direct injection of natural gas in a hydrogen assisted jet ignition engine*. PhD Thesis, The University of Melbourne.
9. Stone C., Mendis K., Daragheh M. (1996) Measurements and modelling of a lean burn gas engine. *Proceedings of the Institution of Mechanical Engineers, Part A: Journal of Power and Energy*, 210.
10. Takagaki S., Raine R. (1997) The effects of compression ratio on nitric oxide and hydrocarbon emissions from a spark-ignition natural gas fuelled engine. SAE technical paper series, No. 970506.
11. Amsden A. (1989) *KIVA-II: A Computer Program for Chemically Reactive Flows with Sprays*. Los Alamos National Laboratory.
12. Amsden A. (1993) *KIVA-3: A KIVA Program with Block-Structured Mesh for Complex Geometries*. Los Alamos National Laboratory.
13. Amsden A. (1997) *KIVA-3V: A Block-Structured KIVA Program for Engines with Vertical or Canted Valves*. Los Alamos National Laboratory.
14. Wang D., Watson H. Modelling the direct injection and mixing of natural gas in an engine. Submitted to Journal of Gas Turbines and Power, 2004.
15. Wang D., Watson H. Modelling enhanced combustion and hydrocarbons production of stratified natural gas mixtures. *To be published*.
16. Boudier P., Henriot S., Poinsot T., Baritaud T. (1992) A model for turbulent flame ignition and propagation in spark ignition engines. *Twenty-fourth Symposium (International) on Combustion*, 503-510.
17. Dryer F., Glassman I. (1972) High-temperature oxidation of CO and CH4. *Fourteenth Symposium (International) on Combustion*.

Contact

Deanna Wang, deannawang@fastmail.fm

ABBREVIATIONS

DI: Direct injection

EOI: End of injection

EVO: Exhaust valve open

Free: Modelled exhaust hydrocarbons not originating from any crevice or wall layer source

Gasket C: Head gasket crevice hydrocarbons

HC: Hydrocarbons

IDUR: Injection duration

Pup: Upstream injection pressure

Ring C: Top land piston ring crevice hydrocarbons

SOI: Start of injection

Spark C: Prechamber crevice hydrocarbons

Wall C: Top land piston ring crevice wall layer hydrocarbons

Wall Q: Wall quench hydrocarbon

Which Fuels For Low CO_2 Engines?
© P. Duret, X. Montagne (Eds.) and Editions Technip, Paris, 2004, pp. 181 – 188
27 rue Ginoux, 75015 Paris

POSSIBLE SHORT-TERM INTRODUCTION OF HYDROGEN AS VEHICLE FUEL / FUEL ADDITIVE

Per TuneståI, Patrik Einewall, Ola Stenlåås, Bengt Johansson
Lund Institute of Technology, Department of Heat & Power Engineering, Lund, Sweden

ABSTRACT – *The infrastructure for natural gas distribution can provide the possibility of an evolutionary as opposed to revolutionary introduction of hydrogen as an additive to natural gas (Hythane). In this study, pure natural gas and a hythane blend with 24.8% (vol.) hydrogen have been used to fuel a multi-cylinder heavy-duty natural gas engine.. Some comparisons with pure hydrogen operation are also presented which indicate severely limited load range due to excessive burn rate causing high heat losses. Hythane increases efficiency, reduces CO_2 emission and improves the HC-NO_X trade-off.*

INTRODUCTION

Carbon dioxide (CO_2) has been recognized as the most important green house gas responsible for the global warming problem. In order to reduce the amount of CO_2 in the atmosphere, the combustion of fossil carbon has to be reduced. This can be achieved by replacing fossil fuels by bio fuels. The plants that the bio fuel is made from absorb as much CO_2 when they grow as is subsequently released through combustion. Thus, the net emission of CO_2 is zero. The problem with bio fuels is that the plants require a lot of space when they grow and space is a scarce resource in densely populated areas. Another way to reduce the combustion of fossil carbon is by decreasing the total use of energy from combustion. Since the world's energy consumption is steadily growing, the only way to achieve this is by increasing the efficiency of the devices that utilize combustion. It is also possible to use fossil fuels with a low carbon content. Natural gas e.g. contains about 25% less carbon per heat energy unit compared to gasoline. In [1], natural gas fuelled Internal Combustion (IC) engines are recognized as a viable intermediate step towards hydrogen powered fuel cells. Finally, it is possible to produce fuels from solar energy, e.g. hydrogen through electrolysis.

There are however problems associated with using hydrogen as a fuel for vehicles. Storing enough hydrogen onboard vehicles is not an easy task. Furthermore, although hydrogen powered fuel cells have a potential for very high fuel effi-ciency, the cost per kW is prohibitive. Obviously, part of this cost can be reduced by mass production but the cost associated with precious metal content is not reduced by an increasing number of units. There is also the possibility of running an IC engine with hydrogen as fuel, but the combustion properties of hydrogen make it less than ideal. Very promising results are however shown in [2] with both high load and high efficiency achieved using direct injection of hydrogen for an IC engine. Finally, there is no existing infrastructure for hydrogen distribution.

The infrastructure problem can be bridged by gradual introduction of hydrogen into the existing infrastructure for natural gas distribution. Hydrogen then acts as a fuel additive with a certain influence on the combustion process. As the hydrogen fraction in the natural gas increases the properties of the gas will more and more resemble those of hydrogen, and gradual modifications can be made to accommodate these changes. The authors think that this evolutionary approach to hydrogen introduction stands a better chance of success than the revolutionary approach of a completely new system for distribution and energy conversion.

This study investigates the influence of hydrogen addition to natural gas on IC engine combustion, performance and emissions. The measurements are performed on a heavy-duty natural gas engine with various combustion chamber geometries giving different levels of turbulence.

EXTENDING THE LEAN LIMIT OF A NATURAL GAS ENGINE

The flame speed of hydrogen is much higher than that of hydrocarbon fuels [3, 4]. Adding hydrogen to natural gas is thus likely to increase the flame speed of the charge. This could be used to extend the lean limit of a natural gas engine to air/fuel ratios where pure natural gas provides insufficient burn rate for stable combustion. A special case of extending the lean limit with hydrogen is presented in [5]. Hydrogen is injected into a prechamber with a spark plug. Partially burned, reacting charge was then injected from the prechamber into the main combustion chamber and served to ignite the charge. This strategy enabled stable engine operation at λ=5. Since the main combustion was essentially that of pure natural gas, there were enormous emissions of HC due to wall and bulk quenching. This phen-omenon is not expected with a homogeneous mix of natural gas and hydrogen though. Indeed, [6] shows that the trade-off between NO_X and HC for a certain equivalence ratio can be improved by addition of hydrogen.

Another way to increase the flame speed is to introduce charge turbulence. Turbulence of suitable length scale wrinkles the flame front and thus increases its surface area. As an effect, the propagation speed of the reaction zone is increased. The flame speed of a certain charge is essentially proportional to the turbulence intens-ity and redesign of the combustion chamber for increased turbulence is an alternative way to ex-tend the lean limit. In [7] two different combus-tion chamber geometries, one with low turbulence and one with high, are operated on a single cylin-der engine with various hythane blends. It is conc-luded that the effects of turbulence and hydrogen addition are to some extent interchangeable.

EXPERIMENTAL SETUP

THE ENGINE

The Engine (TG103/G10A) was originally devel-oped for turbocharged diesel operation and rede-signed by Volvo for turbocharged natural gas operation, see Table 1 for specifications. The fuel was injected upstream of the turbocharger for en-hanced mixing. Natural gas was supplied from the pipeline whereas hythane was supplied premixed from gas bottles.

Table 1. Engine specifications.

Displaced volume/cyl.	1600 cm^3
Compression ratio	11.8:1
Rated power	184 kW (2000 rpm)
Maximum brake torque	1150 Nm (1150
Bore	120.65 mm
Stroke	140 mm
Ignition sequence	1-5-3-6-2-4

MEASUREMENT SYSTEMS

Cylinder pressure

Each cylinder head is equipped with a piezo electric pressure transducer, Kistler 7061B. The signal from the charge amplifier, Kistler 5017A, is processed by two parallel Datel PCI-416 boards in a PC for on-line pressure measurements. The cylinder pressures are measured 5 times per crank angle degree (CAD) using an external clock from a Leine & Linde crank-angle encoder. The pressures are used for heat-release calculations, the program is described in [8]. Pressures are also measured in the inlet manifold (before and after the throttle) and in the exhaust pipe, before the exhaust throttle.

Emissions

Emissions are measured after the turbocharger using a Pierburg AMA 2000 emission analyzer. The analyzer consists of; a Heated Flame Ionization Detector (HFID/FID) for hydrocarbons, a Heated Chemiluminescence Detector (HCLD/CLD) for nitric oxides and a Paramagnetic Detector (PMD) for oxygen (O_2). The HC emissions are presented as methane equivalent (C_1) in the figures. Three Non Dispersive Infrared Detectors (NDIR) measure carbon monoxide (CO high and low) and carbon dioxide.

In addition to lambda calculations from emission analysis, a lean lambda probe (ETAS) is installed in the exhaust pipe, after the turbocharger.

Flows

The mass flow of natural gas is measured with a Bronkhorst F106A-HC.

Torque

The engine is connected to a Schenk U2-30G water brake, controlled by the engine control system. The torque is measured with a load cell, Nobel Elektronik KRG-4.

All data, except in-cylinder pressure, is collected by a HP 34970A Data Acquisition/Switch unit.

Table 2 Natural gas composition

Natural Gas Constituents	% Volume
CH_4	88.06
C_2H_6	6.49
C_3H_8	2.81
C_4H_{10}	1.00
C_5H_{12}	0.20
C_6H_{14}	0.06
CO_2	1.05
N_2	0.33

RESULTS AND DISCUSSION

The performance and emissions were compared between operation with natural gas and a specific hythane blend (24.8% vol. hydrogen). An intermediate operating point of 7 bar BMEP at 1000 rpm was selected.

TEST MATRIX

Since it is unlikely that the same air/fuel ratio and ignition timing should be optimal for both natural gas and hythane, a whole test matrix was evaluated. The air/fuel ratio was swept between λ=1.0 and λ=1.8 in steps of 0.1, and five different ignition angles from MBT to the stability limit (due to over-retarded ignition) were tested for each air/fuel ratio. The stability limit was defined as the ignition timing where COV(IMEP) is 5%. The design limit on the exhaust temperature is 700° C and served as the limit for the ignition timing in the cases when it was reached before the COV(IMEP) limit. The ignition angles were labeled from *a* to *e* where *a* represents MBT(determined by maximum pressure at 11° ATDC) and *e* represents the combustion stability limit. Ignition angles *b*, *c* and *d* are evenly distributed between *a* and *e*. The ignition alternatives are illustrated with their respective p-V diagrams in Fig. 1 (natural gas) and Fig. 2 (hythane blend).

Fig. 3 shows the selected ignition angles according to ignition alternative *a* (the most advanced ignition alternative). As expected, the hythane blend has retarded ignition angles relative to pure natural gas due to the higher burn rate indicated in Fig. 4. The heat release rates for natural gas and hythane respectively at λ=1.7 and ignition alternative *a* are shown in Fig. 5.

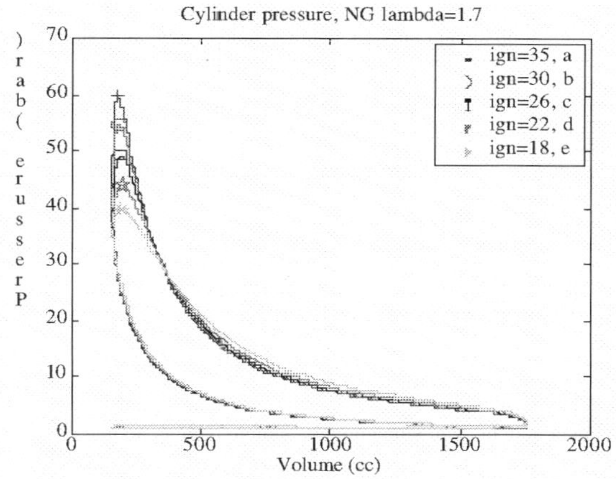

Fig. 1 p-V diagrams for the ignition alternatives *a* – *e* for natural gas at _=1.7

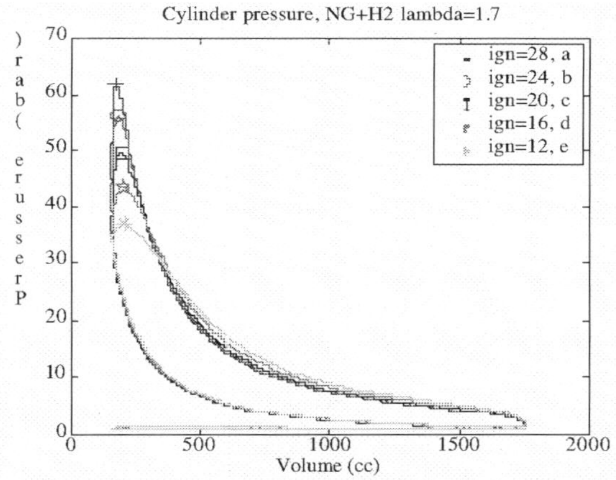

Fig. 2 p-V diagrams for the ignition alternatives *a* – *e* for the hythane blend at _=1.7

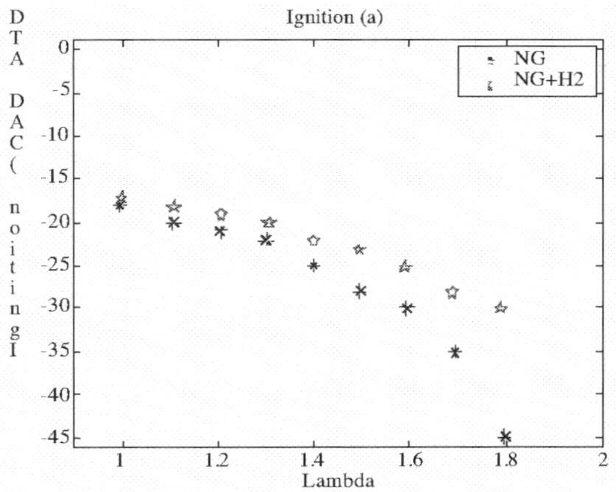

Fig. 3 Ignition angles for ignition alternative *a*

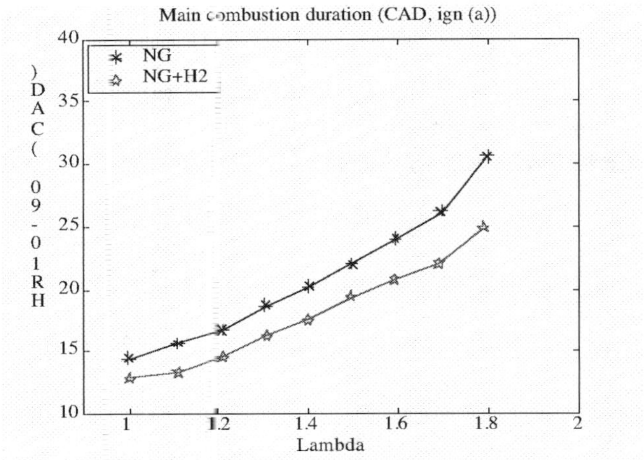

Fig. 4 Combustion duration for ignition alternative *a*

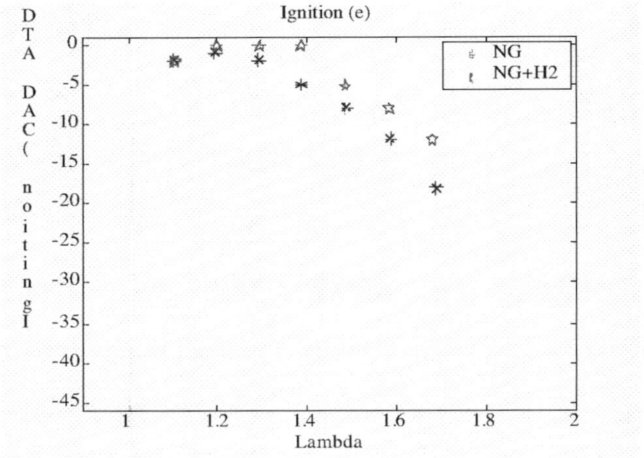

Fig. 6 Ignition angles for ignition alternative *e*

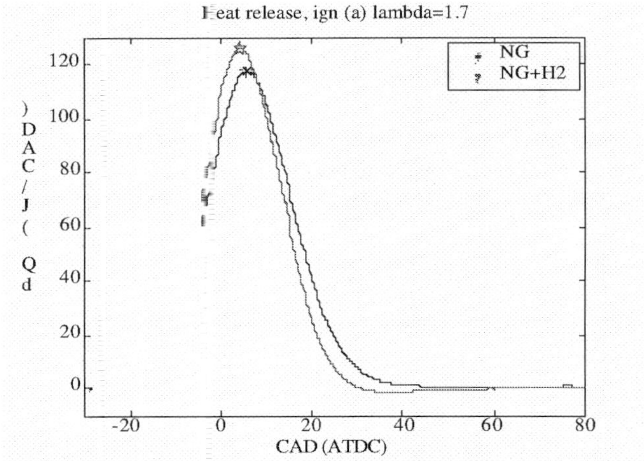

Fig. 5 Heat release rates for ignition alternative *a* at λ=1.7 with natural gas and hythane blend respectively

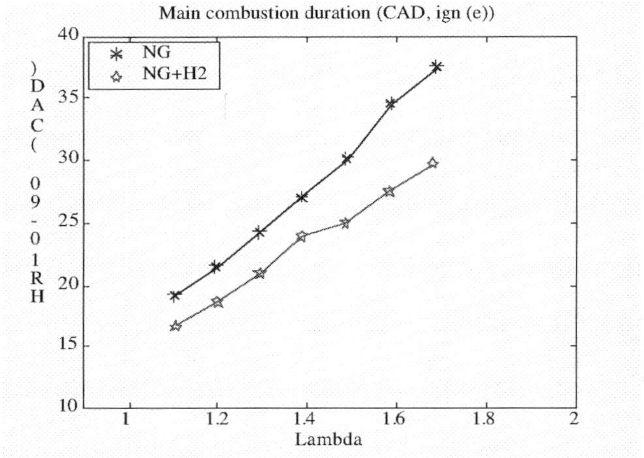

Fig. 7 Combustion duration for ignition alternative *e*

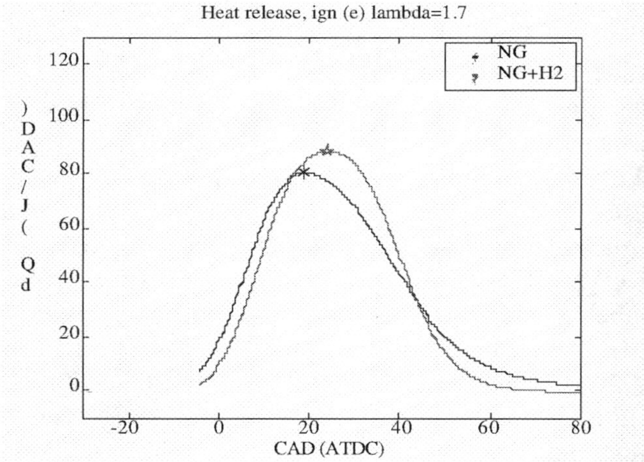

Fig. 8 Heat release rates for ignition alternative *e* at λ=1.7 with natural gas and hythane blend respectively

Fig. 6 shows the corresponding ignition angles according to ignition alternative e (the most retarded ignition alternative). The same behavior is observed with the hythane blend requiring retarded ignition angles relative to pure natural gas. It can be seen here also that the ignition timings for $1.2 \leq \lambda \leq 1.4$ has been limited to 0 (TDC) since a later ignition timing is not relevant from an efficiency point of view.

Fig. 7 shows the combustion duration for ignition alternative *e* which is longer since most of the combustion takes place during the expansion stroke with decreasing pressure and temperature. The heat release rates with natural gas and hythane respectively at λ=1.7 and ignition alternative *e* are shown in Fig. 8. The ignition timings for the three intermediate cases are as mentioned above interpolated between the extremes shown in Fig. 3 and Fig. 6 and the corresponding combustion durations also line up accordingly.

EFFICIENCY AND CO_2

It is reasonable to expect that the brake thermal efficiency should increase with hythane as fuel compared to natural gas since the combustion duration is reduced. With reduced combustion duration the effective expansion ratio increases and more work can be extracted from the gas. This increase in efficiency is likely to be highest where the combustion duration is long with natural gas, i.e. at lean conditions.

Fig. 9 and Fig. 10 show for ignition alternatives *a* and *e* that this is indeed the case. The efficiency is in general higher with hythane and the difference is most pronounced at the leanest conditions. The trend is somewhat broken for ignition alternative *e*, where the efficiency is actually lower with hythane. This can be explained by over-retarded ignition. The hythane blend burns so well even at over-retarded conditions that the combustion stability criterion is sometimes a little too relaxed. The efficiency is (by definition) highest at MBT ignition and drops with retarded ignition.

For comparison a study of operation with pure hydrogen was performed on a single-cylinder version of the same engine. The peak net indicated efficiency for this alternative was 41% at 4 bar net IMEP. Assuming a friction mean effective pressure of 1 bar this corresponds to a brake efficiency of 33%. The lower efficiency with pure hydrogen is due to extremely fast combustion causing increased heat losses. Introduction of EGR could probably reduce the rate of combustion and thus the heat losses.

It should be noted that event though the efficiency gain is fairly small with hythane, the reduction in CO2 emission is substantial. With natural gas composition according to Table 2, the addition of 24.8% hydrogen by volume reduces the energy relative carbon content from 14.8 g C / kJ fuel energy to 10.6 g C / kJ fuel energy. This corresponds to a reduction of CO2 emission by 28% per energy unit. This is in addition to the 25% reduction with natural gas compared to gasoline. Compared to operation with gasoline the reduction in CO_2 emission is 45% assuming the same brake efficiency.

EMISSIONS

When an engine is operated part load with a specific combustion timing, e.g. MBT, there is an infinite number of air/fuel ratios that can be used for each load. The air/fuel ratio that is used will affect the emissions and, specifically, there is a trade-off between HC emisssions and NO_X emissions. With lean operation, HC emissions increase and NO_X emissions decrease with increasing air/fuel ratio. HC emissions increase due to combustion instability. Combustion is too slow to complete before the temperature drops during the expansion stroke. The NO_X emissions decrease because the maximum temperature decreases with more air for essentially the same amount of fuel. It is expected that this trade-off should be affected by the addition of hydrogen to the natural gas since it reduces the combustion duration.

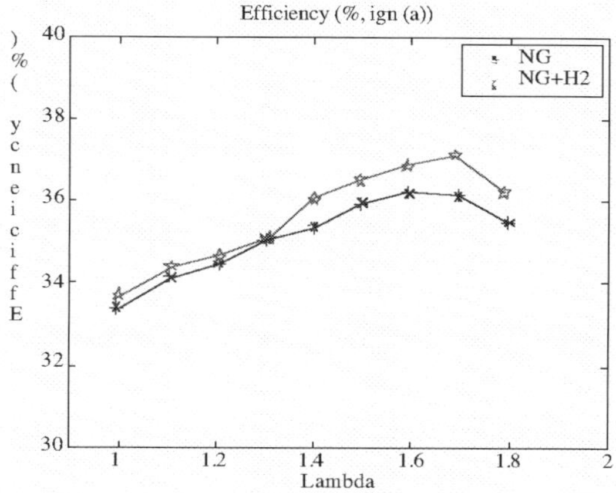

Fig. 9 Brake thermal efficiency for ignition alternative *a*

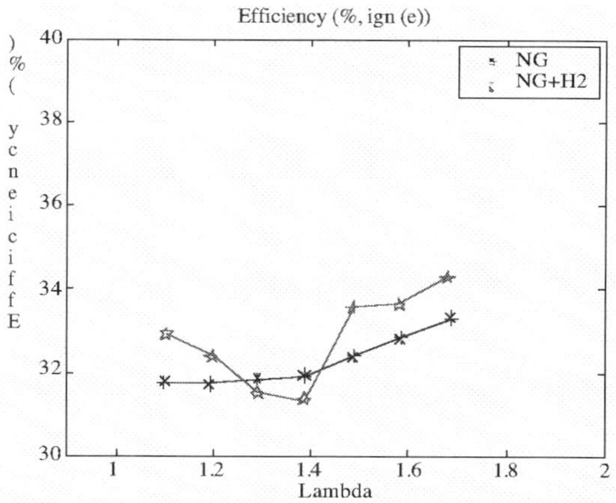

Fig. 10 Brake thermal efficiency for ignition alternative *e*

Fig. 11 - Fig. 15 show that this is indeed the case. For equal NO_X emission levels the reduction in HC emissions is 1-2 g/kWh. All of this can not be attributed to faster combustion. There is also the reduced carbon content in the fuel and the increased efficiency.

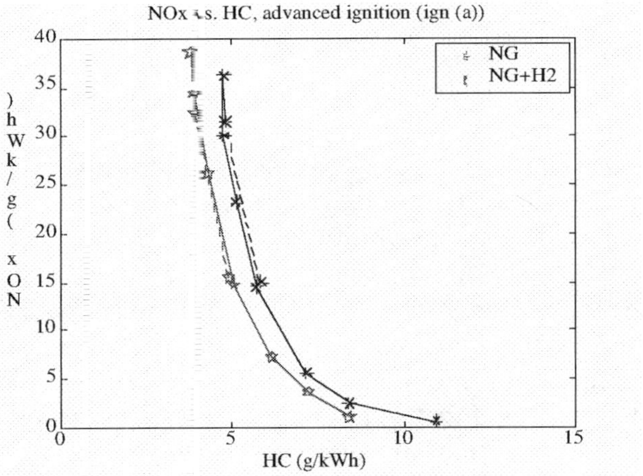

Fig. 11 Comparison between HC-NO_X trade-off between natural gas and hythane for ignition alternative *a*

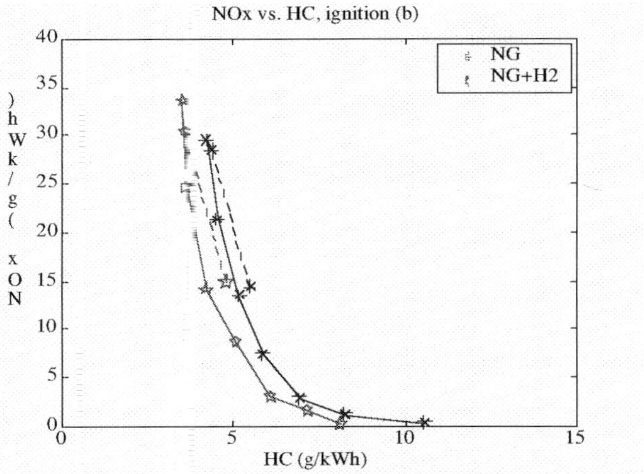

Fig. 12 Comparison between HC-NO_X trade-off between natural gas and hythane for ignition alternative *b*

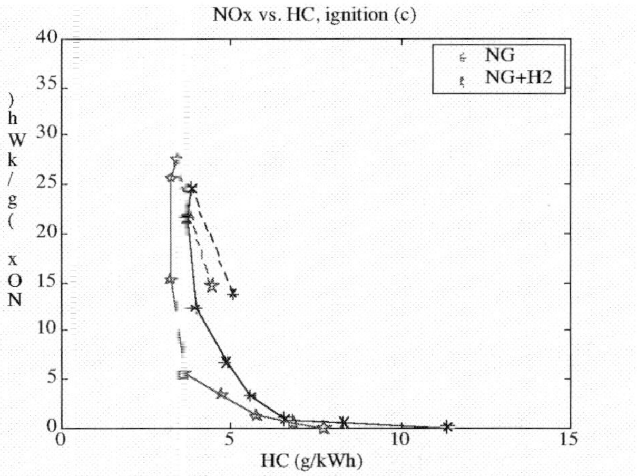

Fig. 13 Comparison between HC-NO_X trade-off between natural gas and hythane for ignition alternative *c*

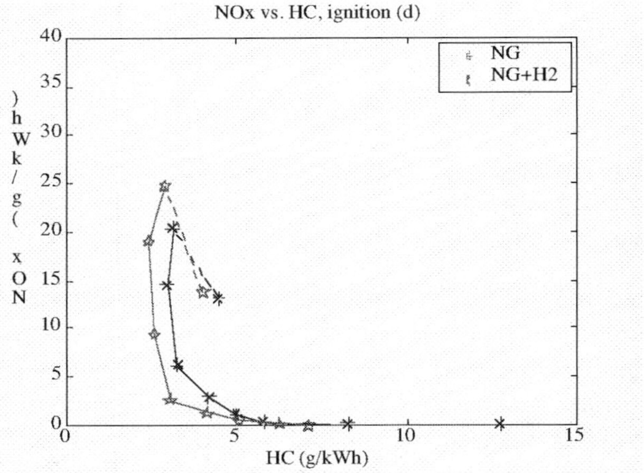

Fig. 14 Comparison between HC-NO_X trade-off between natural gas and hythane for ignition alternative d.

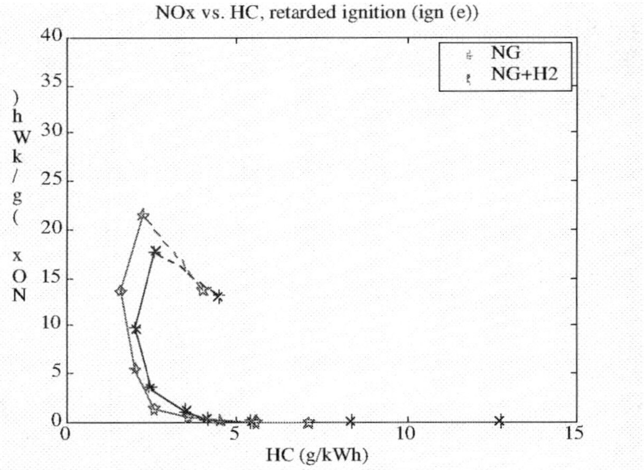

Fig. 15 Comparison between HC-NO_X trade-off between natural gas and hythane for ignition alternative *e*.

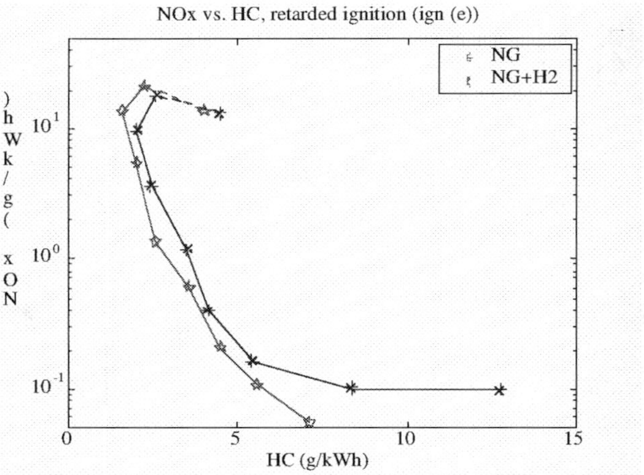

Fig. 16 Comparison between HC-NO_X trade-off between natural gas and hythane for ignition alternative *e*. Logarithmic representation of NO_X emissions

The linear representation of the NO_X emissions for ignition alternative *e* does not reveal the whole truth. In Fig. 16 the NO_X emissions are represented logarithmically instead which reveals that the specific NO_X emissions are merely 0.06 g/kWh at the leanest operating point. This is well below any NO_X regulations and thus NO_X after treatment is not necessary.

CONCLUSION

Operation with hythane (24.8% vol. hydrogen) and natural gas have been compared for a heavy-duty natural gas engine. The study reveals a small increase in efficiency with hythane. The increase is probably due to faster combustion which increases the effective expansion ratio which allows more work to be extracted. The reduction in CO_2 emission is however substantial, 28% due to the reduced carbon content per energy unit alone. Compared to operation with gasoline the reduction in CO_2 emission is 45%, assuming equal efficiency. A comparing single-cylinder study with pure hydrogen revealed a significantly lower efficiency due to increased heat losses.

The trade-off between HC and NO_X emissions parameterized by the air/fuel ratio was compared between hythane and natural gas operation for a five different ignition timing alternatives between MBT and the ignition retard limit. The study shows that the trade-off is improved with hythane. For equal NO_X emission levels hythane operation, in general, emits 1-2 g/kWh less HC. It is also noted that the specific NO_X at the leanest operating point is as low as 0.06 g/kWh which is well below regulations without after-treatment.

It should be noted that the hydrogen content presented in this paper is high enough that some existing natural gas engines may experience decreased performance / efficiency due to the fact that they have been optimized for the existing natural gas quality. Some modification of the engine management may be necessary to take advantage of the change in fuel properties. Bearing this in mind it is the view of the authors that a gradual introduction of hydrogen inte the existing natural gas distribution is a sound way of evolutionary as opposed to revolutionary introduction of hydrogen as fuel.

Acknowledgments. This research was funded by the Swedish Energy Agency (STEM) through the Hythane bus project in the city of Malmö, Sweden. The partners in this project are: Sydkraft, Sydkraft Gas, Swedish Gas Center, Skånetrafiken and Lund Institute of Technology.

REFERENCES

1. Hekkert, M. P., Faaij, A. P. C., Hendriks, F., Neelis, M. L. (2003), Natural gas as an alternative to crude oil in automotive fuel chains well-to-wheel analysis and transition strategy development, Journal of Energy Policy, *In Press*, doi:10.1016/j.enpol.2003.08 018
2. Berckmüller, M. (2003), Potential of a Charged SI-Hydrogen Engine, SAE Paper No. 2003-01-3210
3. Strahle, W. C. (1993), An Introduction to Combustion, Combustion Science and Technology Book Series, Vol. 1, Gordon and Breach Publishers, Amsterdam
4. Apostolescu, N., Chiriac, R. (1996), A Study of Hydrogen-Enriched Gasoline in a Spark Ignition Engine, SAE Paper No. 960603
5. Lumsden, G., Watson, H. C. (1995), Optimum Control of an S.I. Engine with a λ=5 Capability, SAE Paper No. 950689
6. Collier, K., Hoekstra, R. L., Mulligan, N., Jones, C., Hahn, D. (1996), Untreated Exhaust Emissions of a Hydrogen Enriched CNG Production Engine Conversion, SAE Paper No. 960858
7. Tunestål, P., Christensen, M., Einewall, P., Andersson, T., Johansson, B., Jönsson, O. (2002), Hydrogen Addition for Improved Lean Burn Capability of Slow and Fast Burning Natural Gas Combustion Chambers
8. Johansson, B. (1995), On Cycle to Cycle Variations in Spark Ignition Engines, Doctoral Thesis, Lund Institute of Technology

Contact. Per Tunestål received his PhD 2000 in Mechanical Engineering at the University of California, Berkeley. He currently holds a position as Assistant Professor at the Department of Heat & Power Engineering, Lund Institute of Technology.

e-mail: per.tunestal@vok.lth.se
Web: http://www.vok.lth.se

DEFINITIONS, ACRONYMS, ABBREVIATIONS

ATDC After Top Dead Center
BMEP Brake Mean Effective Pressure
COV Coefficient Of Variation (mean/std. dev)

IMEP	Indicated Mean Effective Pressure
Hythane	Blend of natural gas and hydrogen
HC	Hydrocarbon
IC	Internal Combustion (as in IC Engine)
λ	Air excess ratio
MBT	Maximum Brake Torque (ignition timing)
NO_X	Oxides of nitrogen
TDC	Top Dead Center

Which Fuels For Low CO_2 Engines?
© P. Duret, X. Montagne (Eds.) and Editions Technip, Paris, 2004, pp. 189 – 198
27 rue Ginoux, 75015 Paris

THE MIXTURE FORMATION PROCESS OF AN INTERNAL COMBUSTION ENGINE FOR ZERO CO_2-EMISSION VEHICLES FUELED WITH CRYOGENIC HYDROGEN

Dipl.-Ing. Michael Hallmannsegger, Dipl.-Ing. Hans-Christian Fickel
BMW Group Research and Technology

ABSTRACT – *This research work analyses the possibility to predict the mixture formation process of a hydrogen SI engine by means of 3D-CFD simulation. The validation of the calculation is made by comparison with LIF-measurements of an optical engine. Based on test bench measurements performed on a thermodynamical single-cylinder research engine and accompanying 1D- and 3D-CFD simulations, results are shown for both the cryogenic and warm application of hydrogen. Further progress in the R & D process of hydrogen SI engines is expected with the implementation of reliable combustion models in the CFD-tools used.*

INTRODUCTION

Current research activities related to hydrogen as an energy carrier indicate that the concept of hydrogen economy has gained greatly in credibility in recent years. The motivation for these activities is based on the forthcoming exhaustion of oil reserves and the alarming increase of air pollution and global warming effects. For mobile application the different principles of hydrogen storage are also of major interest. BMW focuses on liquid storage systems as a means for the highest energy density by volume as well as weight [1]. Today's BMW hydrogen cars are propelled by bi-fuel engines with external mixture formation at ambient temperatures [2,3]. For future concepts, BMW Group is investigating optimised monofuel engines with direct injection of hydrogen as well as cryogenic port injection (temperatures below 200 K) [4,5,6]. Thereby the focus lies on highest power density and maximum thermal efficiency without emitting pollutants. In Fig. 1 the theoretical power densities of different hydrogen operation modes are compared to gasoline. With cryogenic hydrogen or internal mixture formation the power output of a gasoline engine can be exceeded.

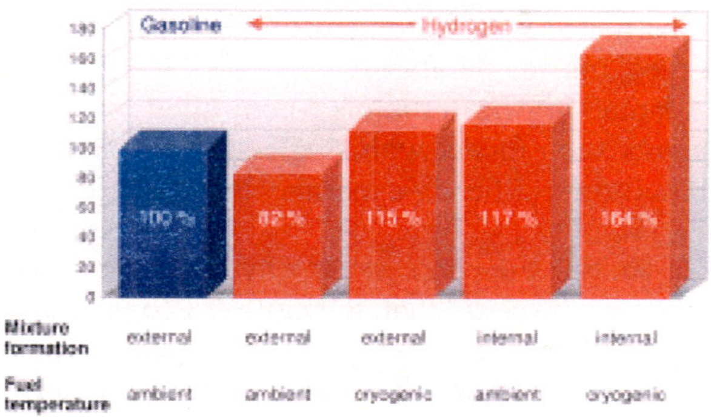

Fig. 1 Theoretical power densities

Hydrogen is the ideal fuel for combustion engines and has distinct physical properties from conventional fuel types. It has a wide flammability range in air (4 ... 75 Vol. %, 10.5 < AFR < 0.5), which allows operation of the engine by quality control. A stoichiometric mixture for maximum indicated mean effective pressure (IMEP) is characterised by a small ignition energy need, a low self-ignition temperature, a short ignition lag and a high burning velocity even due to high diffusivity.

The major steps to compensate for the loss of power at external mixture formation due to the lower volumetric calorific value are supercharging and charge cooling in particular by using cryogenic hydrogen. With cryogenic hydrogen and timed individual port injection, very close to the cylinder intake valves, any problems of backfiring and pre-ignition can be prevented. To achieve further improvements there is a need of developing a computational capability to support the design of high-power-density hydrogen-fueled engines. The simulation tools should cover the hydrogen injection, the mixture formation and the combustion process of an internal combustion engine to meet zero emission requirements. The specific goals of the work are:

- Pointing out the potentials of cryogenic mixture formation.
- Validation of a 3D-CFD-Code to simulate the injection of hydrogen at transonic speeds and its mixture formation in an internal combustion engine.
- Comparison of the engine operation at its full load fueled with hydrogen at ambient and cryogenic temperatures.
- Adaptation of a turbulence-controlled combustion model with single-step chemistry based on test bench experiments.

Fig. 2 shows the integrated hydrogen engine development process that is employed to achieve an improved engine design. After the specification phase the iterative interaction of experiments and simulations and analysis of their results yields an optimised engine design.

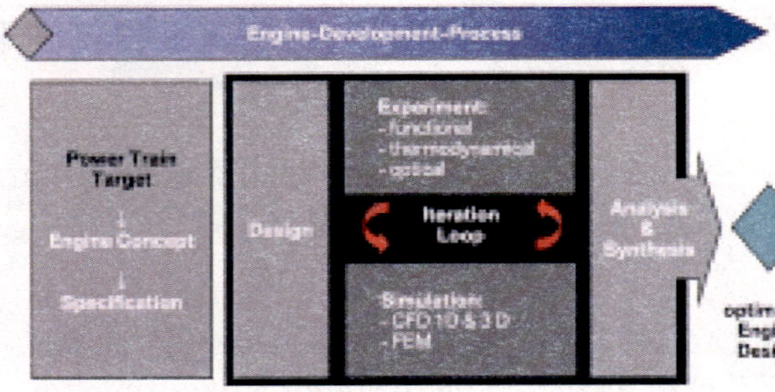

Fig. 2 Integrated engine design process

OPTICAL ENGINE FOR CFD VALIDATION

For the validation of the CFD calculation an optical experiment was considered. Because of safety reasons an optical single-cylinder engine was operated unfired with helium as a substitute for hydrogen. The fluid properties of helium are the closest to those of hydrogen. Its mixture formation

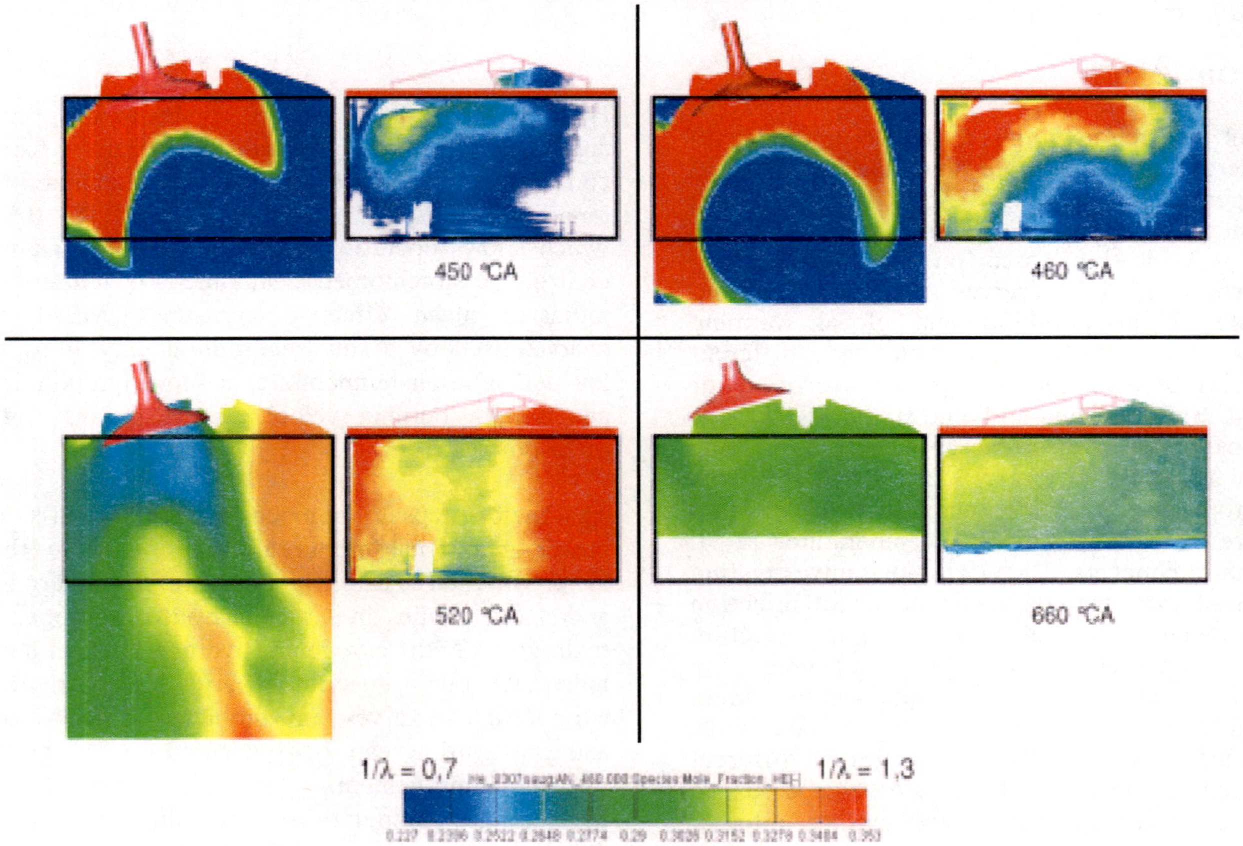

Fig. 3 Comparison of CFD simulation vs. LIF experiment

in the chamber was investigated by laser-induced fluorescence (LIF-technique) of the added tracer triethylamine ($C_6H_{15}N$). In Fig. 4 the experimental set-up of the optical measurement is displayed for the horizontal cut [7]. Without going into detail for both the CFD calculation and measurement background, the comparison between experiment and simulation is briefly shown in Fig. 3. The vertical cutting plane through the symmetric chamber with the inlet valve at different crank angles of the compression stroke is displayed.

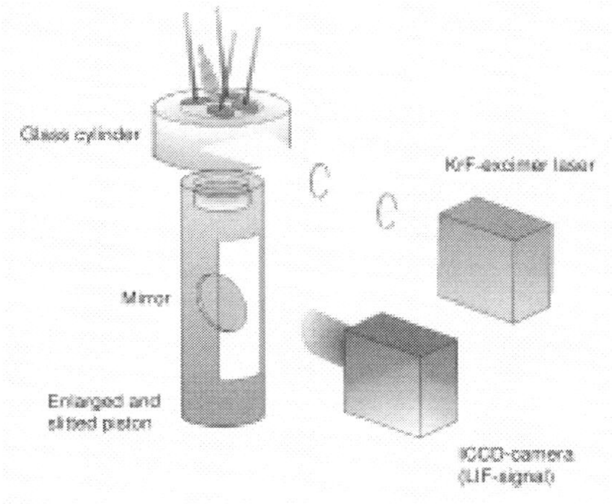

Fig. 4 Set up of optical engine for horizontal cut

The investigated operating point of the engine was 2000 RPM with wide-open throttle and a molar equivalent of injected helium according to a stoichiometric hydrogen air-fuel ratio.

In Fig. 3 the colour bar is scaled in a way that green corresponds to a stoichiometric mixture, red to a rich mixture, and blue and white respectively to pure air. The black frame indicates the viewable window of the glass cylinder. At 450 °CA and 460 °CA the shape of the entering helium and the distribution of the lean and rich regions agree well in simulation and LIF-measurement. As a result, a good homogenisation of the charge is shown in calculation and experiment until the end of the compression stroke.

In this case it could be demonstrated that for the time being CFD simulation has the ability to predict the mixture formation process of an external port-injected hydrogen engine. Because of the large differences in the behaviour of optical and normal engines further results refer to a single cylinder research engine under real operation conditions.

EXPERIMENTAL SETUP

The experimental investigations are conducted using a single-cylinder research engine. The specifications of the engine are summarised in Table 1.

Table 1 Engine properties

Engine Type	1-cylinder, 4-stroke, water cooled, spark ignited
Valves	4, DOHC
Displacement	499.6 cm^3
Bore	86 mm
Stroke	86 mm
Compression ratio	10.5 / 1

Hydrogen is supplied via two electromagnetic port injectors placed in one housing shown in Fig. 5. The ignition system is using a spark plug with high heat conductivity and no remains of residual electric energy. The engine is mounted on an exchangeable frame and connected to a dynamometer.

Fig. 5 Injector unit

The gaseous hydrogen stored in gas bottles outside the engine test bench building is cooled in a heat exchanger guided through liquid nitrogen to a temperature of 77 K. The supplying hydrogen pressure is also controlled separately and a Coriolis flow rate meter measures the mass flow. Cylinder pressure is measured by a water-cooled piezoelectric pressure transducer, and inlet and exhaust pressure with piezo-resistive sensors. Several thermocouples are applied to measure all relevant engine temperatures. To simulate colder hydrogen the intake air was cooled down to 4 °C at the manifold inlet.

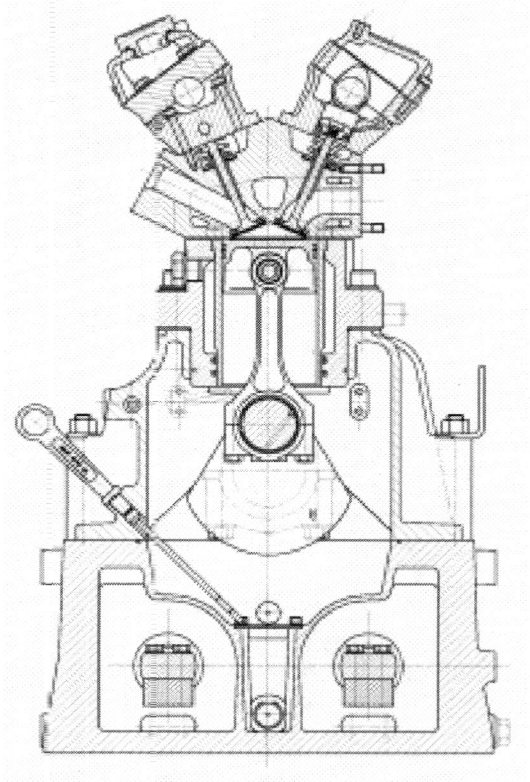

Fig. 6 Cross section of research engine

TESTBED RESULTS

Full-load operation of the engine was conducted to investigate the characteristics in the hydrogen mode at ambient and cryogenic operation. All measurement points were driven at wide-open-throttle and stoichiometric air-fuel ratio.

Fig. 7 shows the variation of the indicated mean effective pressure in terms of the engine speed for the different hydrogen temperature ranges. At 3600 RPM, the maximum volumetric efficiency yields to the highest IMEP. The theoretical gain of 40 % of real cryogenic (~25 K) compared to ambient mixture formation of hydrogen as shown in Fig. 1 was approached by 28.2 % in the quasi-cryogenic set-up with LN_2-cooling.

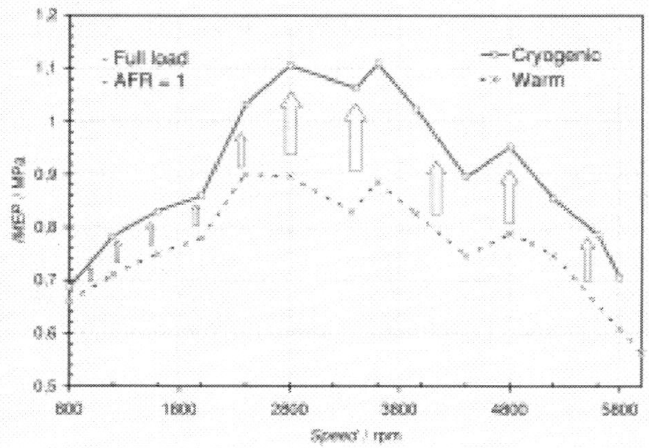

Fig. 7 IMEP over speed

The Indicated Mean Effective Pressure (IMEP) in terms of the air-fuel ratio is shown in Fig. 8 at 3600 RPM and wide-open throttle for the cryogenic operation mode.

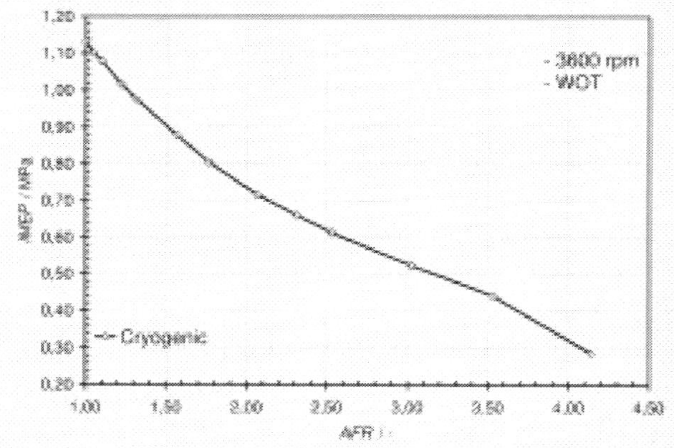

Fig. 8 IMEP over AFR

At the highest IMEP and hydrogen flow rate ($\lambda = 1$) the hydrogen rail temperature achieves a minimum of -145 °C. As the charge becomes leaner the IMEP is reduced monotonically.

In Fig. 9 the plots of cylinder pressure versus crank angle at 2000 RPM are shown for the two fuel temperatures. The maximum pressure of 74 bar is attained for the cryogenic operation. Hydrogen burns steeply after ignition with a maximum pressure gradient of 5.6 bar/°CA. The higher compression pressure in cryogenic mode is due to a raised adiabatic exponent at lower mixture temperatures. At first glance, the power output of the cryogenic mode seems to be strongly increased compared to the ambient mode but crucial for the indicated mean effective pressure is the integral of pressure multiplied by the change in volume. The third curve indicates the CFD-simulated cylinder pressure for the cryogenic combustion.

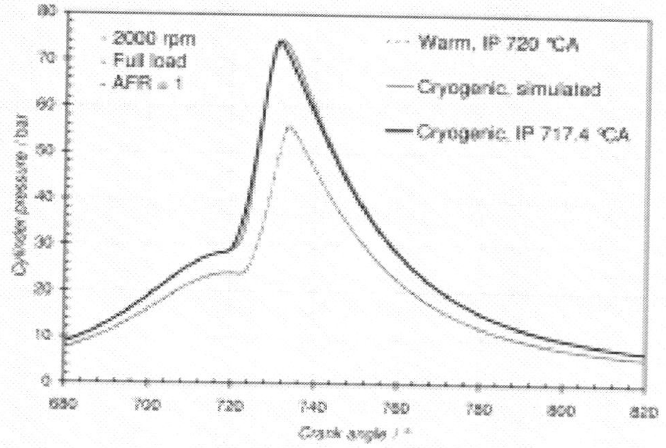

Fig. 9 Cylinder pressure curves

In the next chapter the simulation of the introduced research engine is explained.

COMPUTATIONAL MODELLING

GENERAL TOOL DESCRIPTION

For the numerical investigations the code FIRE™ v8.2, commercialised by AVL, is used. This CFD solver is based on the finite volume approach to discretise spatial dimensions, which rests on the integral form of the general conservation laws applied to polyhedral control volumes. The implicit finite difference procedure is used to discretise time. The solution algorithm employed enables the usage of hexahedral-dominated unstructured moving grids. The coupling of the velocity and pressure fields are effectively attained by the segregated SIMPLE (Semi-Implicit Pressure Linked Equation) algorithm. For the engine application the standard k–ε turbulence model is used. There are several combustion models implemented for premixed, partially-premixed or non-premixed conditions, namely Eddy-Break-Up; Coherent Flamelet; Probability Density Function and others. Because of the fast chemistry and premixed combustion of hydrogen, this simulation work is done with the relatively simple Eddy-Break-Up model by Magnussen to derive the mean reaction rate of combustion [8]:

$$\overline{\rho}\tilde{r}_{fu} = \frac{C_{fu}}{\tau_R}\overline{\rho}\min\left(\tilde{y}_{fu}, \frac{\tilde{y}_{Ox}}{S}, \frac{C_{Pr}\cdot\tilde{y}_{Pr}}{1+S}\right) \quad (1)$$

With adjusted model-parameters (C_{Pr} and C_{fu}) the result of the combustion simulation is shown in Fig. 9 for the cryogenic cylinder pressure curve. The good match between simulation and experiment is obtained without a physical basis, so poor transferability to other geometries or pressure levels is to be expected. Better scalability is expected for more complex combustion models like Coherent Flamelet or Probability Density Function.

As a consequence of averaging, the governing Navier-Stokes equations yield the Reynolds-averaged Navier-Stokes (RANS) equations with unknown correlations. These terms are also called Reynolds stresses or second moments. In order to close the equation set, these variables need to be supplemented before any solution can be achieved. For this, additional algebraic or differential relations are needed, namely what is called a turbulence closure model. The second moments are always vectors of higher order than the basic variables. This complicates the closure problem because six more variables than for a laminar flow of the same complexity must be solved. Two basic levels of modelling are currently used in computational fluid dynamics and transport processes, the first-order and the second-order models. Of the class of first-order eddy-viscosity models the standard k–ε turbulence model is the most widely used turbulence model and is recommended in situations where modelling other physical phenomena, such as chemical reactions or multi-phase interactions, brings in uncertainties that outweigh those inherent in the k–ε turbulence model. The k–ε model consists of the transport equations for the kinetic turbulent energy k and the dissipation ε. The exact k-equation is derived by scalar multiplication of the transport equation for the velocity fluctuation by the velocity fluctuation itself. The ε-equation reflects the composition of the k-equation. One exception is the term which takes into account the fluid compression on the source of ε and which does not have its counterpart in the k-equation. The above model represents the complete and computationally-efficient turbulent closure. However, its shortcomings such as the inability to predict the stress anisotropy or to deal properly with the effects of streamline curvature, rotation, buoyancy and stream wise pressure gradients are well known.

When combustion is considered, the question arises whether the interaction between turbulence and combustion invalidates classical scaling laws known from non-reacting turbulent flows. Hydrogen combustion releases heat and thereby generates flow instability by buoyancy (here neglected) and gas expansion, thus enhancing the transition to turbulence. Turbulence increases the mixing process and thereby enhances combustion to a certain degree. Combustion models must enable the calculation of species transport/mixing phenomena and the simulation of combustion in internal combustion engines under premixed, partially-premixed or non-premixed conditions. Although it is desirable to use detailed reaction mechanisms, available computational resources are inadequate to manage dozens of elementary reactions with many participating species. This is due to the fact that for each species considered in the reaction mechanism, an additional conservation equation must be solved. Chemical kinetic effects are, therefore, accounted for by either a single-step overall reaction for treatment of the oxidation processes in flames, or by separating modelling of low-temperature reactions from modelling of the high temperature kinetics. Mean chemical reaction rates represent a central problem in the numerical simulation, because they appear to be highly non-linear functions of the local values of temperature and species concentrations. Combustion requires that fuel and oxidiser are mixed at the molecular level. This happens mainly due to the turbulent mixing process Once a range of different eddies (turbulent structures) has developed, strain and shear will increase and thereby increase the concentration gradients at the interface between reactants, which in turn enhances their molecular interdiffusion. Molecular mixing of fuel and oxidiser, therefore, takes place at the interface between smaller eddies, as the conduction of heat and the diffusion of radicals out of the reaction zone. Combustion increases gradients even further, since chemical reactions consume the fuel and the oxidiser at the interface. Combustion models thus rely on the empirical evidence that chemical reactions are nearly always fast compared to all turbulent time scales or on the assumption that with the support of molecular diffusion, reactions concentrate in thin layers of a width that is typically smaller than the Kolmogorov scale.

As a first step in modelling the combustion of hydrogen the simple Eddy-Break-Up model introduced 1976 by Magnussen and Hjertager [8] is used. This model assumes that in premixed turbulent flames the reactants are contained in the same eddies and are separated from eddies containing hot combustion products. The chemical reactions usually have time scales that are very short compared to the characteristics of the turbulent transport processes. Thus, it can be assumed that the rate of combustion is determined by the rate of intermixing on a molecular scale of the eddies containing reactants and those containing hot products, in other words by the rate of dissipation of these eddies. The values of the empirical coefficients entering the model have been shown to depend on turbulence and fuel parameters. Adjustment is required with respect to the experimental combustion data of the investigated case. The attractive feature of this model is that it does not call for predictions of fluctuations of the reacting species.

CASE DESCRIPTION

Calculations have been made for a cryogenic and a warm hydrogen operating point at 2000 RPM, wide-open throttle and AFR = 1. The computational domain used is described below:

Moving grid

In Fig. 10 the CAD surface and grid of the computational domain is shown. Meshing was done with the semi-automatic FAME^{TM} Advanced Hybrid meshing tool of AVL. With this grid, consisting of up to 680 000 control volumes, calculations are possible for a whole work cycle over 720 °CA. With the AVL mesher local refinement can easily be conducted in regions where high gradients of the flow variables are expected.

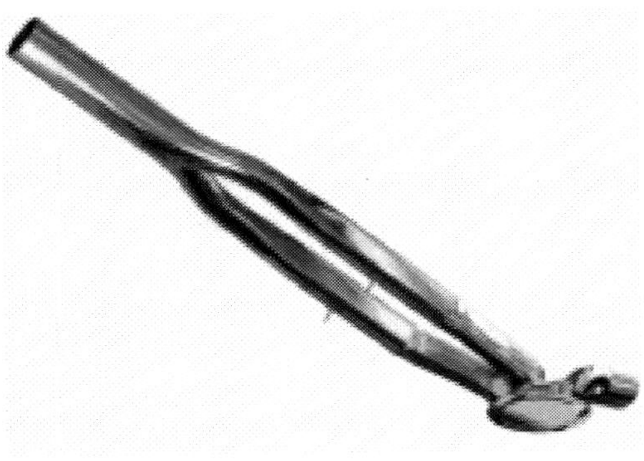

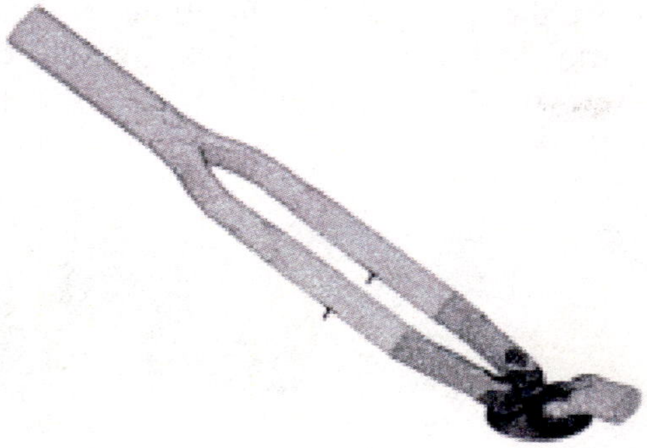

Fig. 10 CAD model and computational grid

For example, a refined grid as shown in Fig. 11 was applied to the orifice region where hydrogen is injected into the fluid domain and shock capturing is needed because of local supersonic flow.

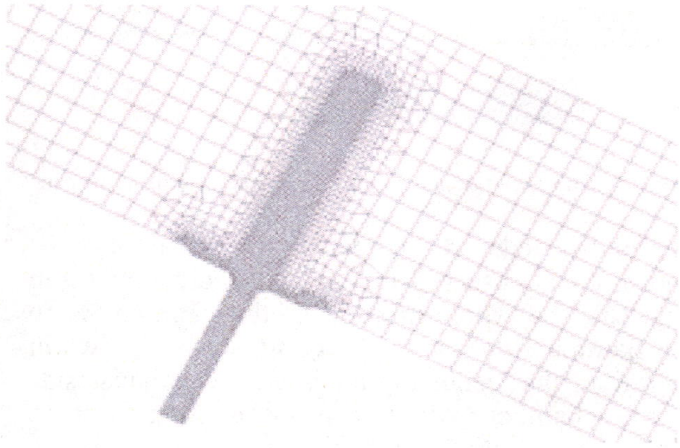

Fig. 11 Refined grid of fuel inlet

Because of the great number of control volumes and the main focus on the hydrogen-mixture formation process, the normally required grid resolution due to near wall treatment was intentionally neglected.

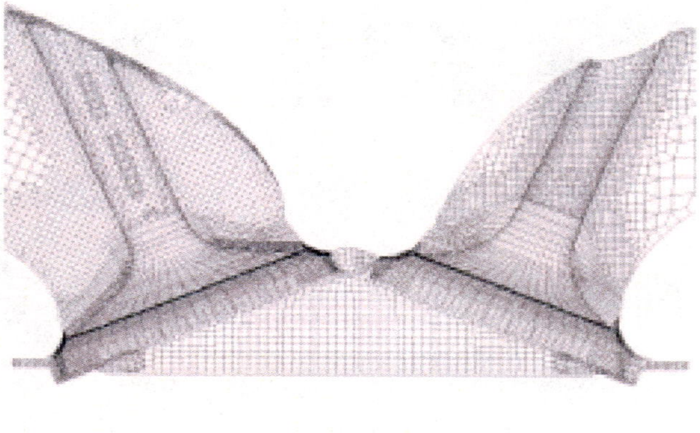

Fig. 12 Moving grid topology

The generation of a moving grid for the whole work cycle is a relatively costly procedure. Fig. 12 shows the separated meshing approach at TDC of the charge exchange. The manifold, intake and exhaust ports, valves and chamber are meshed separately and are topologically connected via an arbitrary interface. At the interface the fluxes are calculated and passed from one side to the other. With this meshing methodology the valve gaps can be meshed with pure hexahedral elements and the cell faces can be aligned perpendicular to the main flow direction for best numerical results. However, this treatment requires up to 30 single mesh pairs and several rezones (mapping of flow field variables between two mesh pairs).

Boundary conditions

Values from the experiments were used to estimate initialisation of the domain and were applied to the interfaces of the model. At the inlet of the manifold and outlet of the exhaust port the transient static air pressures were used as boundary conditions.

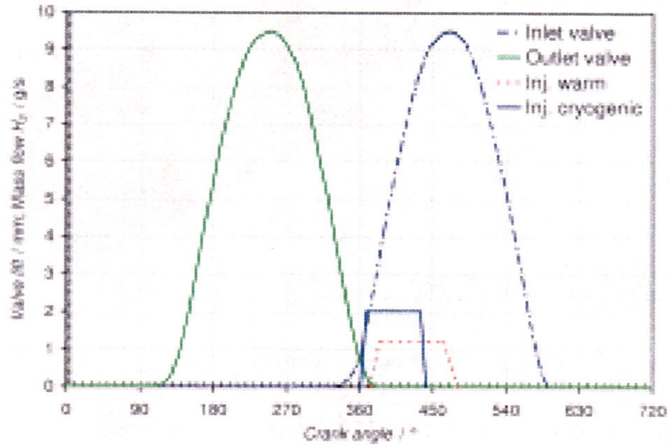

Fig. 13 Valve lift curve and injection timing

Mean temperature values for air, hydrogen and exhaust gas were also taken from measurements. Wall temperatures were estimated and kept constant during calculation. The motion of the grid is prescribed by crank gear relations and valve lift curves as shown in Fig. 13. This graph also shows the injection timing of the chosen warm and cryogenic hydrogen operating points.

SIMULATION RESULTS

For the simulation with hydrogen at ambient and cryogenic temperatures an overview of results is given:

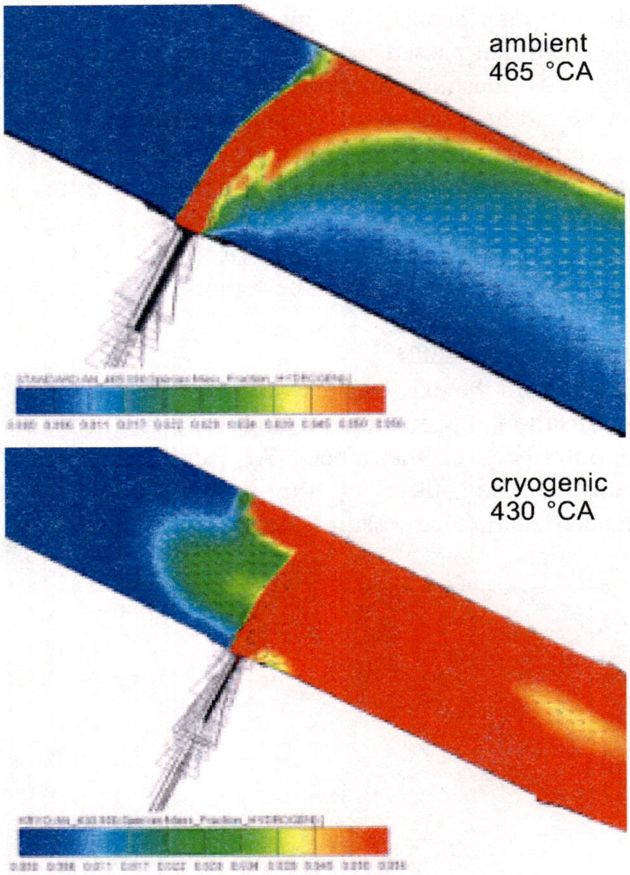

Fig. 14 Injection of hydrogen

Fig. 14 shows the deflected injection of hydrogen into the intake tube for the ambient (upper image) and cryogenic (lower image) cases. The time steps are shown just before the injections collapse. Once again, the green colour corresponds to stoichiometric composition. The velocity vectors indicate the minor influence of the airflow on the hydrogen jet penetration. Crucial to intrusion depth is the momentum ratio of the crossing streams. The maximum velocities are about 1600 m/s for the warm and 1452 m/s for the cryogenic hydrogen at the orifice. The intake air flows at approximately 37 m/s. With cryogenic injection the intake tube is highly filled with hydrogen, in comparison to warm injection, due to higher density. Although the investigated engine has only small charge motion, the vertical cut through the symmetry plane of the chamber in Fig. 15 shows the good homogenisation tendency of the warm hydrogen-air mixture during the compression stroke, including the four-prong spark plug modelled in detail.

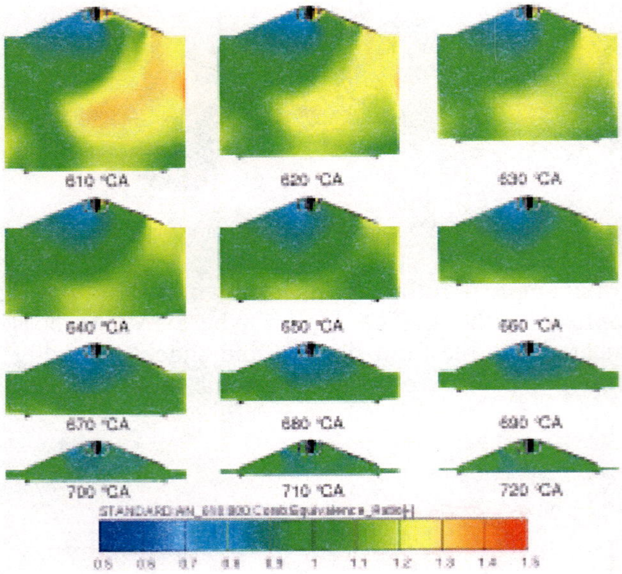

Fig. 15 Charge homogenisation, 2000 rpm, warm hydrogen

Although reduced tendency toward homogenisation is observed in the cryogenic operation (Fig. 16), the whole mixture of the chamber is in the flammability range. In contrast to the warm operation, the leaner region is on the exhaust side of the chamber, which may be a good way to reduce the heating of the exhaust side during combustion.

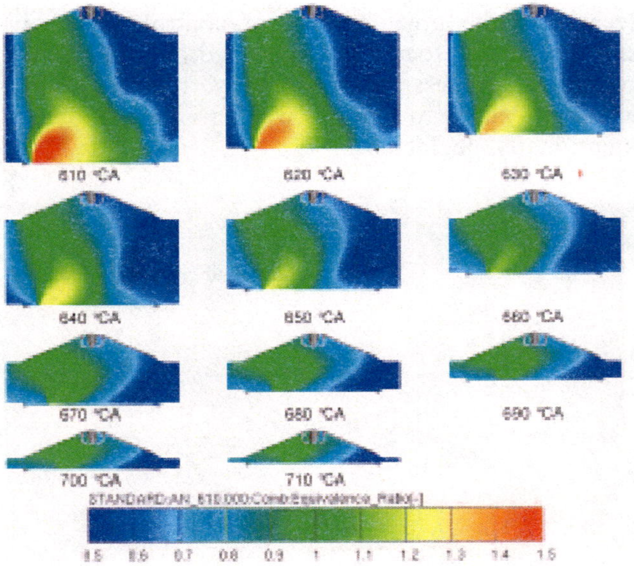

Fig. 16 Charge homogenisation, 2000 rpm, cryogenic hydrogen

As another example for the cryogenic calculation results, Fig. 17 shows the temperature distribution during the intake stroke, where hot residual gas behind the intake valve is enclosed by the cold fresh charge. It also illustrates the increased charge density by using a cryogenic mixture formation.

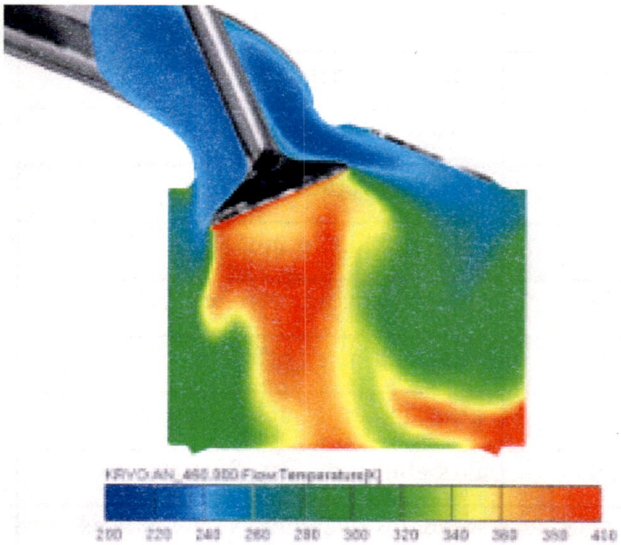

Fig. 17 Temperature distribution at 460 °CA, cryogenic H_2

The improved charge exchange for cryogenic mixture formation is presented in Fig. 18. The curve of the mass flow into the cylinder has a peak when the intake valve is already closing. Before the valve is shut at 588 °CA, in both warm and cryogenic modes the charge is flowing out of the chamber back into the intake ports. Mixture outside the chamber within the flammability limits is the prerequisite for backfire occurrence.

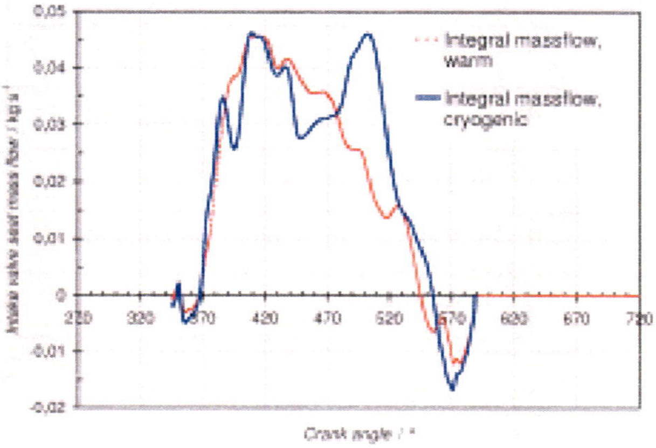

Fig. 18 Mass flow into cylinder

CONCLUSION

Using the CFD-tool FIRE™ v8.2, it was possible at reasonable cost (computational resources) to simulate the hydrogen mixture formation with transonic injection. The validation and proof of the reliability of the simulation results was achieved by LIF-measurements of a helium-operated optical engine. The developed simulation process is already able to predict the operating behaviour even of a virtual hydrogen engine. First promising results in simulating the hydrogen combustion of an internal combustion engine were obtained by the Eddy-Break-Up model. On the simulation side the following conclusions can be drawn:

- Fluid dynamics simulation of hydrogen-driven engine is not a matter of course feature for current CFD codes, due to high density gradients and Mach numbers greater than unity.
- The hydrogen mixture formation of an internal combustion engine can now be simulated within one week of computing time on four CPUs.

On the experimental side the following conclusions can be made for cryogenic hydrogen operation:

- The increased volumetric heat value and improved charge exchange resulting from the use of cryogenic hydrogen clearly increases the power output of the engine.
- No knock borderline exists to limit the optimal choice of spark timing.
- An overall operation without occurrence of pre-ignition or backfire phenomena even at stoichiometric equivalence ratio is possible in cryogenic mode.

Further progress in simulation and experiment will be related to investigation of other combustion models and transferability to different operation conditions.

In January 2004 the research project HyICE (Optimisation of the Hydrogen ICE) has been started. Several partners in this project, which has been funded by the European Union for the next three years, carry out detailed research of hydrogen engines. The HyICE consortium consists of nine partners, from automotive OEMs and suppliers to research institutes.

REFERENCES

1. Pehr, K., Burckhardt, S., Korn, T., Koppi, J., Partsch, P. (2002) Hydrogen - The Fuel of the Future - The BMW 750hL. *Automobiltechnische Zeitschrift* 104, 2, 120-131.
2. Schüers, A., Abel, A., Artmann, R., Fickel, C., Preis, M. (2002) Twelve-Cylinder Hydrogen Engine in the BMW 750hL. *Motortechnische Zeitschrift* 63, 2, 98-105.
3. Freymann, R., Pehr, K., Strobl, W. (2002) Twenty-Five Years of Continuous Hydrogen Research at BMW. *Proc. of the JSAE Annual Congres* 25, 17-20.
4. Göschel, B. (2003) The Hydrogen Combustion Engine as the Drive System for the BMW of the Future. *24th Int. Vienna Engine Symposium.*
5. Schüers, A., Gerbig, F., Wimmer, A., Kovac, K., (2003) Thermodynamic Analysis of the Combustion Process of the Hydrogen ICE with Internal Mixture Formation. *Symposium on the Combustion Process of ICE, Graz.*
6. Berckmüller, M., Rottengruber, H., Eder, A., Brehm, N., Elsässer, G., Müller-Alander, G., Schwarz, C. (2003) Potentials of a Charged SI-Hydrogen Engine. *SAE*, 2003-01-3210.
7. Blotevogel T., Hartmann M. (2003) Anwendung der planaren laserinduzierten Fluoreszenz zur qualitativen und quantitativen Gemischbildungsuntersuchung bei Gasmotoren, speziell bei Wasserstoffmotoren. *Proc. Dessau Gas Engine Conference*, 113-131.
8. Magnussen B.F., Hjertager B.H. (1976) On mathematical modeling of turbulent combustion with special emphasis on soot formation and combustion. *Int. Symp on Combustion* 16, 719-729.
9. AVL LIST GmbH (2003) FIRE™ v8.1.1 user guide.

ACRONYMS, ABBREVIATIONS

AFR	Air-fuel ratio
CA	Crank angle
CAD	Computer-aided design
CFD	Computational fluid dynamics
CFM	Coherent flamelet model
DOHC	Double overhead camshaft
EBU	Eddy-Break-Up model
Eddy	A turbulent structure, e.g., a vortex
IMEP	Indicated mean effective pressure
LIF	Laser induced fluorescence
PDF	Probability density function
RANS	Reynolds averaged Navier-Stokes
SIMPLE	Semi-implicit pressure linked equation
TDC	Top dead center
WOT	Wide open throttle
ε	Dissipation rate of turbulence (m^2/s^3)
ρ	Density
τ_R	Turbulent mixing time scale
C_{fu}	Model constant - fuel
C_{Pr}	Model constant - product
k	Turbulent kinetic energy (m^2/s^2)
r_{fu}	Fuel consumption
S	Stoichiometric oxygen requirement
y	Mass fraction

WHICH FUELS FOR LOW CO_2 ENGINES?

Which Fuels For Low CO_2 Engines?
© P. Duret, X. Montagne (Eds.) and Editions Technip, Paris, 2004, pp. 201 – 208
27 rue Ginoux, 75015 Paris

FURORE – FUTURE ROAD VEHICLE RESEARCH, A ROADMAP FOR THE FUTURE

A CONTRIBUTION TO THE IDENTIFICATION OF KEY TECHNOLOGIES FOR FUELS AND POWERTRAIN FOR A SUSTAINABLE DEVELOPMENT OF EUROPEAN ROAD TRANSPORT

Josef Affenzeller

AVL List GmbH, on behalf of the FURORE Consortium

ABSTRACT – *Individual traffic will be the major pillar of global traffic in the next 30 to 50 years. The road vehicle of the future will obviously incorporate an array of technologies. In consideration of still uncertain basis of social and environmental conditions, the question arises as to how and which technology routes should be followed to solve problems and conflicts faced by society in terms of energy consumption, emissions and safety and to meet the needs of future mobility.*

In June 2002 the European Automotive Research Partners Association EARPA initiated the Thematic Network FURORE – Future Road Vehicle Research. FURORE established a platform of European stakeholders to discuss breakthrough technologies and the corresponding research demand for vehicles of the year 2020 and beyond. The network focused predominantly on road vehicles powered by internal combustion engines, but was also analysing potential breakthrough technologies in advanced fuels and systems such as hybrids and alternative propulsion. The FURORE Network aimed at creating an integrated, long-term research strategy for the European automotive sector, which is based on the specific point of view of research organisations and Universities.

INTRODUCTION

The road vehicle of the future will incorporate an array of new technologies. In consideration of the still-uncertain basis of social and environmental conditions in twenty years, the question arises as to which technology routes should be followed to meet the needs of future mobility and to reduce environmental and economical problems and conflicts faced by society due to increased mobility. This refers in particular to energy consumption and the CO_2 challenge, toxic emissions, noise and safety issues.

The European Thematic Network *FURORE – Future Road Vehicle Research* – established a platform of European stakeholders to discuss breakthrough technologies and the corresponding research demand for vehicles of the year 2020 and beyond[1]. The network focused predominantly on the research demand for road vehicles powered by internal combustion engines, but was also analysing the potential of advanced fuels and alternative propulsion systems. Note: the challenges of production and infrastructure are certainly important issues too but they are not a special FURORE project topic.

Automotive research and development has become a more international, collaborative effort within and between research organisations, universities and industry. The FURORE Network aimed at improving this collaboration by creating an integrated, long-term research strategy for the European automotive sector, based on the specific point of view of research organisations and universities, providing broad independent expertise for future research programmes regarding sustainable road transport.

FURORE was initiated by EARPA, the European Automotive Research Partners Association. For the realisation of this ambitious project, several leading European research organisations and universities formed a task force.

[1] For more information, see: http://www.furore-network.com

The results of the FURORE activities were published as an automotive R&D technology roadmap and the main findings for energy, fuels and powertrain are summarised in this paper. The roadmap is intended to aid co-operation between industry, research institutions and universities based on a system approach that integrates the leading stakeholders from the relevant technological areas.

VISION FOR 2020 AND BEYOND

The breakthrough technologies and the derived research demand are based on a comprehensive analysis of future road transport scenarios, voting exercises, workshops and individual discussions. In general technological targets can be split up into research targets, engineering targets and commercial targets. It is essential to mention that FURORE focuses primarily on research targets. This means that the potential of a technology to realise intended technological objectives is examined but the aspects of feasibility and large-scale production are not investigated in detail.

VIEW ON FUTURE EUROPEAN TRAFFIC SCENARIOS

Due to its enlargement to include Central and Eastern European Countries, the European Union is expected to be the world's strongest economic area in 2020. More than 70% will live in urban and suburban agglomerations and more than 25% of the population will be older than 60 years. These are selected criteria which have to be considered when discussing future technologies. Future mobility will be focused predominantly on the citizens and their needs, which are basically individual mobility and the availability of goods. By 2020 there will already be positive results of today's efforts in optimising intermodality and interoperability among the different transportation systems. Nevertheless, the road will still be the dominant mode of travel and transportation and person-kilometres will further increase. It is possible that there will be subscription systems with the possibility to use several means of individual and collective transportation depending on safety, effectiveness, cost, time and comfort. Citizens will expect the freedom to choose among the different means of transportation with seamless transfer from one to the other and with a minimum of concern for the vehicle used. Goods transport will increase as well, due to EU enlargement and integration, just-in-time production and distribution (both world-wide) and freight and fleet deregulation to decrease costs and increase competition. An adequate infrastructure (including physical network and traffic management systems) will be required to keep travel time to a minimum, reduce congestion and realise efficient intermodality. As the establishment of trans-European transport networks is a prerequisite for future mobility, much improvement will have to be realised by 2020.

VIEW ON FUTURE EUROPEAN ENERGY SCENARIOS

Traffic forecasts for the next 10-20 years show that the increase of traffic volume and congestion will counteract the significant improvement of the vehicle technologies and the negative consequences of energy consumption, emissions, noise and safety will further increase. To enhance the quality of air, future legislation will impose even lower pollutant emission targets, approaching zero-impact level, but recognising the diminishing return as this zero-impact is approached. The EC aims at a substantial substitution of the oil-based fuel market by alternatives (in particular natural gas, bio-fuels and hydrogen) by 2020. A further objective is the reduction of the European Union's energy dependency which will also improve the security of energy supply. Today, transport is 98% dependent on crude oil, representing 67% of final oil demand. There is consensus that, if the use of alternatives grows until 2020 in line with EU targets, the majority of alternative fuels will be supplements and blends to existing liquid fuel types. Hydrogen - often seen as the ultimate goal - is another option. H_2-penetration until and also beyond 2020 depends greatly on progress in solving the challenges of production and storage. Apart from that it requires political decisions on prime energy sources which have not yet been made.

For certain alternative fuels there are still major hurdles - technological as well as cost-related -which need to be solved before a major increase in market share can be expected. However, given the continuing pressure on CO_2-emissions, voluntary agreements are likely to spread to all sectors. In order to meet EU greenhouse gas emission targets, as well as to decrease energy dependency, moving towards alternative fuels alone will not be sufficient. In addition, overall energy consumption in transportation must be reduced significantly. Together with considerable efficiency gains it is believed that this is the only measure which will lead to success.

ENERGY AND FUELS: OBJECTIVES AND IDENTIFIED BREAKTHROUGH TECHNOLOGIES

Encouraged by the European Commission's above-mentioned target for alternative fuels, the portfolio of available fuels will become more heterogeneous by 2020. The introduction of several new fuels is a challenge due to the costly proliferation of many different engine and emission control technologies, and a parallel proliferation of supporting fuel infrastructures. In some cases these new fuels or blends may be synergistic with new combustion modes, but this remains to be verified.

Conventional fuels:

The main fuels for transportation, i.e. gasoline and diesel fuel, will not change dramatically but fuel specifications will further evolve. Both will become practically sulphur-free (10 ppm sulphur achieved in 2009), and other property improvements should be evaluated strictly with regard to well-to-wheel efficiency and CO_2 impacts.

Advanced conventional fuels:

These are mainly mineraloil-based, such as syn-fuels and blended fuels. They are directly linked to future combustion requirements and/or optimisation constraints. Pure or blended fuels must be considered. For liquids from natural gas or coal, cost and CO_2 have to be tracked.

Alternative fuels:

Natural gas, biomass to liquid fuels, bio-fuels and hydrogen are the main candidates of the very ambitious alternative fuel proposal. Large-scale production facilities are still missing (apart from natural gas) and it can be estimated that there also is an optimisation potential in the production processes when building up those production facilities. Nevertheless the major issue with these fuels is how to produce, store and distribute them at a competitive price.

It is assumed that pure hydrogen could only attain a non-negligible market share beyond the considered time period, by 2030 and later, for example also as a valuable 'add-in' for biomass-based renewable fuels. Production of hydrogen from hydrocarbons either in the plant or in the vehicle is not efficient with regard to CO_2 unless development of large centralised systems with CO_2 sequestration and the security of supply objectives, and does not seem to be a feasible long term solution, even if the Natural gas option is today the best cost efficient solution. The Production of hydrogen from electricity via electrolysis is needed in large scale quantities and not only for niche applications. This seems not feasible from renewable sources only. Therefore the use of nuclear power is one option but would definitely lead to a general political discussion on energy supply. The Furore Network carried out a study about the technical potential of biomass for different future scenarios for an enlarged EU 30. This potential results basically from residues (logging, straw, used wood) and cultivated energy plants. The outcome shows that the maximum technical potential for biofuels (not considering the constraints, which reduce this max. technical potential to "real" figures for fuels) could cover between 30% and 45% of the inland energy demand for road transport.[2]

Hydrogen could serve as an intermediate vector for the possible use of electricity in road transportation. As the electricity distribution grid is largely existing and competitive, research on efficient and cost-effective on-board electricity storage development should not be overlooked as an interesting solution, especially when combined with CO_2 control of power generation (CO_2 centralised sequestration process potential). Focussed basic research is needed to solve the problems of energy storage, especially batteries.

Major concerns with most of all these alternative solutions are production costs, and for gaseous fuels the distribution and improvement of storage solutions; if there is no breakthrough improvement in this area, alternative fuels will remain marginal and will serve only niche markets. All future

[2] The study is available at: http://www.furore-network.com

research initiatives must consider fuels, engines and aftertreatment as interactive systems.

FURORE Roadmap for Energy and Fuel:

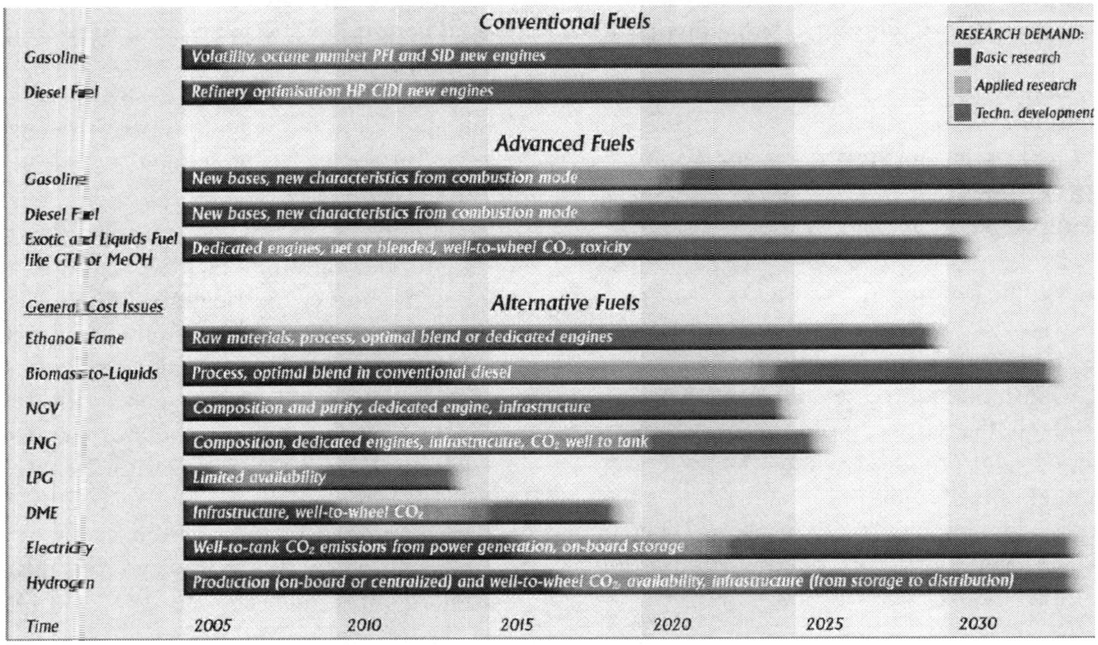

POWERTRAIN INCL. EMISSION AND AFTERTREATMENT: OBJECTIVES AND IDENTIFIED BREAKTHROUGH

Vehicle propulsion system technology development is an evolutionary process, influenced by the requirements of customers, the legislative boundary conditions like emissions and safety standards, the energy resources and prices influenced by production/distribution costs and not in the least by excise duties and taxes. The general targets for technology trends and visions are energy consumption reduction, near zero emissions and alternative fuel compatible power systems.

SPARK IGNITION ENGINES-STATE-OF-THE-ART TECHNOLOGY UNTIL 2007

In the near future, the importance of down-sized spark ignition engines will increase considerably. Boosting (supercharging and turbocharging) will be combined with displacement reduction, sometimes with new, redesigned engines. The feasible reduction in cylinder displacement is up to 40 percent, with a corresponding benefit in fuel consumption and carbon dioxide emissions of up to 20 percent. The adoption of direct injection will be another means of improving engine efficiency, but in terms of costs, these engines with stratified combustion are more expensive than conventional spark ignition engines due to the need for advanced injection technology and additional nitrogen oxide aftertreatment. As an interim step, direct injection with homogeneous combustion (enabling conventional emission control) in combination with supercharging will take a growing market share. Variable valve trains enable a significant fuel economy increase and CO_2 reduction up to 20%.

Another option are concepts with a variable compression ratio (VCR) in combination with downsizing, which offer (for supercharged engines) a potential up to 25% CO_2 reduction. Finally, first low cost belt driven starter generators operating at 14V will reduce fuel consumption by enabling automatic stop/start, but without re-generative braking and torque support which require 42V systems. Naturally, many of these technologies can be effective in combination but research work must be carried out to find a cost-effective solution for high fuel economy for each vehicle type.

SPARK IGNITION ENGINES-TECHNOLOGY TRENDS UNTIL 2020

For the time span to 2020 it is most likely that above-mentioned technologies will further evolve and will feature advanced, adaptive control of these flexible sub-systems, using new sensor technologies. New combustion systems such as controlled auto-ignition with very low engine out emissions of Nox and particulates will be realised to be effective in a wide engine map area and without full load penalties. Some evolution of fuel properties may support this new combustion philosophy. Spark ignition engine technology is fully compatible with both natural gas and hydrogen, and when logistical problems are solved and if energy policy promotes these fuels into the marketplace, combustion systems tailored to these fuels will emerge. Extreme-charged lean burn spark ignited combustion processes are promising in this respect too. This combustion technology has extremely low Nox emissions and may only need oxidation or three way-catalytic converters.

LIGHT-DUTY DIESEL ENGINES-STATE-OF-THE-ART TECHNOLOGY UNTIL 2007

The significance of down-sized diesel engines will also grow in the next decade, because of increased supercharging rates, intercooling as well as the possibility of electrically assisted turbochargers and variable valve train concepts, which enable an optimised torque and transient characteristic. A specific power output up to 70 kilowatt per litre seems to be reachable. Compared with current diesel engine technology, a fuel consumption benefit of up to 25 percent is predicted as a result of down-sizing measures and further reduction of friction losses.

Advanced fuel injection systems, which allow an adapted injection characteristic such as pilot, split and post injection as well as rate shaping, will reduce the local exhaust emissions substantially, perhaps avoiding the need for additional particulate and Nox control devices in small to medium vehicles. Also injection nozzles with variable injection-hole size will be a part of these advanced diesel engine concepts.

HEAVY DUTY DIESEL ENGINES – STATE-OF-THE-ART TECHNOLOGY UNTIL 2007

The challenges for the heavy duty engine are similar, with greater emphasis on emission control but a stronger market desire for efficiency in order to reduce operating cost. Fuel consumption and Nox/PM emission will be addressed by combustion process improvements, including application of flexible high pressure fuel injection, four valves per cylinder, improved boosting, electronic control and low oil consumption. There are a variety of combustion process philosophies with corresponding emission control needs. Exhaust gas recirculation (EGR), particulate traps, selective catalytic reduction (SCR) based on urea in combination with oxidation catalyst and lean Nox traps will be used.

DIESEL ENGINES: TECHNOLOGY TRENDS UNTIL 2020

To obtain further reductions in emissions and fuel consumption, sub-system variability will increase for both heavy duty and light duty diesel engines, enabled by advanced control systems. To increase the degree of down-sizing, in-cylinder peak pressure will increase, calling for new engine design concepts including supercharging and improved materials. With single cycle control strategies, emissions will decrease further. A greater step in Nox emission reduction in a wide mapping area will be expected for homogeneous charge compression ignition (HCCI) in the part load range without full load fuel consumption penalties.

HYBRID AND AUXILIARY POWER UNITS

Increasing on-board electric power requirements, together with a desire for powertrain efficiency improvement, creates a need for hybrid-electric technology and auxiliary power units. Hybrids with an Integrated Starter Generator and "42 Volt board-net", in combination with a down-sized engine, are set to become a widespread measure for reducing fuel consumption and emissions, especially in city traffic. This "mild hybrid" technology provides good fuel saving potential by applying engine shut off and brake energy recovery, and by using the electric motor to assist for improved launch of vehicles with down-sized engines. Costs, weight, reliability and increased complexity were identified as the hurdles and barriers for this technology. Research has to be done on electric machines, the integration of the Starter Generator , the control system and the reduction of weight. Nevertheless hybrid technology with a high share of electrical power calls for a greater change to the powertrain of a conventional vehicle. New transmissions, clutches and big electric machines are needed, together with high power batteries, which remain very expensive and not sufficiently developed. Research is needed on the components in a hybrid powertrain, like new transmissions, electric motors, batteries and supercaps. Also the integration and control of all the components is an important issue. Despite the additional complexity and components compared with the conventional powertrains, full hybrids will slowly penetrate the market. Particularly the cost/fuel & emission saving ratio has to be considered. A different approach, which reduces the many issues related to batteries and electrical energy storage in general, is the use of small, efficient auxiliary power units (APUs) to supply basic power needs, supplemented by a main internal combustion engine providing high power.

Fuel cells or combustion engines (perhaps innovative linear, or rotary types) are suitable as an auxiliary power unit. Fuel cell systems as APUs require significant reduction in cost, weight and size, both of the stack itself and also for the fuel and air supply systems and the heat management. Reforming of liquid fuels, or feasible hydrogen storage will be required. The reformer adds complexity and costs to the fuel cell system. Reformers for gasoline and diesel are in a very early technological stage and a lot of research is needed to guarantee the hydrogen generation during the transient operation in a vehicle.

TECHNOLOGY VISIONS BEYOND 2020

It is likely that the "variable" combustion engine, with highly variable subsystems under the control of a sophisticated overall powertrain management system, will continue to evolve in combination with improved, partly renewably sourced, fuels. The diesel homogeneous charge compression ignition (HCCI) and gasoline controlled auto ignition (CAI) combustion processes, which have many similarities, may merge along with fuel properties into a "combined combustion system". In combination with the improved aftertreatment technologies, "zero impact" emission levels will be reached.

In combination with advanced transmissions (probably dual-clutch or infinitely variable type) with high efficiency and flexibility the energy consumption of vehicle power trains will be reduced further. Together with new materials and new design rules a specific engine power of 150 kW/litre for gasoline and 75 kW/litre for diesel engines can be expected. Electric hybridisation may be commonplace beyond 2020, both as an efficiency improvement enabler and provider of extra power needed for x-by-wire systems. Significant improvements, and possibly a breakthrough in energy storage technology are required to overcome the major hurdle, provided that the research effort has been supported in the intervening years. Thus, if there is more CO_2-free electrical energy available from renewable or nuclear sources, battery-electric vehicles have to be discussed again. Pure battery vehicles already exist in niche applications and have been under research for decades. The main research field is consequently the battery, which has to be improved regarding energy and power density, reliability as well as with regard to cost reduction. For the electric powertrain itself new, more cost-effective and material-saving production methods for electric motors are an issue.

Fuel cell vehicles running on hydrogen have the advantage of being real zero-emission vehicles. Technologically, they can be considered as electric vehicles in which the battery has been replaced or supplemented by a fuel cell. In many applications a secondary, conventional battery will be needed for start-up and for conditioning of the fuel cell. Fuel cell stacks show best efficiency at part load which is favourable for inner city driving. The stack is only one part of the fuel cell power train efficiency – it is necessary to take into account the air system, the thermal management, the cabin heating/cooling, the cooled climate start (freezing of combustion water in the cell), the electric machine and the power converter, which decrease the overall powertrain efficiency. On top of this, the efficiency of the reformer (if used) and well-to-tank efficiency of fuel production have to be considered. Combining these factors, today's fuel cell vehicles using non-renewable hydrogen do not offer an improvement in well-to-wheels efficiency over the best IC-engine hybrid technologies. As the fuel cell technology is relatively new, high basic and applied research efforts are necessary to reduce production cost, increase overall energy efficiency and to reduce size and weight. To overcome these problems, real breakthroughs in the stack and fuel storage technology are necessary. The market chances of fuel cell vehicles are generally assessed more pessimistically than a few years ago expecting it mainly in niche applications also for the period after 2020.

FURORE Powertrain Roadmap

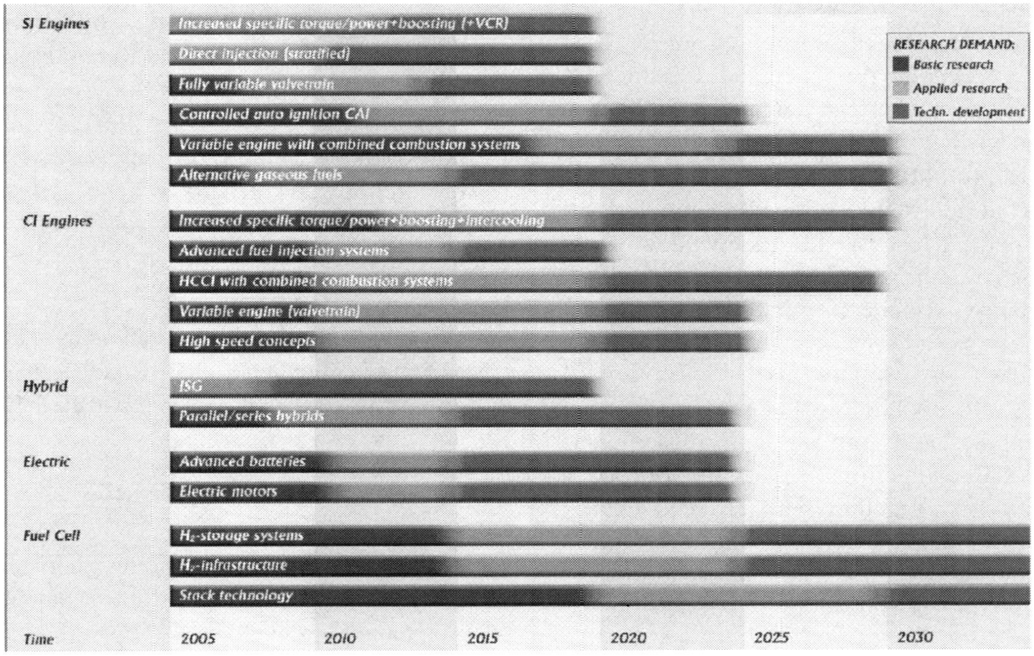

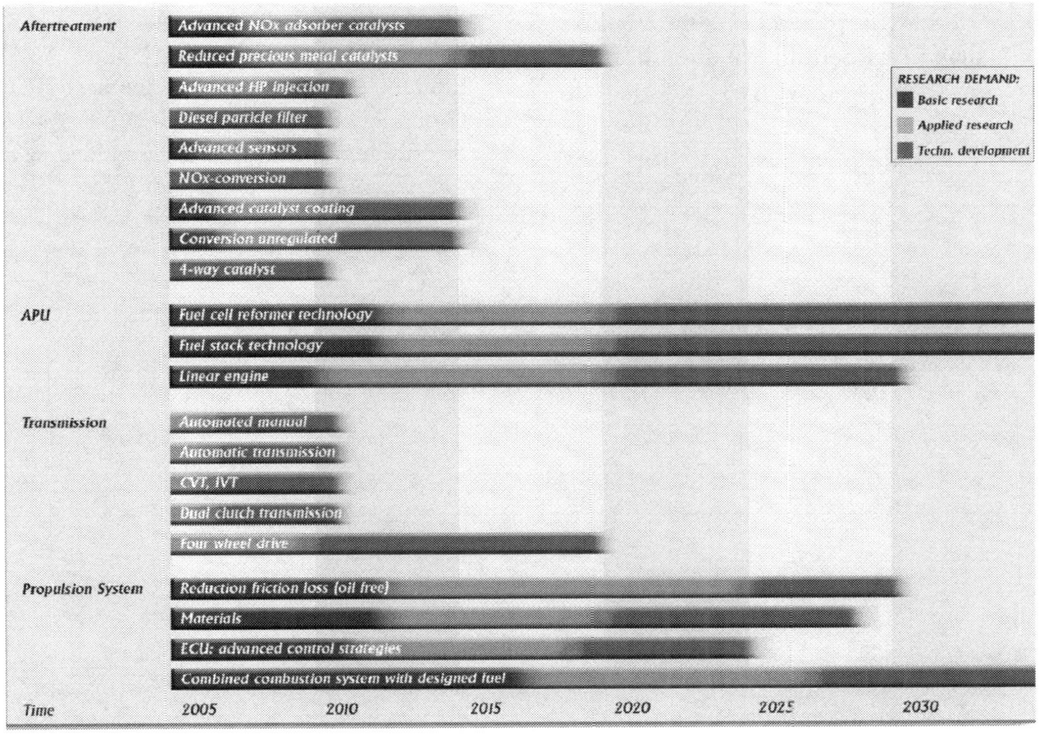

CONCLUSIONS

Road transport technology has historically followed an evolutionary path, with many small steps combining to achieve significant and remarkable progress in reliability, safety, comfort, performance and environmental impact. This study concludes that this evolutionary approach, promoted to a "fast track" pace by strategically focussed research, is well placed to meet the challenges of 2020 and beyond. In the year 2020 and beyond we will have an increased plurality of different propulsion technologies, but internal combustion engines will still be dominant. Most of all, independent from future scenarios on propulsion systems or fuels, it will be essential to save energy wherever possible.

Adequate research investment in the evolution of proven powertrain technology based on internal combustion engines and conventionally-based fuels guarantees global competitiveness of the European Automotive industry together with reduced energy dependency and improved environment. Research in completely new technologies is an additional must to promote sustainable advances in environment and energy security.

Apart from the needed technical research, only improvements in terms of production, distribution and storage will make fuel cell and hydrogen based powertrain systems a competitive alternative. Vehicle weight is crucial for both fuel consumption and safety issues. Intensive research for new materials and production processes including recycling technologies is needed.

A holistic system approach which integrates the concerned stakeholders (infrastructure, vehicle manufacturers, research providers etc.) and the concerned scientific areas (materials, electronics, telematics etc.) will lead to remarkable technological progress. The integration of specific and generic technologies as well as development tools and platforms will lead to better, faster and cheaper research results, finally crucially strengthening the sustainable development of the European road transport sector.

Contact. further information about the FURORE network and the entire FURORE R&D technology roadmap document are available on the furore website: www.furore-network.com

Acknowledgment. Have contributed to this work:
AVL List (Co-ordination), FEV Motorentechnik, Idiada Automotive Technology, IFP Engines & Energy, Ricardo Consulting Engineers, TNO Automotive, RWTH Aachen/ika, VTT Processes, LMS International, Le Moteur Moderne, Fraunhofer Gesellschaft/LBF, University Sheffield/EMD, Istituto Motori CNR, TRL Limited, Additional Contributions by EUCAR/Working Group Powertrain, CONCAWE, IEA, and other organisations dealing with automotive research.

Which Fuels For Low CO_2 Engines?
© P. Duret, X. Montagne (Eds.) and Editions Technip, Paris, 2004, pp. 209 –216
27 rue Ginoux, 75015 Paris

GREENHOUSE GAS EFFECT IN REAL USES DRIVING CONDITIONS FOR CONVENTIONAL AND ALTERNATIVE FUELS

Gabriel Plassat
ADEME

ABSTRACT ñ *Greenhouse gas effects on climate are become more and more obvious, visible and demonstrate by international experts. First step for France is stabilization of emissions between 1990 and 2008-10 (Kyoto Protocol). Next step is a factor 4 reduction for 2050. For transport, this reduction has to be done in a growing sector, entirely petrol dependant with diffuse consumption unit. The factor 4 reduction will operate by traffic reduction and unit emissions reduction choosing best technology. For leading these choices, ADEME evaluate emissions on passengers cars and heavy duty vehicle on real cycle. For HDV, buses represent a laboratory for new technologies. This paper present pollutants and greenhouse gas results on buses for different technologies.*

INTRODUCTION

Considering present standard economy, greenhouse gas emissions are linked with PIB for almost all the countries except USA ("way of life"), Japon and France (nuclear).

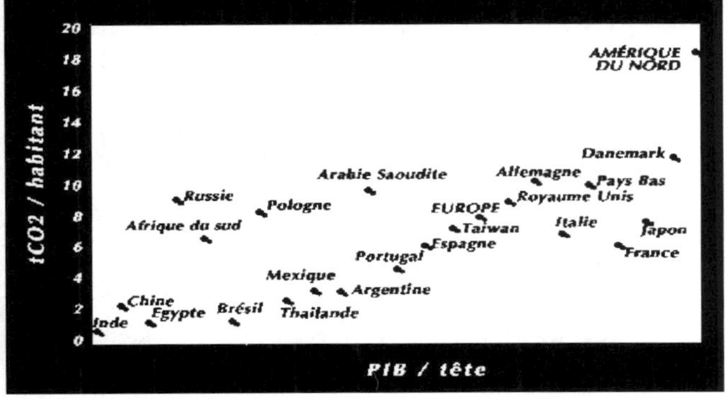

Fig. 1 Greenhouse gas in tCO_2 function PIB, Enerdata 1998

But these emissions will evaluate, especially by increasing of asia emissions. Climate stabilization is possible by a reduction of greenhouse gas emissions around 450 ppm, which induce "only" 2∞C increase.

For an earth population around 10 billions people, unit emission has to be 500 kg CO_2 / year / person (for 450ppm). Kyoto protocol is the first step to stabilize emissions (for France) between 1990 and 2008-10. Next step is a factor 4 reduction for 2050.

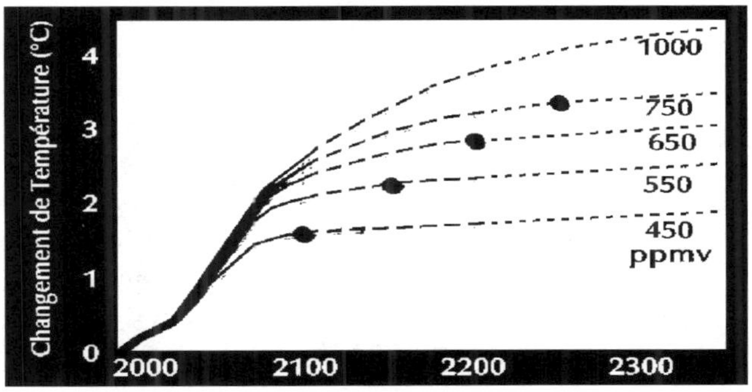

Fig. 2 Temperature increase function CO_2 ppm, GIEC 2000

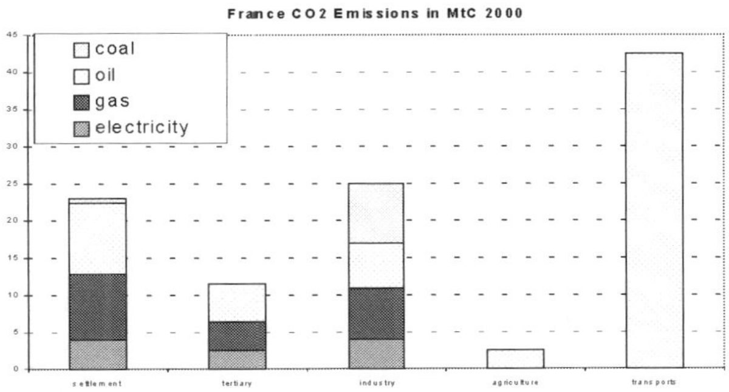

Fig. 3 Emissions CO_2 France, 2000

For transport, this reduction has to be done in a growing sector, entirely petrol dependant with single diffuse consumption unit. The factor 4 reduction will operate by traffic reduction and unit emissions reduction acting choosing best technology

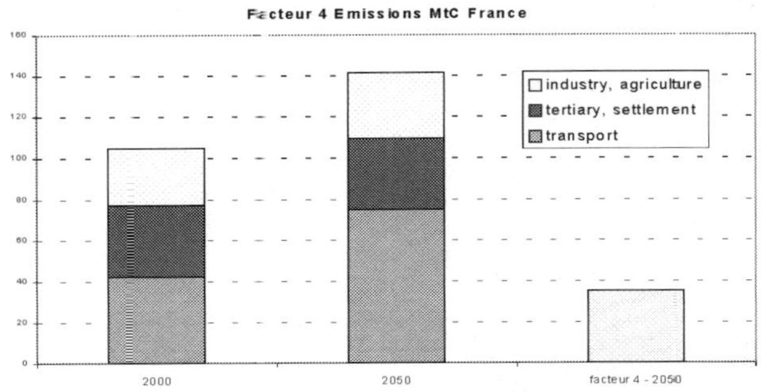

Fig. 4 Factor 4 reduction emissions France 2050

In parallel, pollutants emissions have to be reduced on vehicles especially on particulates, NOx but also CO and HC. Public health, ozone pollution are also under consideration. Emissions are regulated with specific norms in Europe, Japan and USA.

In 1999, ADEME started a national program relating to a comprehensive study on urban transports. Buses represent a laboratory for new technologies before application on truck industries. These evaluations will enable technologies forecast and tax/incentive to promote best technologies.

Several technical solutions were evaluated to quantify on a real driving condition for buses, on one hand pollutants emissions and fuel consumption, and on the other hand reliability and cost in real existing fleet.

There are two different actions types:

ï Actions to implement for reducing the pollution caused by existing fleet (retrofitting) using fuel modifications, Diesel particulates filters or NOx reduction system. Also these technologies were evaluated: Diesel with low sulphur content, Diesel with water emulsion, Diesel with biofuel content, five different DPF, and a DeNOx system are under test (an estimation of NOx reduction is done).

ï New measures to be adopted when renovating a fleet, opting for vehicles whose emissions are lower than Diesel, using either of NGV (stoechiometric and lean combustion solutions with Euro II engines), or of LPG, or of hybrid bus (Neoplan, a real hybrid and Mercedes CITO, a Diesel electric engine were tested) or of electric buses.

First of all, representative driving cycles for bus were established by ADEME and industrials partners. On this cycle, pollutants were measured as well as fuel consumption, at the beginning of a program and one year after to quantify reliability and increase/decrease of pollutants emissions.

At the same time, under real conditions, each technology was tested on buses fleet in use in a city. Information such as fuel consumption, lubricant analysis, problem on the technology was followed during a one year program.

BUSES EMISSIONS

The reduction of atmospheric and noise pollution has particular importance in the sector of urban buses. In many cases, they illustrate the city's environmental image and contribute to reinforcing the attractiveness of public transports. France's fleet in service, presently put at about 15,000 units, consumes about 2 per cent of the total energy of city transport. It causes about 2 per cent of the Hydrocarbons emissions and from 4 to 6 per cent of the NO_x emissions and particles. These vehicles typically have a long life span (about 15 years) and are relatively expensive to buy, about 200.000 euros per unit.

With all these conditions, buses can be a kind of laboratory vehicle for new technologies before application on truck industry.

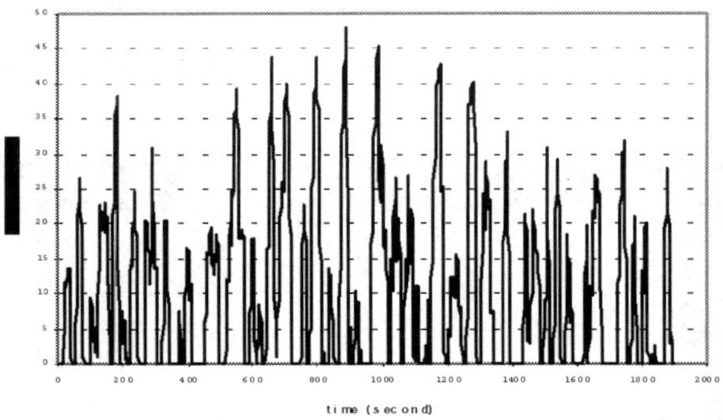

Fig. 5 ADEME/RATP cycle speed function time

ADEME METHODOLOGY

At first, the following cycle was established by ADEME and RATP using an urban line in Paris.

This cycle [speed = f (time)] is then considered representative for urban buses by professionals.

On this cycle, pollutants (regulated and unregulated) were measured as well as fuel consumption, at the beginning of a program and one year after to quantify reliability and increase/decrease of pollutants emissions.

At the same time, in real condition, each technology was tested on buses fleet in use in a city. Information such as fuel consumption, lubricant analysis, problem on the technology and cost associated were following during a one year program.

TECHNOLOGIES TESTED

Reducing urban transport emissions and fuel consumption can be achieved through two different ways described below: improve existing fleet by adding post treatments system or modified fuels or renew fleet.

ACTIONS TO IMPLEMENT FOR THE EXISTING FLEET

The technological solutions in this category have the common characteristic of being able to be applied quickly and on a large number of vehicles. They therefore offer an immediate impact on the fleet's emissions and consumption.

FUELS

Diesel formulae can be modified. This means modifying the chemical composition of the fuel without however affecting its main characteristics (cetane, viscosity). Two kinds of solution stand out:

- Commonly called "city fuels" or VLS (very low sulphur) Diesel in allusion to their low sulphur content.
- Added oxygenated compounds, enabling improved combustion. Biofuels and Diesel Water Emulsion (DWE) are included in this category.

LOW SULPHUR CONTENT:

This is Diesel fuel made with traditional refining procedures (distillation, hydro-processing, crackingÖ) but for which specifications have been made more stringent so as to eliminate pollution-generating compounds or to improve certain performances. The most representative parameter is the sulphur content. This concentration will be reduced from the present level of 350 ppm (or 0.035 wt %) for the entire fleet to 50 ppm by 2005. Other parameters such as the poly-aromatic content and the distillation curve are also being reworked. Using these products makes it possible to moderately reduce particulate emissions on the order of 10 per cent as well as reduce the emission of certain toxic compounds such as the poly aromatic hydrocarbons (PAH).

Using only these fuels is not really as optimal solution: their use requires a certain number of operations that have to be done on the level of fixed infrastructures (tanks). However used with post-treatment (see below), they offer very attractive performance.

OXYGENATED DIESEL FUELS

To date the following products are available:

- The Diesel Water Emulsions (DWE), such as AquazoleÆ, the principle of which is a lowering of combustion temperature by adding water. This fuel was the subject of technical, economic and environmental testing on the public transport systems in the Paris area. Emissions measurements concerned emulsions are presented in this study (average of 3 DWE present on French market) considering 10% water, over consumption of 10%, so GWP is equal to classic Diesel.
- The bioDiesels are composed of a blend of vegetable oil esters (in France, methyl ester of rapeseed ñRME- is generally chosen) and Diesel. This product, traditionally called Diester, is an oxygenated compound of agricultural origin. It can go into the composition of traditional Diesel up to 5 per cent. Higher concentrations (commonly 30 per cent) are useable and remain compatible with present engines. The main environmental attractivity of this technology lies on the one hand in incorporating renewable energy in road fuels (which leads to limiting CO_2 emissions) and, on the other, in the outlet it provides the

farm production of oil-producing plants. Specific tests were done to verify compatibility of bioDiesel and particulates traps. Emissions results of 30% RME mixed with Diesel are presented in this study, coupled with particulates filters.

POST-TREATMENT

Apart from action related to fuels, it is also advisable to implement effective post-treatment systems to clean up buses in service. These systems make it possible to complete combustion after the chamber and thereby to reduce the emissions of unburned products (CO, HC and particles).

There are three main possibilities:

The oxidation catalyst

The oxidation catalyst is a system that makes it possible to treat unburned chemicals in the exhaust (CO, HC) thanks both to the catalytic action of the metals, the bases of which are impregnated and the presence of oxygen in the exhaust. The present offer in second installation is weak, for these products, offered as standard on new cars, have been supplanted in the second-installation market by particulate traps providing action on solid pollutants.

Particulate traps

Although the perfecting of particulate traps for years has run up against a problem of mechanical stress and thermal overheating during regeneration, satisfactory technical solutions have occurred over the past few years and are now being used on a more or less large scale in several fleets. ADEME tested different technologies in partnership with operators. They are differentiated by the regeneration strategy and by the nature of the platform used:

- the CRT (a combined oxidation catalyst and particulate trap) the regeneration principle of which is the formation upstream of the NO_2 filter that facilitates oxidation of soot. The companies market this kind of device (HJS, EMINOX). The base is in cordierite. Very low sulphur content Diesel is necessary for this filter in order to function properly.
- the ENGELHARD particulate trap whose substrate is directly washcoated with catalytic metal. A VLS fuel is recommended for this system.
- the AIRMEEX particulate trap which uses a fuel additive (Iron Strontium) to regenerate the base filter in silicon carbide. This system is compatible with Diesel at present-day standards with 350ppm sulphur fuel.
- the COMELA particulate trap which is a CRT with 2 options: Diesel fuel injection in the particulate trap piloted by pressure drop and exhaust temperature, and valves in the particulate trap which can seal up part of the filter to maintain a certain temperature level in a part of the filter. VLS fuel is needed.
- the RECYCL'AIR particulate trap which is an electrostatic filter working with 350ppm sulphur fuel.

The NOx catalyst

There remains the delicate problem of reducing nitrogen oxides, the subject of considerable research for Diesel. For post treatment, two solutions are possible:

- The catalytic reduction of this product that can only be done by introducing a reducing agent such as urea in the exhaust gas.
- The implementation of NO_x-trapping systems on bases such as barium. These systems absorb the NO_x during the motor's normal operations (poor mixture), then release them during the rich-functioning phase (caused by a temporary change of the motor's settings). In these conditions it is possible to reduce the NO_x. Yet these devices call for a complicated monitoring of the engine and cannot be seriously planned except on complex engines (of the common rail type).

On engine, system like EGR or Urea DeNOx is also integrated. In 2004, reduction of NOx with EGR and urea will be evaluated on buses in retrofitting. This system will also be validated on a fleet. In this study, **an estimation** of NOx reduction was done using urea solution: -50% on NOx and +10% on GWP for fuel due to urea production.

NEW SOLUTIONS TO ADOPT WHEN RENEWING FLEETS BUSES

At the same time as action on fuels and post-treatment, it appears desirable to begin looking for answers aimed at improving the conditions of use of buses in service. We might mention the optimisation of use and assignment of vehicles according to their performance, training drivers to drive economically, and rigorous maintenance of the fleet with an increase in technical checks.

It is necessary to accelerate the renewal of the fleet. This makes it possible to replace old vehicles with generally reduced efficiency by new vehicles (in conformity with the latest standards).

Traditional Diesel engine types are at present undergoing many modifications intended to reduce polluting emissions (EGR, electronic injectionÖ). Emissions standards are tending to be more severe. It will force testing the performance of vehicles according to cycles nearer real operating conditions. The effect of improving fuels (mentioned above) will contribute to the technological development.

Although enjoying a great deal of technological expertise, the Diesel sector has inconveniences in terms of polluting emissions:

- w/o DPF, it generates more particles emissions that those of other road fuels,
- with Diesel's being a heavy product, its combustion creates heavy unburned hydrocarbons,
- Diesel engines emit NO_x when carrying heavy loads.

Even though the devices described above give rise to hope for reducing polluting emissions (especially particles), it is nonetheless advisable to call for new technology with an eye to protecting the environment and encouraging **energy diversification**. Because of their great lightness, gaseous fuels (NGVs or LPGs) in relation to classic fuels demonstrate undeniable environmental benefits due to their composition. They offer characteristics that enable good compatibility with command ignition engine types.

NATURAL GAS VEHICLE (NGV)

NGVs make a considerable reduction in sound and smell emissions and do away with black smoke. Stored as gas at 200 bars pressure, it gives a bus about 300 kms' worth of autonomy.

Two strategies exist at present:

- the stochiometric solution (without excess air or fuel), enabling the adoption of a 3-ways catalyst reducing a strong proportion of the CO, HC and NO_x emissions,
- the "poor" solution (strong air excess), very promising in terms of consumption and emissions, it is more complicated to implement. For the moment it allows only the use of an oxidation catalyst.

A pilot programme was conducted on six sites by ADEME and its partners (GART, GDF, UTP) so as to draw environmental, technical and economic lessons from the NGV solution.

LIQUEFIED PETROLEUM GAS (LPG)

Like the NGVs, the LPGs have undeniable environmental advantages. And like the NGVs, LPGs are presently being assessed in a multi-site programme run by ADEME, GART and UTP. It began in 2000. At present there is but one engine-size. It uses technology derived from modern petrol engines (liquid injection, stochiometric mixture and a 3-ways catalyst).

ELECTRIC BUSES

As far as electricity is concerned, it presents a real environmental interest for applications such as the minibus or midibus in town centres:

- the absence of exhaust gases, thus of emissions near where it is driven,
- suppression of sound emissions,
- a more flexible and safe drive, reducing the risk of accident.

Wide-scale distribution of this technology runs up against questions of cost and technological problems (associated with battery autonomy).

Choosing this solution is wise when it is associated with specific transportation services such as:

- town-centre services or tourist sites
- parking shuttles from city rims to town centres.

HYBRID VEHICLES

Hybridisation consists of associating thermal engines and electric engines in a vehicle. Through the derivation of a fraction of the supplied power and the storing of energy in batteries, electric traction makes it possible to decouple the power asked for by the vehicle from that supplied by the thermal engine. Operating in the purely electric mode is sometimes possible in dense urban areas.

The implementation of hybrid systems has the advantage of not requiring specific supply infrastructure. Neoplan (MAN) solution is tested in south of France (town of Avignon) in comparison to Diesel. CITO (Mercedes) was also tested. Hybrid vehicles are presently being tested on an operational basis in France. Neoplan program will go on in 2004 with new kind of battery. No emission measurements are presented in this study.

RESULTS

GENERAL

Graphic in Appendix present a synthesis of mains pollutants (particulates and NOx) on axes in gram per km on ADEME/RATP cycle. Greenhouse gas emissions (from well to wheel) are also indicated by the area of the circle and value next to the circle.

Different groups can be identified:

- Diesel "standard" with fuel 350ppm, with Diesel with water, with bioDiesel (30%)
- Diesel with particulates filter and same fuels
- Gas fuel (NGV, biogas and LPG)
- Hybrid technology with piston engine and turbine.

Estimation for Diesel with [Biodiesel, Particulates filter and DeNOx], for Hybrid with particulates filter, and for impact on increasing speed cycle (+5 km/h)is presented

ANALYSIS

Greenhouse gas emission can not be separated from pollutants emissions for technology choice.

In general, the higher benefit in greenhouse gas reduction, the higher investment is needed. For example, biogas which present the smallest emission need NGV buses and purification station to clean biogas.

<u>Hybrid</u>

Hybrid technology presents small benefit for greenhouse gas due to a lack of optimization in energy management. Meanwhile, this bus can run in ZEV mode in center. In the future, this technology will be privileged for urban conditions (buses, garbage truck, goods delivery in town). Engines from passagers cars will be used for cost reasons.

Turbine engine (Capstone turbine) present very good result in pollutants reduction but greenhouse gas emissions have to be improved.

<u>NGV and LPG</u>

Gas fuel using spark ignition engine confirm particulates and NOx advantages but also confirm low efficiency of this engine with low load and speed conditions.

For HDV, NGV engines use always Diesel base engine with present disadvantages for optimization. Progress directions are direct injection to reduce pumping losses, hybridizing with more optimized car engines.

Biogas presents best results for greenhouse gas but this solution will be limited due to cost and refueling station.

<u>Diesel</u>

Diesel engines present good efficiency and cost are reduced with truck industry volume.

Impact of Diesel with water is limited on greenhouse gas.

BioDiesel (30%) presents an interesting solution without modification on engine. Reduction on greenhouse gas is important and can be quickly spread for all vehicles. Moreover, Biodiesel is compatible with particulates filter.

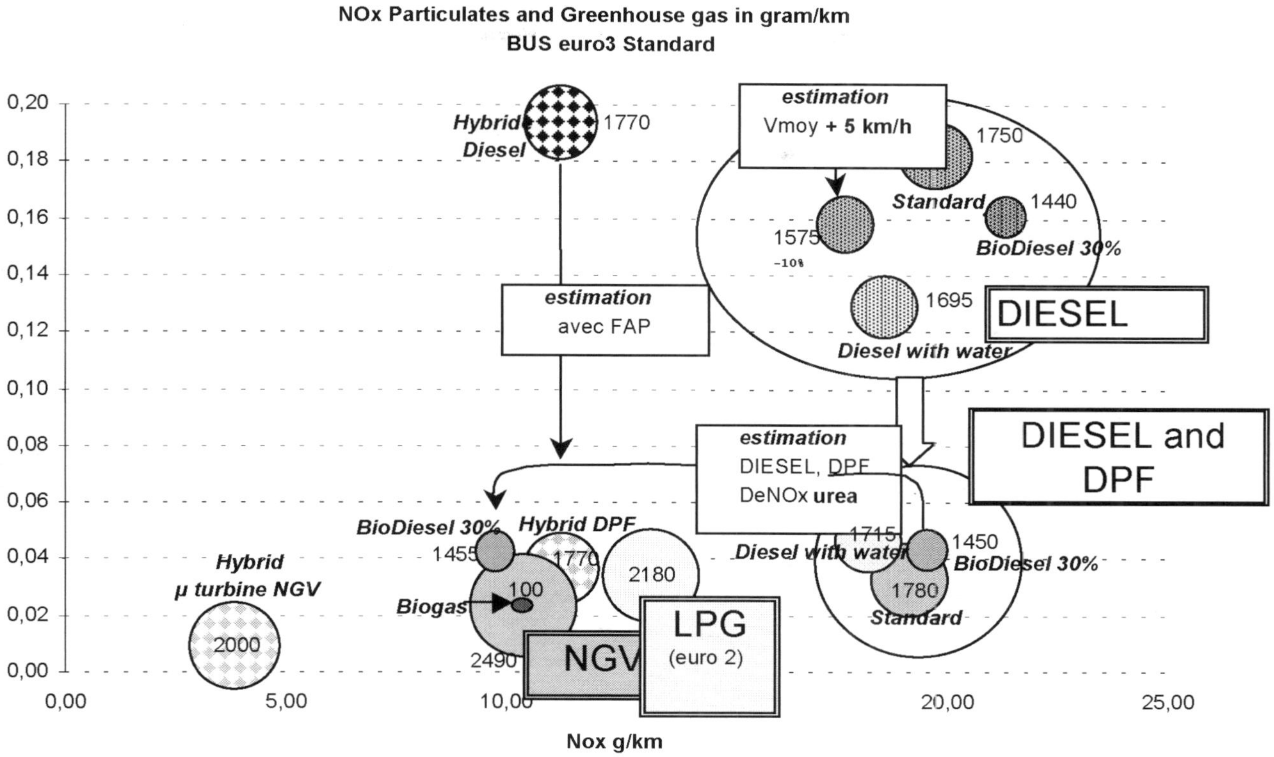

An estimation with DeNOx system (-50%) on NOx, allows Diesel with particulates filter and BioDiesel to reach same pollutants emissions as NGV with 40% reduction in greenhouse gas.

Traffic condition

Buses have to efficient on duration journey to replace passenger car. Special line for buses reduces time and also allow engine to work in more efficient area. An increase of 5 km/h on average speed reduces greenhouse gas emissions by 10% and NOx by 15%.

CONCLUSION

Technologies for Heavy Duty Vehicles present potential for greenhouse gas reduction. In general, the higher benefit in greenhouse gas reduction, the higher investment or fuel cost is needed. Heavy duty industries propose product for two different kind of economy.

On one hand, persons transportation depend of a national economy. So, persons transportation can use energy with a national specific price. Best technologies: using massively electricity with train and renewable fuel (biofuel) with buses (and passengers car).

On the other hand, goods transportation is connected to a international economy. Energy price have to be world price. Rail and water transport have to increased to the maximum feasible. For road, technologies depend on the cycle use of the vehicle:

Long road: energy density is the main point, so fuels have to be liquid. Diesel will stay with biofuel at maximum achievable with area. Fuel cell pay appear for auxiliary production unit.

Urban condition (buses, garbage truck and goods delivery truck): various of energy and technologies can be used as ADEME evaluation showed. Hybrid with NGV, Biofuel, hythane, Ö and some fuel cell will be used.

With mixed conditions (road and urban), hybrid will take a good part using Biodiesel, NGV, Diesel.

Evolutions will take time for vehicle renewal (around 10-15 years), infrastructure for fuel, for train, roadmaps have to be established with all the partners. Technologies won't be sufficient and traffic reduction will be necessary.

Acknowledgments. French partners for all the different studies are RATP, Connex, Transdev, Keolis, UTP, GART and ElectricitÈ and Gaz de France.

REFERENCES

1. Radanne Pierre ñ MIES (2004), La division par 4 des Èmissions de dioxyde decarbone en France d'ici 2050, introduction au dÈbat.
2. Prevot Henri (2004) Politique ÈnergÈtique nationale et lutte contre l'effet de serre.

Contact. Gabriel.plassat@ademe.fr

Which Fuels For Low CO_2 Engines?
© P. Duret, X. Montagne (Eds.) and Editions Technip, Paris, 2004, pp. 217 – 221
27 rue Ginoux, 75015 Paris

ALTERNATIVE FUELS FOR ROAD TRANSPORT

Franz X. Söldner, Jorgan Henningsen
European Commission, Brussels

ABSTRACT – *Alternative fuels substituting oil are supported by the European Commission with the target of achieving a 20% market share of road transport fuels by 2020. The main objectives are improvement in the security of energy supply and reduction of greenhouse gas emissions. The three major lines with an attractive long-term potential are biofuels, natural gas and hydrogen.*

INTRODUCTION

The European Commission has set out a possible target of 20% market share of road transport fuels for alternative fuels in its two framework documents for EU transport and energy policy, the Green Paper on security of energy supply [1] and the White Paper on a common transport policy [2]. The main objectives of introducing alternative fuels are improvement in security of energy supply by source diversification and climate protection by reducing greenhouse gas emissions.

A Communication of the European Commission on alternative fuels [3] has identified biofuels, natural gas and hydrogen as three main candidates of alternative fuels presenting a sufficiently large long-term market potential to ensure a reasonable perspective for infrastructure and product development.

A biofuels directive [4] introduced together with the Communication on alternative fuels sets reference targets for the market share of biofuels, gradually increasing from 2 to 5.75% over the period 2005-2010. A competitive market position of biofuels can be obtained by means of reduced fuel taxation as provided by the EU directive on the taxation of energy products [5].

The market development scenario of a staged introduction of the three main alternative fuels as projected in the Communication is shown in Table 1.

	Biofuel	Natural Gas	Hydrogen	Total
2005	2			2
2010	5.75	2		8
2015	(7)	5	2	14
2020	(8)	10	5	(23)

Table 1. Development scenario for the market share of alternative motor fuels (in percent).

An expert group with participants from the main stakeholders in Europe has been established by the European Commission to analyse technical and economic prospects of alternative fuels, with the focus on natural gas and hydrogen, but also including new developments on biofuels, and LPG. This Alternative Fuels Contact Group used as a reference for their assessment a well-to-wheel analysis of fuel pathways undertaken by a consortium formed by European oil industry and car manufacturer associations and the Joint Research Centre of the European Commission (JRC Ispra) [6].

The Alternative Fuels Contact Group published its report with recommendations on a possible approach to the market development of alternative fuels in December 2003 [7].

Hydrogen as possible future energy carrier has been investigated by a High Level Group called by the European Commission [8]. To facilitate development and deployment of hydrogen and fuel cell based energy systems, the European Commission recently has launched the European Hydrogen and Fuel Cells Technology Platform.

This paper presents the technological and economic perspectives of the main alternative motor fuels natural gas, hydrogen and biomass derived fuels, as appearing from these activities.

NATURAL GAS

Natural gas potentially could reach a market share well above 5% by 2020. The main long-distance transport and distribution system for natural gas could fully support the extra demand from transport, as 10% share in the EU motor fuel market would only correspond to about 5% of the overall demand for natural gas in the EU, with the situation projected in energy/transport scenarios for 2020.

Main requirements for a broader market penetration of natural gas as motor fuel are the build-up of a network of re-fuelling stations and a wide and attractive natural gas vehicle choice to be offered by the car manufacturers. Recent initiatives in several Member States of the European Union on infrastructure build-up for natural gas supply and new natural gas vehicle models from major car manufacturers provide the required signals to the markets.

Development of the re-fuelling infrastructure is key for a possible market broadening of natural gas in the motor vehicle sector. A network of filling stations is a pre-condition to build consumer acceptance and encourage vehicle production by industry.

Investment into an expansion of the infrastructure for natural gas fuel supply could become economic if vehicle numbers increase as expected in market development scenario studies in conditions of improving services. Effectively utilised natural gas re-fuelling stations can be competitive with gasoline or diesel filling stations, since the lower commodity cost of natural gas compensates for the higher cost of the re-fuelling installation.

Mature vehicle technology is available, and an existing range of gasoline backed bi-fuelled and increasingly also dedicated mono-fuel natural gas vehicle models would allow a broader market introduction.

CO_2 emissions from natural gas vehicles are about 20% lower than from corresponding gasoline vehicles and comparable to diesel vehicles, with present technology. Future vehicle technology developments are expected to further improve mainly the efficiency of spark ignition engines and bring them close to diesel engines. As a result, natural gas vehicles are expected to have about 15% lower CO_2 emissions than both, gasoline and diesel vehicles by 2010. Optimised mono-fuel natural gas engine technology might allow even higher gains.

Local air quality can profit from the lower pollutant emissions of natural gas vehicles compared to diesel vehicles, in particular with regard to particulates and NOx. No additional equipment would be required to meet future tighter emission standards presently under discussion.

Substitution of part of the high demand for diesel in Europe by natural gas could also re-balance the stretched ratio between diesel and gasoline production in refineries and improve their energy efficiency and greenhouse gas intensity.

In the early stages of market development, fleet and local markets, such as urban transport, offer the potential for high utilisation rates of refuelling stations, providing an economic case for further investment and expansion of the network. Drivers with high annual mileage would also be preferred first phase customers, as they would boost the share of natural gas fuel consumption with a proportionally low share of vehicles.

Economic incentives compensating for temporary economic disadvantages could support a broader market penetration of natural gas vehicles. The EU directive on the taxation of energy products [5] gives the possibility of reduced fuel taxation for natural gas as a motor fuel. Long-term guarantees on such incentives are essential to build consumer confidence in the early market opening phase. After a transition period, a self-supportive mass market could develop, the need for economic incentives would decrease.

In summary, natural gas as motor fuel offers an important fuel diversification, a significant CO_2 emission reduction in road transport, and contributions to improvement in local air quality. Natural gas could become economically competitive with gasoline and diesel, without tax incentives, in a sufficiently large market through cost reductions for vehicle production and infrastructure utilisation. A motor fuel market share of 2% would already provide the conditions for economies of scale in a mass market.

HYDROGEN

Hydrogen is a potential future universal energy carrier. It thereby could provide access to any primary energy source for nearly any energy consuming appliances. This capability of a very wide diversification of energy supply is of particular interest for the road transport sector, which almost entirely depends on mineral oil based fuels.

No greenhouse gases are released from hydrogen applications in the final stage of energy consumption. The hydrogen production pathway, however, is decisive for the greenhouse gas emissions and energy use overall. Hydrogen production from natural gas steam reforming or coal gasification could become an environmentally acceptable option by including CO_2 sequestration.

Hydrogen production is well established. The largest part of industrially produced hydrogen by far comes from natural gas steam reforming. Electrolysis of water is commercially fully developed as well and is mostly used for smaller volume supply and on-site production, in the absence of convenient access to central production facilities. Cost for hydrogen from electrolysis, however, is at least a factor 2 higher than for natural gas steam reforming. Hydrogen production from biomass gasification could offer an efficient pathway for the utilisation of renewable resources. Co-production of hydrogen and liquid hydrocarbon fuels in large biomass gasification/ liquefaction plants would allow building up high-volume clean hydrogen production capacity in a flexible and cost-effective way.

In the long term, hydrogen offers the potential for full reliance on non-fossil energy resources. Key factors are cost of the energy input and the efficiency of energy conversion.

Scenarios for establishing a hydrogen infrastructure in Europe have been investigated in the EU funded research project HyNet. The cost for a standardised hydrogen re-fuelling station in conditions of a beginning mass market has been estimated to the amount of 250,000-300,000 €. With a customer base of 500 vehicles per station, as required for economic operation, the installation cost would be of order 500-600 € per vehicle. Cost for fuel supply infrastructure provisions therefore a small compared to the cost of vehicles. Investment could be recovered already in the medium term if markets are gradually built up from clusters of hydrogen applications, with a view on closely matching growing hydrogen demand and expansion of hydrogen supply.

Hydrogen vehicles are developed along two different technology systems. Hydrogen combustion engines are based on conventional spark ignition engines modified for hydrogen injection. Fuel cell systems use the electricity produced in the chemical reaction of hydrogen with oxygen to water to drive an electrical motor. Fuel cell vehicles are expected to achieve significantly higher energy efficiency than internal combustion engine vehicles, particularly in part-load operation typical for urban traffic. At high-load running, in particular of heavy duty vehicles on long distance transport, possible advantages of fuel cell systems over diesel engines are less obvious.

Cost for hydrogen as a fuel depends on the production and distribution pathway. Overall fuel cost per km could become comparable with present fuel costs for petrol vehicles, with mass market conditions for refuelling infrastructure and high energy efficiency fuel cell vehicles.

Cost of vehicles is a major obstacle for the introduction of hydrogen in the transport sector. It is mainly determined by the power train and the hydrogen on-board storage/supply system. The considerable cost reductions required for fuel cell vehicles would need to be achieved through technological improvements and economy of scale in large production series. Hydrogen vehicles with internal combustion engines could more rapidly approach prices of conventional petrol vehicles. Development efforts aim at bringing down cost of hydrogen vehicles and ensuring durability and reliability.

The share of hydrogen in road transport fuels could reach a few percent by 2020, particularly if faster market opening with less expensive hydrogen internal combustion engines and acceleration of fuel cell development by stationary applications can prepare the ground for a wide introduction of high efficiency full cell vehicles, which still require longer technological development times.

BIOMASS BASED FUELS

Key issues determining the future development of these fuels will be the availability of biomass and possibilities of cost reductions. The use of woody feedstock could increase the resource potential and reduce costs. This feedstock would be either waste products from agriculture, forestry or wood-based industries, or energy plants specifically grown for the purpose.

Two new pathways for fuels derived from biomass have recently emerged, i.e. bioethanol production from woody feedstock via enzymes, and synthetic biofuels production through biomass gasification and subsequent liquefaction in a Fischer-Tropsch process

Biomass-To-Liquid (BTL) fuels are particularly attractive for the following reasons:

- BTL-diesel has excellent diesel properties.
- BTL-diesel can be blended at any mixing ratio with mineral oil derived diesel, without need for engine modifications.
- Substitution of mineral oil based diesel by BTL diesel could help re-balancing mineral oil refining and improve its energy efficiency.

The maximum technical potential for all biomass-derived fuels produced in the EU should significantly exceed the level of 7-8% market share envisaged in the Communication on alternative fuels for biofuels [3]. First demonstrations of BTL fuels on pilot plant scale have just started, while technology development for biofuels produced through the breaking down of woody biomass by enzymes is more advanced.

Fuels can be purpose-tailored in the BTL process. A wider availability of these fuels could support improved engine technology with better energy efficiency and lower CO_2 emissions. Further development is required to improve the economics of the production process and the logistics of feedstock supply. Co-production of BTL biofuels and hydrogen could provide a cost-efficient pathway to large-scale renewable hydrogen production.

SUMMARY

Alternative motor fuels have the potential to gain significant market share within the next decades, even exceeding the 20% substitution target set out by the Commission for 2020. Cost for alternative fuels could come down already at low market shares due to economy of scale. Economics of alternative fuels, in particular natural gas, is likely to gain in future from increases of oil prices due to supply shortages. Sustained market broadening of alternative motor fuels could improve the environmental performance of road transport and start a transition away from the preponderant oil dependency of the transport sector, thereby improving security of energy supply.

Substitution of 20% market share of oil derived fuels by alternative fuels in the scenario discussed in this paper could result in a saving in oil consumption of about 70 Mtoe annually by 2020. This is of the same order as the projected entire EU oil production in the North Sea by that time. For CO_2 emissions, this substitution scenario could provide a saving of about 80 Mt $CO_{2,eq}$/year in 2020. This corresponds to nearly 10% of present CO_2 emissions from transport.

Acknowledgments. The work of the Alternative Fuels Contact Group provides an important basis for this paper.

REFERENCES

1. Communication of the European Commission European transport policy for 2010: time to decide , COM(2000)769.
2. Communication of the European Commission: Towards a European strategy for the security of energy supply, COM(2001)370.
3. Communication of the European Commission on alternative fuels for road transportation and on a set of measures to promote the use of biofuels, COM(2001)547.
4. Directive on the market share of biofuels, 2003/30/EC.

5. Directive on the taxation of energy products, 2003/96/EC.
6. Well-to-Wheels Analysis of Future Automotive Fuels and Powertrains in the European Context, CONCAWE/EUCAR/JRC, http://ies.jrc.cec.eu.int/Download/eh/
7. Market Development of Alternative Fuels, Report of the Alternative Fuels Contact Group, 2003, http://europa.eu.int/comm/energy_transport/en/envir_en.html
8. Hydrogen Energy and Fuel Cells, A vision of our future, EUR 20719, Brussels, 2003.

Which Fuels For Low CO_2 Engines?
© P. Duret, X. Montagne (Eds.) and Editions Technip, Paris, 2004, pp. 223 – 233
27 rue Ginoux, 75015 Paris

WHICH TRANSPORT FUELS FOR THE FUTURE? THE OIL INDUSTRY PERSPECTIVE

Jean-François Larivé[1], Neville Thompson[1], Bruno Célard[2]

1. Concawe 2. Europia

ABSTRACT – *Transport policies in Europe have for many years been mainly driven by the need to reduce the adverse effect of vehicle emissions on air quality, particularly in urban areas. The Auto-Oil process explored the various options available with the objective to determine the most cost-effective way to attain the air quality objectives while recognising that contributions from both fuels and vehicles were needed.*

Resulting measures in terms of vehicle emission limits (the "Euro" standards) as well as fuel specifications (the fuels directive), have resulted in an impressive reduction of the contribution of road transport to air pollution, this in the context of a continuous increase in the volume of transport. most air quality issues are now well under control although some areas of concern remain, such as ground level ozone and particulates.

Climate change concerns have now shifted the focus towards greenhouse gas emissions. While air pollutant emissions have been successfully abated, GHG and particularly CO_2 emissions keep increasing under the combined influence of increasing transport volume and the extra energy required to produce fuels with tighter quality requirements. Efforts by car manufacturers to improve fuel efficiency are also constrained by the need to achieve very low pollutant emissions.

Another aspect that is currently receiving much attention is security of energy supply, transport being seen by some as particularly vulnerable as it relies virtually exclusively on oil.

GHG emission reduction and security of energy supply are the two main drivers for the development of alternative fuels in Europe.

WHAT HAS BEEN ACHIEVED IN TERMS OF AIR QUALITY AND GHG EMISSIONS?

Over the past decade, vehicle emissions limits and fuel specifications have been tightened considerably. As an example, **Figure 1** illustrates the reduction of the emission limits for light-duty diesel vehicles.

Figure 2 shows the evolution of the European road fuel specifications.

These measures, coupled with legislation on stationary sources are resulting in impressive air quality improvements as illustrated for in **Figure 3 & 4** for PM_{10} and VOCs.

Given the very low emissions from advanced gasoline and diesel vehicles operating on sulphur-free fuels, there is little remaining potential for further improvements to air quality through changes to other fuel properties. Moreover, such changes would require additional refinery processing with consequently higher overall CO_2 emissions, as well as potentially constraining future fuel supply volumes, in particular for diesel.

Much work is in progress on new combustion concepts such as HCCI (homogeneous charge compression ignition) and CAI (controlled auto-ignition) with a view to reducing engine-out emissions and hence the cost of exhaust gas after-treatment. In the near term such combustion is only likely to operate at part load conditions, so these systems will have to be able to operate on conventional gasoline and diesel fuels. Longer term the objective is to extend the envelope of homogeneous operation and in this regard the importance of fuel quality needs to be studied. Early indications are that, for gasoline based systems, the

RON-MON balance may change and for diesel, more volatile fuel with lower cetane number may help. Some workers have even suggested that gasoline and diesel engine technologies may eventually combine with consequent need for a single fuel somewhere between today's gasoline and diesel fuels. These issues need to be carefully researched as these new engine technologies evolve.

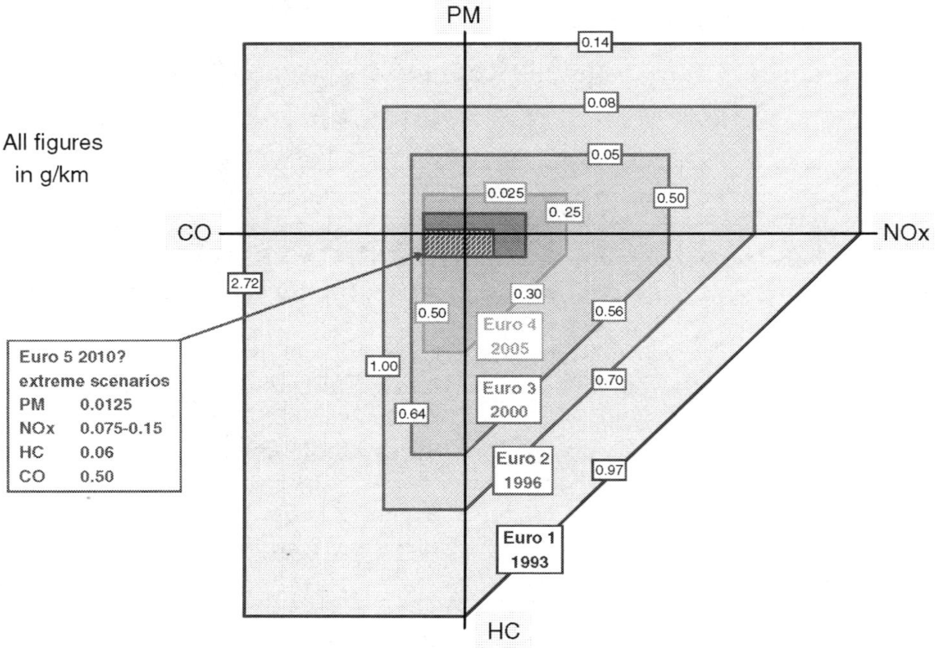

Fig. 1 European emission standards for light-duty diesel vehicles

Year			1993	1995	1996	2000	2005	2009
Gasoline Unleaded 95/85			**EN228**					
Sulphur	ppm m/m	max	1000	500		150	50/10	10
Benzene	% v/v	max	5			1		
Aromatics	% v/v	max				42	35	
Olefins	% v/v	max				18		
Oxygen	% m/m	max	2.5 (1)			2.7		
RVP (summer)	kPa	max	up to 80			60 (2)		
E100	% v/v	min	40(s)/43(w)			46		
FBP	°C	max	215			210		
Year			**1993**	**1995**	**1996**	**2000**	**2005**	**2009**
Diesel (standard grade)			**EN590**					
CI		min	46					
CN		min	49			51		
Sulphur	ppm m/m	max	2000		500	350	50/10	10 (3)
Density	kg/m³	min	820					
		max	860			845		
T95	deg C	max	370			360		
PAH	% m/m	max				11		
Lubricity	µm @ 60°C	max			460			

(1) Up to 3.7% at Member State discretion. Individual limits apply to specific compounds
(2) 70 kPa max allowed in Member States with arctic or severe winter conditions
(3) End date for full introduction of 10 mg/kg S max diesel remains subject to further review

Fig. 2 Evolution of European road fuels specifications in the last decade

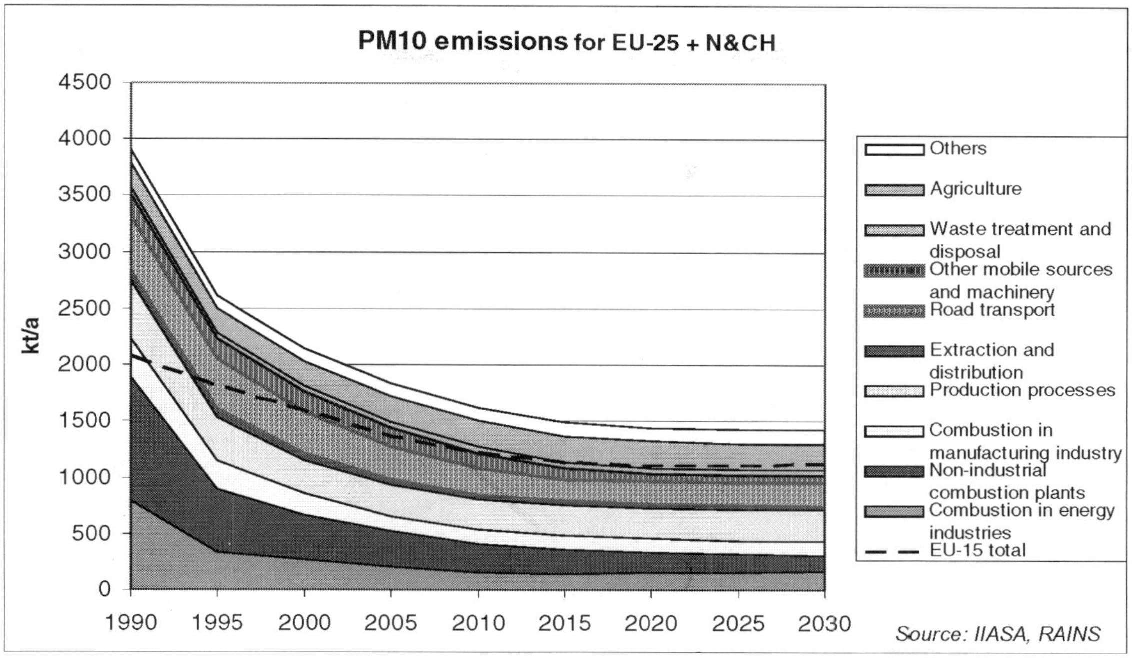

Fig. 3 Evolution of PM10 emissions in Europe

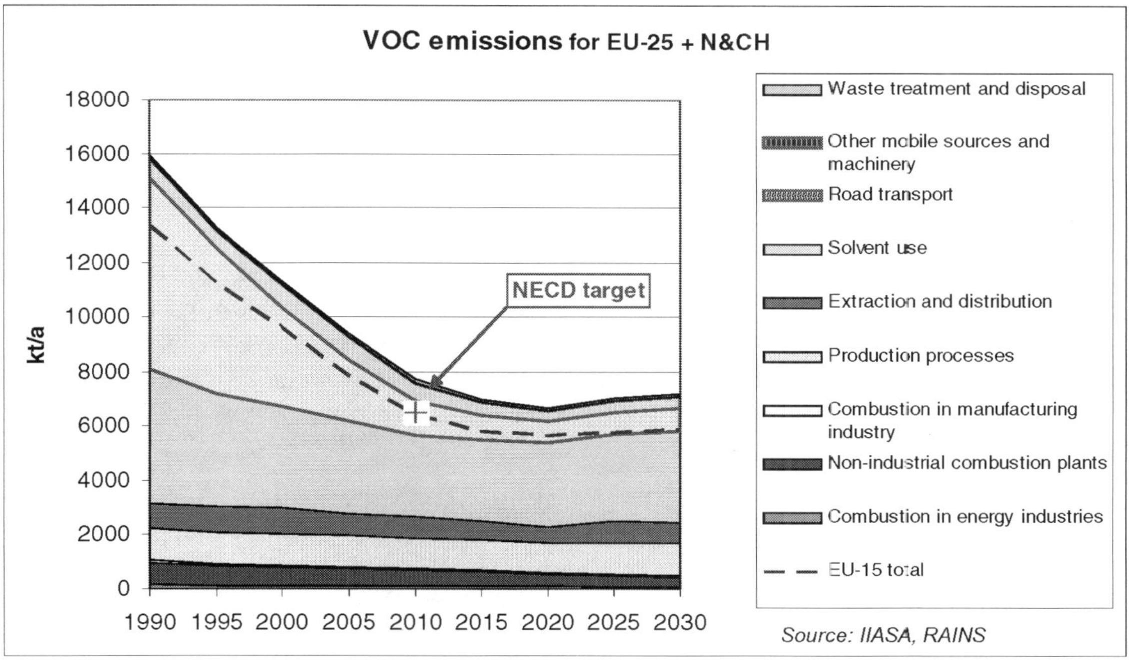

Fig. 4 Evolution of VOC emissions in Europe

However, CO_2 is an even bigger challenge than conventional pollutant emissions. The rapid increase of transport demand as well as the additional energy required for producing vehicles that emit less and fuels with tighter specifications result in a steady increase of CO_2 emissions, this in spite of efforts by car manufacturers to increase the fuel efficiency of vehicles and by the Oil Industry to improve energy efficiency in the refineries.

ALTERNATIVE FUELS: WHAT ARE THE OPTIONS?

The quest for alternative transport fuels in Europe is driven by two main issues:

- The desire to reduce GHG emissions particularly from the transport sector which relies virtually exclusively on fossil fuels and is forecast to continue on a strong growth path (transport contributes some 30% of CO_2 emissions in Europe)
- Concerns about the continued availability of transport fuels solely derived from oil.

At the EU level, the alternative fuel drive was launched by the Green paper "Towards a European strategy for the security of energy supply" published in 2001, proposing a target of 20% of alternative fuels in 2020. This was followed by the "Biofuels Promotion" Directive introducing an indicative target for Member States for the inclusion of 5.75%[1] of biofuels in the road fuel pool by 2010. DG-TREN's Alternative Fuels Contact Group, established in 2002, took up the task of advising the Commission on which other options could be available to reach the original 2020 objective. The group's work focussed primarily on CNG and hydrogen although advanced biofuels such as "Biomass-to-Liquids" (BTL) were also considered. Subsequently the Commission used the results of the Contact Group work to define their strategic guidelines in the "Marketing Development of Alternative Fuels" report published in December 2003.Ordered list, using numbers.

The WTW study jointly carried out by EUCAR, CONCAWE and the EU Commission's Joint Research Centre provided the basis to evaluate a large number of options from the point of view of energy use and GHG emissions while also giving information on costs and potential volumes. The results of the study are used in the following sections to discuss the various options.

[1] Energy content basis

THE OVERALL PICTURE: ENERGY VS. GHG

Figure 5 shows, for all pathways, the GHG emissions against the total energy used, regardless of its origin, be it fossil or renewable.

From this analysis, it clearly appears that the majority of alternative fuel options are more energy-intensive than the conventional, oil-based ones. GHG emissions can indeed be reduced but in most cases at the expense of energy.

If fossil energy resources are clearly limited, this is in practice also the case for renewable resources such as land, solar and wind for which only a limited number of sites are available. Efficient use of energy, whatever its origin, should therefore remain a major objective.

CONVENTIONAL FUELS (Fig. 6)

Developments in engine and vehicle technologies will continue to contribute to the reduction of energy use and GHG emissions. In the timeframe 2002 to 2010, higher energy efficiency improvements are predicted for the spark ignition engine technology (gasoline and CNG) than for the Diesel engine technology. Hybridization of the conventional engine technologies can provide further energy and GHG emission benefits, although it will increase the complexity and cost of the vehicles.

CNG (Fig. 7)

The origin of the natural gas and the supply route are critical to the overall WTW energy use and GHG emissions of CNG pathways. Longer supply routes will become more prevalent in the future as Europe's demand for natural gas grows (mainly for electricity production).

Today the WTW GHG emissions for CNG lie between gasoline and diesel, approaching diesel in the best case. Beyond 2010, greater engine efficiency gains are predicted for CNG vehicles, especially with hybridization. WTW GHG emissions become marginally lower than those of diesel. WTW energy use remains higher than for conventional fuels.

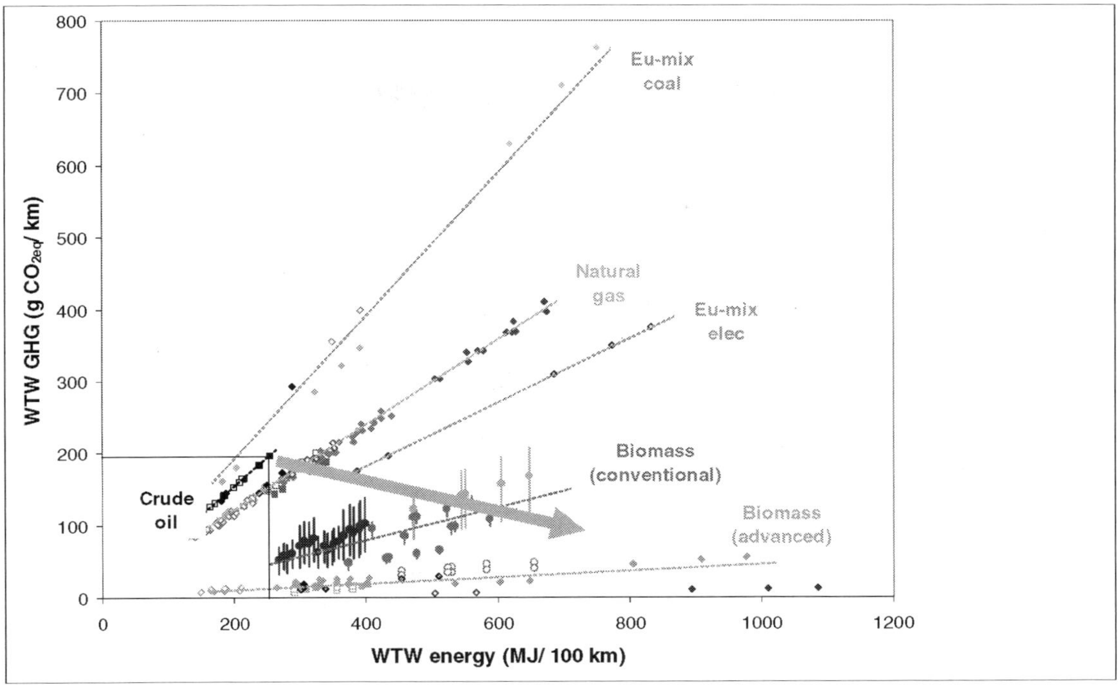

Fig. 5 Less GHG, more energy?

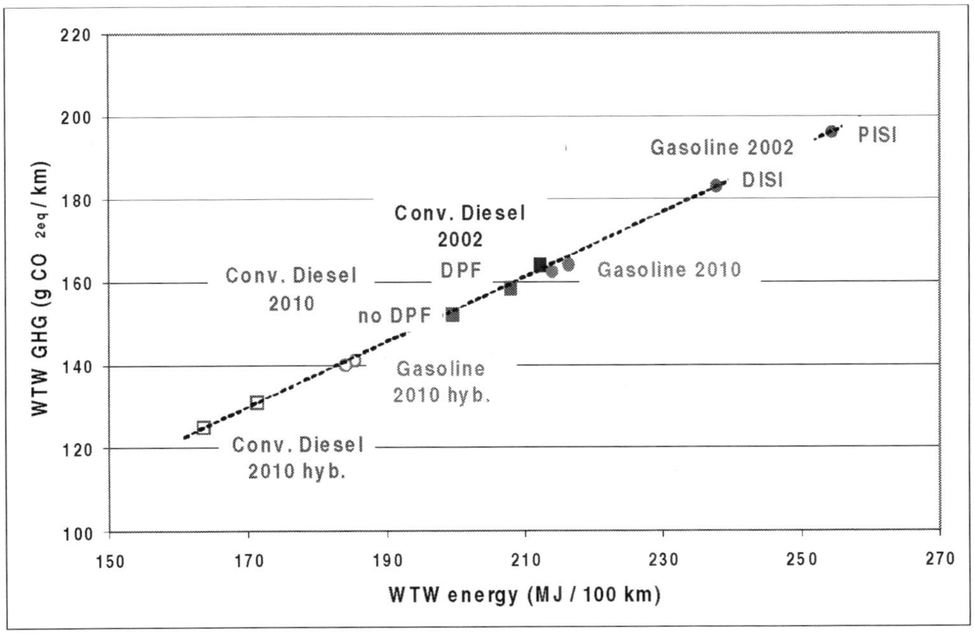

Fig. 6 Advanced powertrains operated with conventional fuels still have potential for improvement

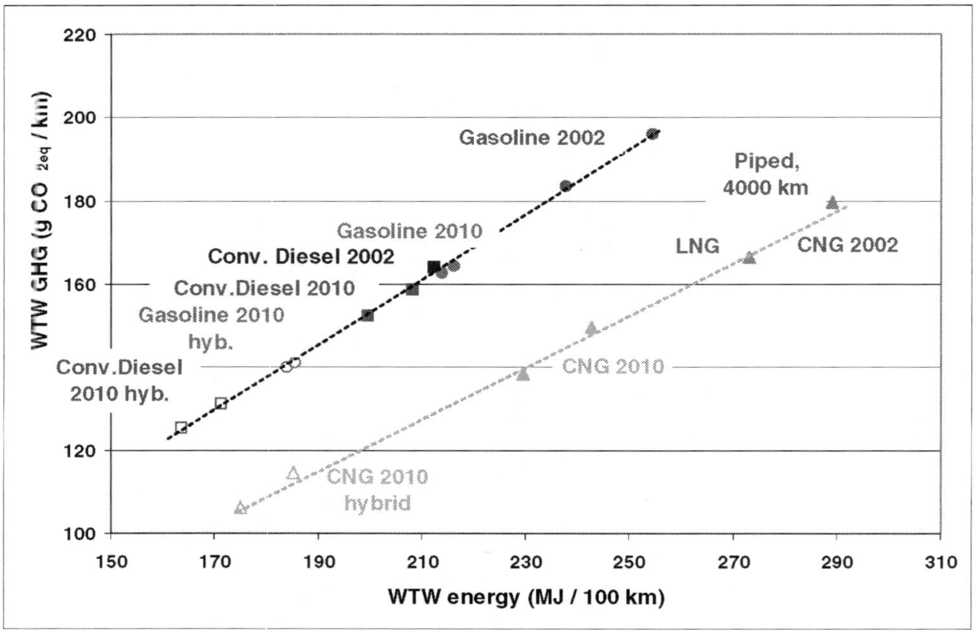

Fig. 7 CNG offer limited GHG gains, at the cost of extra energy

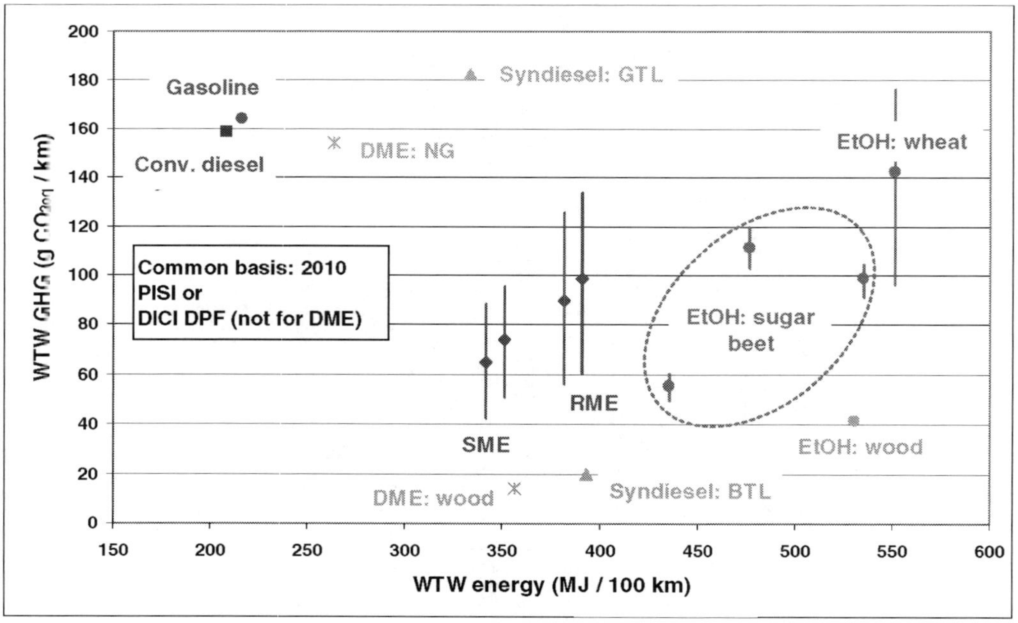

Fig. 8 GTL, conventional and advanced biofuels

The cost of CO_2 avoided is relatively high as CNG requires specially adapted vehicles and a dedicated distribution and refuelling infrastructure. Targeted application in fleet markets may be more effective than widespread use in personal cars.

The established natural gas supply routes to Europe are from domestic sources as well as imports principally from Algeria and Russia. European gas production is set to decline in the coming decades. Europe will have to increasingly rely on imports from a small number of large suppliers, (Russia being by far the largest,) through long-term contracts implying rigid commitments on the customer side. The bulk of these imports will reach Europe through a small number of pipelines. This is in the context of a forecast strong growth of demand principally for the power sector. From this point of view development of natural gas use for the transport sector represents a diversification but hardly an improvement of the security of supply as gas transport infrastructures are heavy, inflexible and vulnerable.

ALTERNATIVE LIGUID FUELS **(Fig. 8)**

A number of routes are available to produce alternative liquid fuels that can be used neat or in blends with conventional fuels in the existing infrastructure and vehicles.

Conventionally produced bio-fuels such as ethanol and FAME provide some GHG benefits but are energy-intensive compared to conventional crude oil based fuels. The GHG balance of conventional biofuels is particularly uncertain because of nitrous oxide emissions from agriculture. Potential volumes of ethanol and FAME are limited. The cost/benefit crucially depends on the specific pathway, by-product usage and N_2O emissions.

GTL (Gas-to-Liquids) processes enable high quality diesel fuel to be produced from natural gas. However, the WTW GHG emissions are somewhat higher than for conventional diesel fuel. Only limited GTL volumes can be expected to be available by 2010 and beyond.

New BTL (Biomass-to-Liquids) processes are being developed to produce synthetic fuels with lower overall GHG emissions, though still high energy use. BTL processes have the potential to save substantially more GHG emissions than current bio-fuel options at comparable cost and merit further study. Issues such as land and biomass resources, material collection, plant size, efficiency and costs, may limit the application of these processes.

DME (Di-Methyl Ether) can be produced from natural gas or biomass at lower energy use and GHG emissions than other GTL or BTL fuels. Implementation costs would be much higher as DME would require specifically designed engines and a dedicated distribution and refuelling infrastructure.

HYDROGEN **(Fig. 9)**

Hydrogen can be produced from virtually any energy source, and this is one of its attractions. The benefits in terms of GHG and the total energy required vary, however, considerably.

In the short term, natural gas is the only viable and cheapest source of large scale hydrogen. If hydrogen is produced from natural gas, the WTW energy use and GHG emissions are higher for hydrogen ICE vehicles than for conventional and CNG vehicles. WTW GHG emissions savings can only be achieved if hydrogen is used in fuel cell vehicles.

Large scale central hydrogen production (e.g. from coal or gas) offers the potential for CO_2 capture and sequestration and this merits further study.

The hydrogen production routes via electrolysis of water are generally very energy-intensive, combining the inefficiencies of electricity generation and of electrolysis as such. The GHG balance critically depends on the origin of the electricity. Using EU-mix electricity results in higher GHG emissions than producing hydrogen directly from NG.

Hydrogen from non-fossil sources (biomass, wind, nuclear) offers low overall GHG emissions. Renewable resources have, however, a limited potential and can be used to produce e.g. electricity, often with higher GHG benefits.

Indirect hydrogen through on-board reformers offers little GHG benefit compared to advanced conventional powertrains or hybrids.

The future of hydrogen is inextricably linked to the successful commercial deployment of the fuel cell, which holds the promise of much higher energy efficiency than conventional powertrains. There are many technical and economic hurdles to be overcome before hydrogen can be produced, stored, distributed and used at the scale required for road transport applications. The required vehicle technologies are complex and expensive while the issue of customer acceptance for a completely new type of vehicle should not be underestimated.

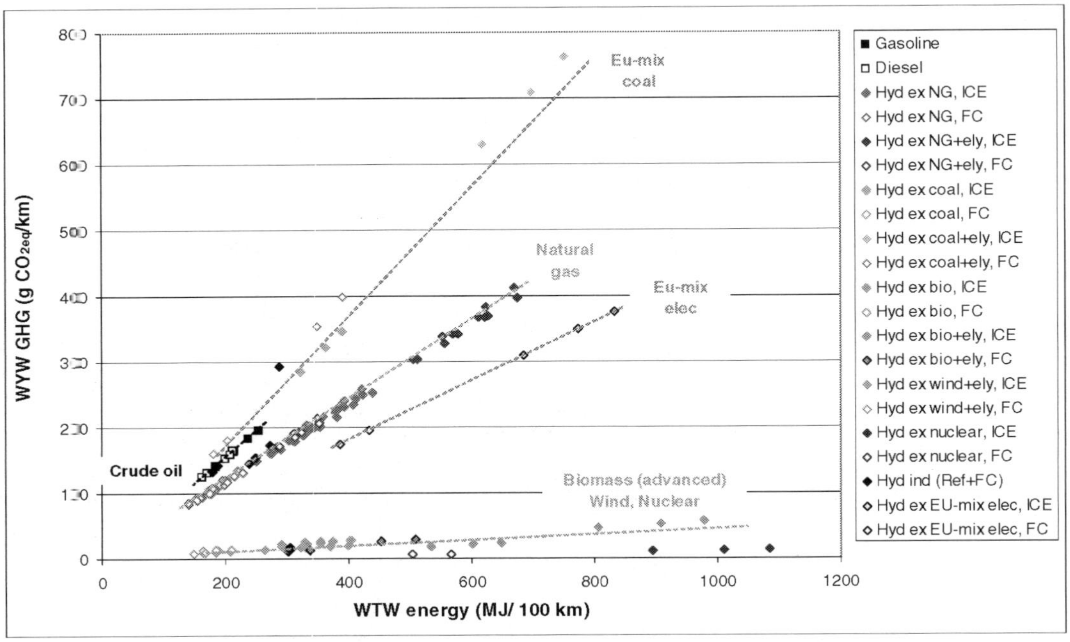

Fig. 9 There are many routes to hydrogen

COST

The way to alternative fuels is generally costly inasmuch as it requires new production facilities, transport infrastructure for raw materials and, particularly in the case of gaseous fuels, entire distribution infrastructures as well as specially adapted vehicles.

GHG reduction being one of the main objective of alternative fuels it is convenient to express the costs per tonne of CO_2 avoided. **Figure 10** shows the results for an arbitrary 5% fleet substitution scenario.

Costs start around the 200 € /t mark for the best biofuel pathways and can go up to several thousand €/t e.g. in the case of on-board reformers / fuel cells using fossil fuels. To put these figures in perspective, the threshold value considered in the Kyoto protocol discussions was 20 €/t.

CNG offers limited savings at a comparatively high cost, mainly due to the distribution infrastructure and additional vehicle cost. Hydrogen from natural gas offers more savings but at a much higher cost.

The least expensive options are biofuels and particularly advanced biomass conversion processes which offer a high percentage of savings with costs towards the lower end of the scale.

AVAILABILITY

In order to complete the picture, the issue of the potential volumes of each type of fuel that could reasonably be produced, must be considered.

Availability of the quantity of natural gas that would be required to substitute 5-10% of the European fuel market would not represent more than about 5% of the total EU gas consumption and would therefore not be a serious issue. We have seen, however, that the expected benefits in terms of GHG avoidance are also limited.

Biomass carries many hopes but the potential is limited. Land use must be balanced between food production, energy crops and other uses. Residues and wastes do offer significant resources but their efficient collection is often fraught with difficulties and there are, in many cases, competing alternative uses (particularly for wood waste). **Figure 11** presents an estimate of the maximum amounts of liquid fuels that could be produced from biomass in Europe in different scenarios. The maximum potential is about 10% of the expected 2010 demand. The actual amount is critically dependent on the type of crop and final fuel, advanced biofuels offering a higher scope.

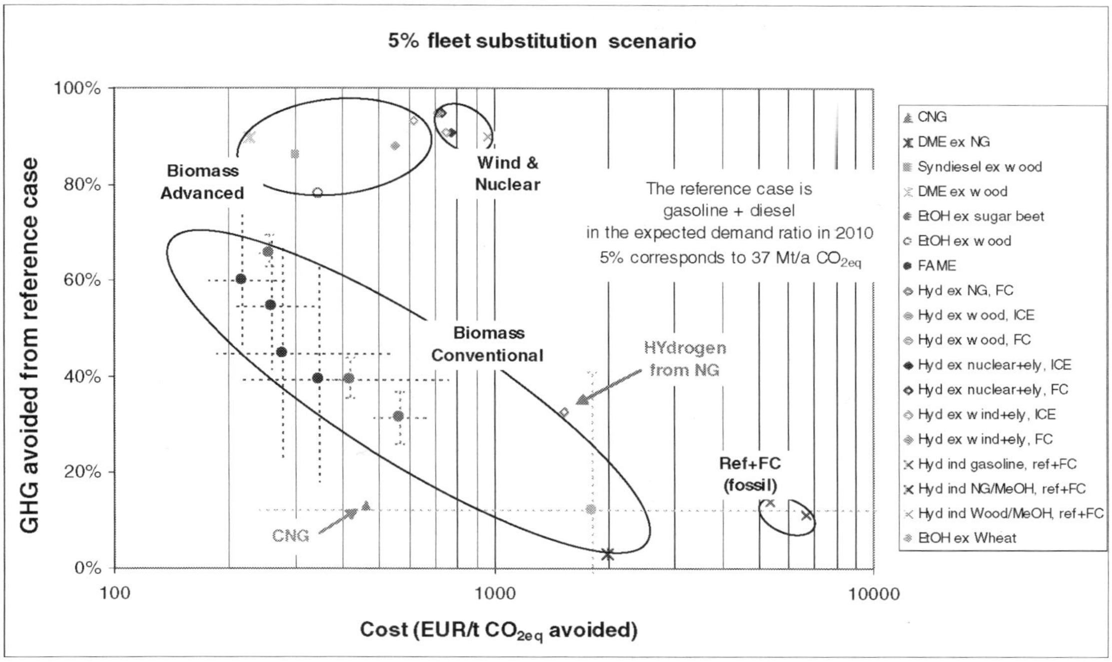

Fig. 10 Cost of CO_2 avoidance

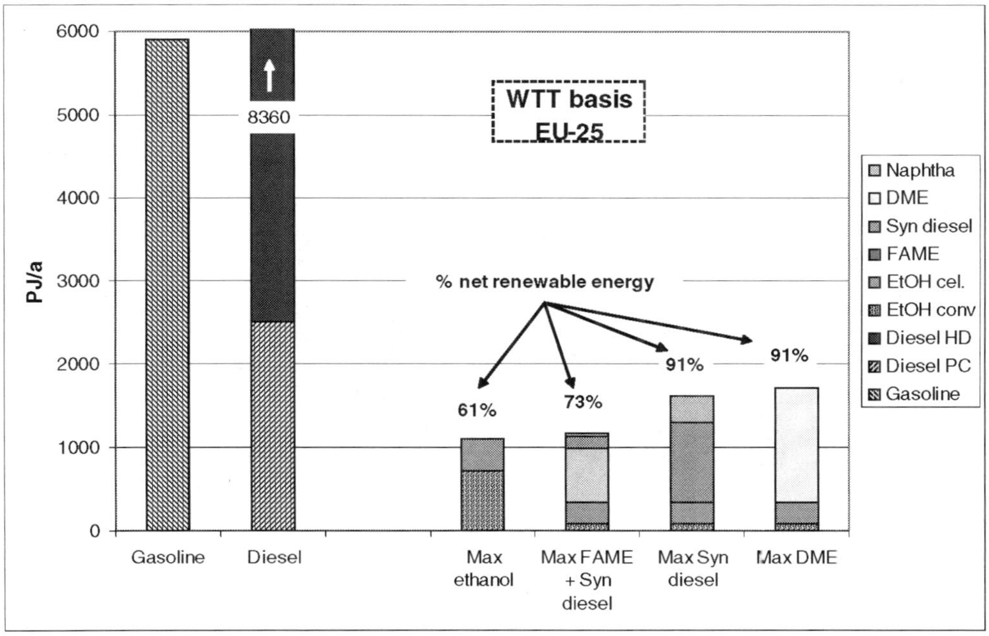

Fig. 11 Potential of biomass-based liquid fuels in Europe

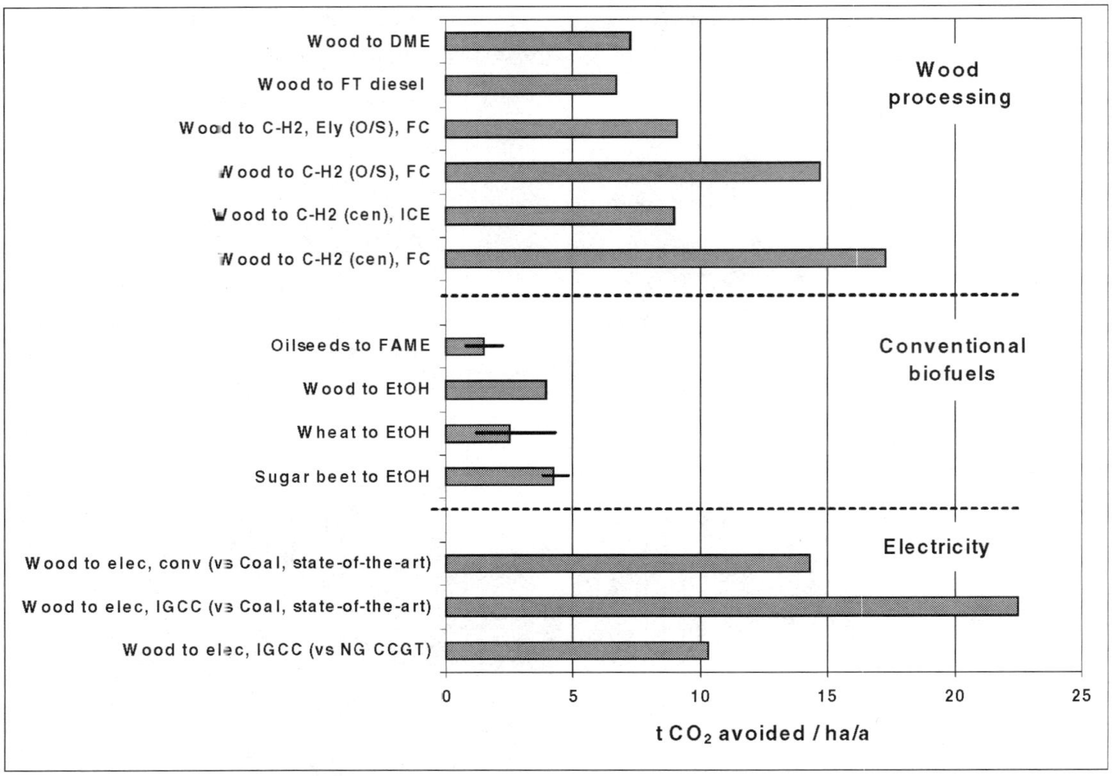

Fig. 12 Potential CO_2 avoidance from 1ha of land

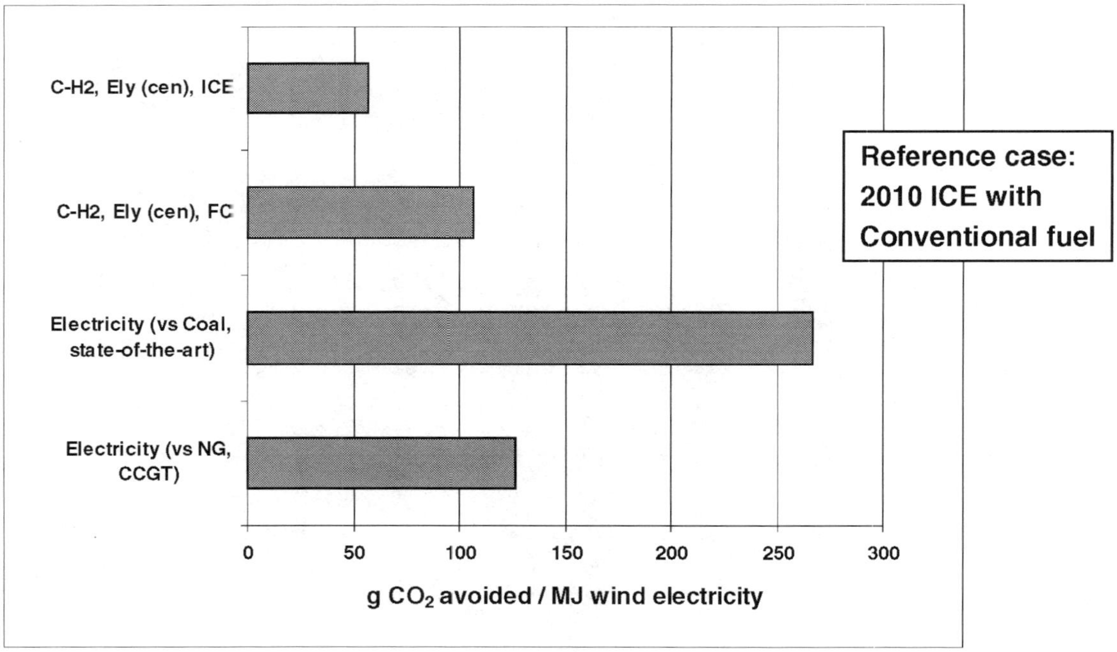

Fig. 13 Potential CO_2 avoidance from 1 MJ of wind electricity

In the context of limited land availability, **Figure 12** illustrates the potential CO_2 avoidance from various uses of one hectare of land.

Because of relatively low yields and often partial use of the biomass for fuel production, conventional biofuels offer only limited potential. Advanced biofuels based on woody biomass fare better, hydrogen offering the best options when associated with a fuel cell. Straight electricity production, particularly when substituting coal gives high CO_2 avoidance.

Other renewable resources such as wind are generally suitable to produce hydrogen via electricity and electrolysis. These hydrogen routes are therefore in direct competition with renewable electricity as illustrated in **Figure 13**.

WHAT PLACE FOR CONVENTIONA OIL-BASED FUELS?

From the foregoing, it is clear that, although alternative fuels and vehicles will make some contribution, conventional fuels and powertrains will continue to provide the bulk of road transport demand for the foreseeable future.

Powertrains will further evolve to meet the very low European emission standards while steadily improving fuel efficiency. This will require increasingly sophisticated engine and after-treatment technologies e.g.

- Direct injection, variable valve actuation, downsizing for spark ignited engines
- Multiple high pressure injections, exhaust gas recirculation for diesel
- Improved TWC, PM traps, lean NOx converters
- Hybrid configurations.

The refining industry is currently engaged in a major effort to introduce sulphur-free fuels (10 mg/kg max S) that will meet the needs of all advanced vehicle technologies that can be expected in the 2010 time-frame. Changes to other fuel properties offer little or no additional air quality benefit, would increase CO_2 emissions and add to security of supply concerns, especially for diesel, where meeting the future demand is already a challenge.

Novel combustion systems e.g. HCCI, CAI may require fuels with different specifications. This needs further research as these systems evolve.

Achevé d'imprimer par Corlet Numérique - 14110 Condé-sur-Noireau
N° d'Imprimeur : 18872 - Dépôt légal : septembre 2004 - Imprimé en France